Industrial Design in Engineering
a marriage of techniques

Edited by
Charles H. Flurscheim
BA FEng FIMechE FIEE FeIIEEE

 Springer-Verlag Berlin Heidelberg GmbH

Industrial Design in Engineering
a marriage of techniques
Edited by Charles H. Flurscheim

First edition published 1983 by
The Design Council
28 Haymarket
London SW1Y 4SU

Typeset in the United Kingdom by
Santype International Ltd
Salisbury

All rights reserved. No part of this publication
may be reproduced, stored in a retrieval
system, or transmitted, in any form or by any
means, electronic, mechanical, photocopying,
recording or otherwise, without the prior
permission of The Design Council.

© Springer-Verlag Berlin Heidelberg 1983
Originally published by The Desgin Council in 1983.
Softcover reprint of the hardcover 1st edition 1983
ISBN 978-3-662-12059-0   ISBN 978-3-662-12057-6 (eBook)
DOI 10.1007/978-3-662-12057-6

British Library CIP Data

Industrial design in engineering
   1. Design, industrial
   I. Flurscheim, Charles H.
   745.2   TS171

Contents

Acknowledgements   v

Authors   vii

1   Objectives and techniques of industrial design in engineering   1
Charles H. Flurscheim

2   Introduction to ergonomics   19
John Wood

3   Ergonomics—display and control applications   55
David G. Davies

4   Ergonomics—environmental factors   81
Rod J. Graves

5   Form   105
Peter Murdoch and Charles H. Flurscheim

6   Colour   133
Allan Whitfield and Tom Wiltshire

7   Machine graphics   159
Ronald Easterby

8   Selection of materials for the man/machine interface   175
Alec B. Kirkbride

9   Selection of finishes for the man/machine interface   195
Noël London and Howard Upjohn

10	Design for maintenance Antony Gibbs and Charles H. Flurscheim	205
11	Psychological factors in man/machine interface design Peter Murdoch	213
12	Design of control components Noël London and Howard Upjohn	227
13	Industrial design and safety Professor Richard T. Booth	245
14	Models as an aid to design Jack Hilton	263
15	Standardisation and quality in relation to industrial design W. H. Mayall	277
16	Specification for the development of engineering products Charles H. Flurscheim	289
17	The integration of industrial design and engineering Antony Gibbs and Charles H. Flurscheim	297
18	Engineering product design—some solutions in practice Charles H. Flurscheim	303

Subject Index 377

Name Index 387

Acknowledgements

At the period when I was designing engineering equipment there was little information on how to optimise the man/machine interface. Later, when I became responsible for technical policy for a wide range of products, the importance of the area covered by Industrial Design, and the difficulties inherent in its integration with engineering, became more sharply emphasised; and even today, although there are now excellent specialised industrial design text books, this difficulty still persists. This book was therefore conceived with the objective of providing engineers with a brief picture of the philosophies embraced within the total spectrum of Industrial Design, together with some practical guidance, in order to assist in the marriage of these techniques with engineering.

I am therefore grateful to the Authors who, from their wide experience, have contributed chapters in support of this objective. When design problems involve several techniques, these have been discussed here in the respective chapters from the appropriate technological viewpoints, and this overlap seems to me to be constructive. Editorial discontinuities also occur in a multi-author concept, and here I hope these will be accepted as a small price to pay in return for the authority this approach permits.

A work of this character generates many technical and editorial problems, and I am especially indebted to Mr Terry Bishop of The Design Council without whose contribution and help this project would never have been completed.

I am also once again indebted to the Partners, Merz and McLellan, Consulting Engineers, who have provided me with the services needed in preparation of this book, and, for her continued assistance, to Miss Emily Reeve.

Charles H. Flurscheim

Authors

Booth, Richard T. JP PhD DIC FIMechE CEng MIOSH
Professor and Head of Department of
Occupational Health and Safety
University of Aston in Birmingham

Davies, David G. BTech
Managing Director
Alpha Industrial Ergonomics Consultants Ltd

Easterby, Ronald S. BEE MSc CEng
Senior Lecturer in Applied Psychology
University of Aston in Birmingham

Flurscheim, Charles H. BA FEng FIMechE FIEE FellEEE
Formerly
Chairman, Associated Nuclear Services
Consultant to Merz and McLellan
Director of Engineering, Associated Electrical Industries Ltd
Member of Council, Royal College of Art

Gibbs, Antony FSIAD
Joint Senior Partner
Helix/Hop Studios, Industrial Design Consultancy

Graves, Rod J. MSC MA DipED AMBPS
Group Head, Environmental Ergonomics Branch
National Coal Board

Hilton, Jack H. MSIAD
Manager, Central Purchasing and Industrial Design Departments
Mather and Platt Ltd

Kirkbride, Alec B. BA(Cantab) FSIAD FRSA
Senior Partner
Kirkbride Associates, Industrial Designers

London, Noël RDI FSIAD NDD
Joint Senior Partner
London Associates, Industrial Design Consultants

Mayall, William H. CEng AFRAeS
Design Consultant

Murdoch, Peter DesRCA FSIAD
Partner
Murdoch Design Associates Ltd, Industrial Design Consultants

Upjohn, Howard RDI FSIAD NDD (*d. 1980*)
Joint Senior Partner
London and Upjohn, Industrial Design Consultants
and Tutor, School of Industrial Design
Royal College of Art

Whitfield, T. W. A. BA PhD
Senior Lecturer, School of Industrial Design
Leicester Polytechnic

Wiltshire, T. J. BSc PhD MCIBS
Lecturer, Building Science Section, School of Architecture
The University of Newcastle upon Tyne

Wood, John BTech MDesRCA
Director
Communications Complex Design Ltd, Industrial Design Consultants

1 Objectives and techniques of industrial design in engineering

Charles H. Flurscheim

1.1 Objectives of industrial design in engineering

It is useful at the start of any project to define its scope, its objectives and the inventory of tools available. This book is concerned with the design of all classes of engineering-based machines and systems, from heavy engineering through electronics to domestic appliances, and specifically with the aspects of their design that affect their relationships with man. The objective of all engineers is to optimise fitness for purpose, based on market requirements and the criteria of quality, performance and cost effectiveness. Industrial design, which embraces aesthetic, ergonomic and graphic techniques—or arts, for they can be considered as either—provides tools that can assist in the specification of what is needed by the market and in design for the man/machine interface. These techniques are used by the designer to analyse the interfaces through which the two-way traffic of information, control instructions, manual operations, environmental influences and impressions pass between machines and those who use them, or are affected by them, and to guide their detail design requirements. These techniques may not be considered 'engineering' but, as their title of 'human factors engineering', [1.1] generally used in America, reminds us, they are an essential part of the wide spectrum of techniques drawn on by the engineer, including specialist expertise when appropriate. The problems to which they can be applied can therefore be considered in terms of the engineer's criteria of performance, quality and cost effectiveness.

The term performance is used here in relation to function; the machine's functional performance may lend itself to quantitative assessment, or it may be that it can only be assessed in comparison with similar devices. The objective of industrial design in either case is to improve in-service performance through improvements in the man/machine relationship.

The term quality is used here to include a number of important attributes dependent on design: appearance and visual coherence, which are significant factors in achieving and demonstrating quality; the compatibility of machines with their human and physical environment; reliability, especially as affected by maintenance facilities; safety, as influenced by control considerations; and the less specific quality of 'wholeness', which is dependent on the designer's ability to optimise every detail in its relationship to the unit as a whole, and in which the synthesising approach of industrial design makes a major contribution.

Cost effectiveness is an essential aspect of all product development, for without it no product can be commercially successful. The objectives in this area are both to improve effectiveness through the industrial design contribution to performance and quality, and also to reduce the actual costs of production and operation.

Although the objectives of industrial design techniques used in engineering are therefore very wide, this has really only been recognised quite recently as systematic ergonomics and design procedures have been added to existing and long-standing visual techniques. It is therefore relevant to examine the historical background to this development, the nature of the three techniques, and the classes of problem to which they can be applied.

1.2 Historical background of industrial design in engineering

Aesthetic design is by far the oldest of the industrial design techniques, and has influenced architecture and domestic design since the earliest civilisations. Examples of aesthetic solutions to practical problems are to be seen in ancient pottery, and in the proportions, ornamentation and siting of Greek temples.

The convex stone pedestal and vertical perspective of the columns of the Parthenon, for instance, show evidence of an analytical design approach to offset optical illusions, the principles of which are still in use today. The collapse of the Roman Empire in the third century AD ushered in a darker period, in which art and science were submerged for a thousand years, penetrated in medieval times by occasional shafts of light shining on structures such as the Gothic cathedrals, which were based on design principles of elasticity and equilibrium, usually sponsored by ecclesiastical patronage.

The revival of humanism in the fifteenth century, known as the Renaissance, included a renewed appreciation of aesthetic ideas, not only in art and architecture but also in the emerging field of machine design. The Renaissance also heralded a revived interest in science and a growing recognition of the value of invention to society, as established in the fifteenth-century Venetian patent laws. Encouraged by this, and supported by advances in mechanics, materials science and fabrication processes, an increasing number of engineering-based products began to appear, such as the universal joint invented by the Italian, Cardano. These owed as much to the revival in appreciation of aesthetic design as they did to the renewed interest in science and, perhaps because new scientific ideas sometimes clashed dangerously with establishment beliefs, it is not surprising that many of them were conceived or disguised as works of art.

As the centuries flowed by a gradual change took place, at first towards restriction and refinement of ornamentation, and then towards the use of form and detail design to create aesthetic satisfaction through expression of function. Examples of the artistic approach are the fifteenth-century German Astrolabe (Figure 1.1), the French Medallion lathe of 1750 (Figure 1.2), the working components of which appear to float in a confection of ormolu, and the English microscope by George Adams (1750/1800) (Figure 1.3). Even as late as the 1830s this style persisted to some extent, as shown by the English Columbian printing press (Figure 1.4). The trend towards restricted ornamentation is shown in the beam engine of 1830 with its Grecian detail embellishment on an otherwise functional layout (Figure 1.5), or the Anglo-American Arab printing press of 1881 (Figure 1.6).

The concept of functional machine design seems to have originated around 1800, but it should not be overlooked that, three centuries earlier, Leonardo da Vinci recognised that ornamentation in the idiom of his time did not always contribute to his designs. This is demonstrated in his graceful flying machine, in which he

1.2 Historical background of industrial design in engineering

Figure 1.1 (right) German astrolabe of 1462. (*National Maritime Museum*)

Figure 1.2 (below) French medallion lathe of 1750. (*Science Museum*)

1.2 Historical background of industrial design in engineering

Figure 1.3 (opposite) Silver double microscope by George Adams, London, second half of the eighteenth century. (*Crown Copyright, Science Museum*)

Figure 1.4 (right) Columbian printing press of 1830. (*Science Museum*)

Figure 1.5 (above) Factory beam engine of 1830. (*Science Museum*)

envisaged a fixed wing structure with ornithopter wing tips (Figure 1.7). Examples of developed products were the Maudslay treadle lathe made in England about 1810 (Figure 1.8), the German microscope by Frauenhofer of 1817 (Figure 1.9), the Berrel handbrace of 1850 (Figure 1.10), and the double expansion steam engine of the SS Britannic of 1874 (Figure 1.11), all of which exhibit their functional purpose with impressive simplicity, anticipating modern design by a century or more.

A nineteenth-century milestone in the history of engineering design was the Great Exhibition of 1851 instigated by Prince Albert, the Prince Consort. This brought together under one roof machines from all parts of the industrial world and, by encouraging comparison and competition, laid the foundations for a step forward in design. New materials were another stimulus, as basic components were fabricated first in bronze, then cast iron, wrought iron and, later, steel. Then in 1879 the collapse of the Tay Bridge, supported on cast-iron columns, marked another milestone, as it stimulated engineers to look afresh at the problem, relatively new

Figure 1.6 Anglo-American 'Arab' clamshell pattern printing press of 1881. (*Science Museum*)

Figure 1.7 (above) Model of Leonardo da Vinci's ornithopter design, dating from the sixteenth century. (*Scala/Vision International*)

Figure 1.8 (right) Headstock of Maudslay's treadle bench lathe of 1810. (*Science Museum*)

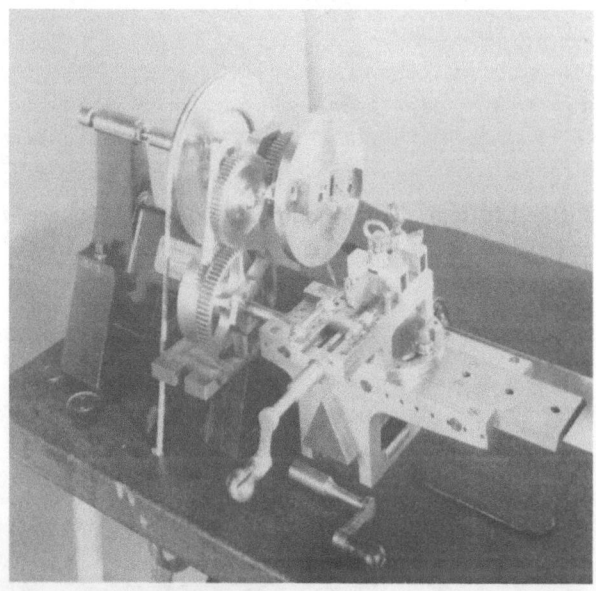

Figure 1.9 (left) German achromatic microscope by Josef von Frauenhofer, about 1817. (*Deutches Museum, Munich*)

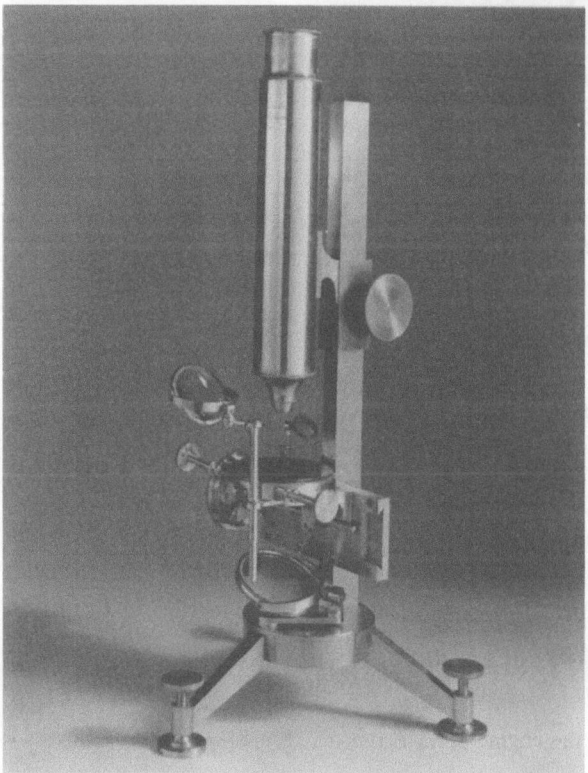

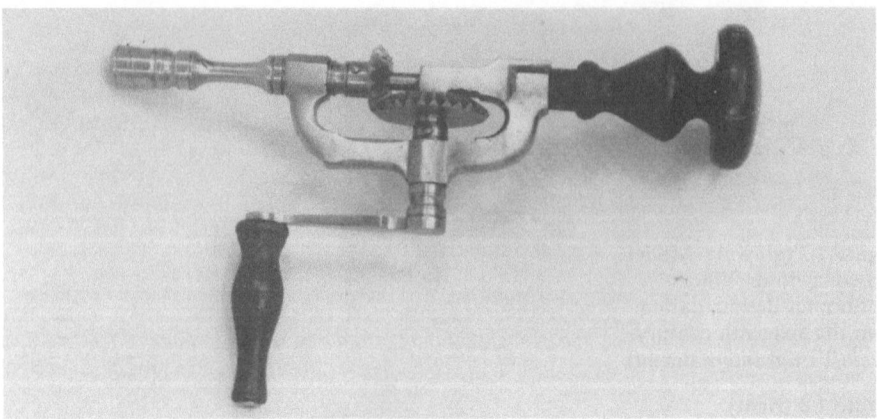

Figure 1.10 (top)   Berrel handbrace, about 1850. (*Science Museum*)

Figure 1.11 (above)   Model of the engine of *SS Britannic*, 1874. (*Science Museum*)

and simple in those times, of material specification, not only for bridges, but for all products. With the vastly increased range of materials that are now available, materials selection has today become a most difficult design problem, not only in relation to performance but also in terms of the man/machine interface, and is thus a critical part of the designer's responsibility.

At the turn of the century the introduction of new forms of power once more accelerated the rate of expansion in variety and performance of engineering products. This rapid growth, and the disorientation it caused in public taste, may explain the stagnation, and even deterioration, in aesthetic design that seems to have followed. There were notable exceptions; in the 1920s Ettore Bugatti, for example, was concerned not only with the performance of his road-racing cars, but with their aesthetic affect, which combined a sense of proportion with extreme attention to detail (Figure 1.12). In the same period W. O. Bentley exploited a different, but equally valid, approach; his robust-looking cars (Figure 1.13), described by Bugatti as '*les plus vite camions du monde*', contrasted strongly with

Figure 1.12 Bugatti car, 1928. (*Hugh Conway*)

Figure 1.13 Bentley car, 1928. (*Hugh Conway*)

the sensitivity of Bugatti's elegant creations.

Some products are developed in stages over many years, and this process tends to refine their visual appearance hand in hand with their functional design. The locomotives of the 1840s (Figure 1.14) can be compared with those of a century later (Figure 1.15) which, as the result of continuous development, reached a peak in impressive appearance before the fashion for streamlining, which contributed virtually nothing to performance, yet spoiled accessibility and increased cost, detracted from their true form. This improvement in appearance can be attributed very largely to the cumulative result of repeated redesign, but also—and especially with British locomotives—to the commercial competition that existed in that period between different railway companies. This encouraged the creation of a

Figure 1.14 (top)   LNWR locomotive, about 1844.

Figure 1.15 (above)   LMS Class 7P locomotive, 1938. (*National Railway Museum*)

1.2 Historical background of industrial design in engineering

Figure 1.16 Cab of the LMS locomotive shown in Figure 1.15. (*National Railway Museum*)

company image through, among other things, impressive rolling stock, typified by attention to the design of components and the orderly arrangement of exposed details as well as the use of house colour schemes.

Locomotive cabins also benefited from this process as a result of increased attention to the working conditions of the drivers, although the control system itself generally remained a rather crude array of valves and levers, differing from its predecessors more in complication than in quality of presentation (Figure 1.16).

In another field, a comparison of the early V-type 50hp Antoinette aircraft engine of 1906 (Figure 1.17) with the 2750hp Rolls-Royce engine of 1931 (Figure 1.18) again shows how the intensive development of performance can be reflected in functional appearance, even in machines with little aesthetic content. Nevertheless this gradual method of improving appearance is rarely possible, and more extensive and organised design is usually needed to achieve good results.

Throughout the centuries, machines had of course been designed to be used by men and women, but their manual operational and control requirements were elementary, and were met by ad hoc arrangements and by skilled craftsmanship rather than by ergonomic analysis of need. The results were often crude, but were adequate for their purpose. The need for a more systematic approach began to be realised early in the twentieth century, influenced to a large extent initially by a few outstanding contemporary bridges and buildings which, decades before their time, had been designed to use steel, glass and concrete in undisguised forms appropriate to the materials. Between the wars this approach was reinforced by a flow of ideas affecting all aspects of industrial design that poured out from the Bauhaus at Dessau, inspired by its director Walter Gropius. In England an Industrial Fatigue Research Board was set up to study relevant ergonomic problems. Nevertheless, prior to 1939, the impact of industrial design in the engineering world was virtually confined to equipment such as car bodies or domestic appliances, where the market value of form and style had been established by the work of designers such as Raymond Loewy in the USA. A wider recognition of the importance of industrial design techniques had to wait for the upsurge in control problems that occurred during and after the 1939–45 war. In this period, and in parallel with the tremendous increase in complication of modern machines and systems, the value of ergonomics as well as aesthetics and graphics began to be realised and their application expanded. This trend is still accelerating, and it had led by the 1960s to a new era in machine design, exemplified by a comparison of old and new Colchester lathes (Figure 1.19), or by the integrated and balanced form of the Goulder gear measuring machine (Figure 1.20). The intensive study of human physical and mental capabilities and limitations has made it possible to apply techniques to the more onerous present-day man/machine interface problems that are better informed in visual terms and are more logical and systematic in terms of ergonomic and graphic criteria.

Figure 1.17   50hp Antoinette aircraft engine, 1906. (*Science Museum*)

1.3 Industrial design techniques as applied in engineering

The age-long preoccupation with aesthetic design as compared with the brief history of ergonomic and graphic techniques is perhaps the reason why industrial design is so often thought to be concerned only with appearance, or even with the limited aspect of this subject commonly referred to as style. This is quite unrepresentative of the present-day position, in which aesthetic, ergonomic and graphic techniques all contribute to some extent, depending on the type of product and the nature of the interface problems.

Aesthetic design is concerned with form, colour, style and the compatibility of an object with its visual environment. Form includes shape, proportion, balance, texture and finish, and good design in terms of form is the result of trained analysis and appreciation, combined with the application of a few basic rules, which in engineering should be looked on as guides, not laws. In the areas of texture and finish form overlaps with ergonomics and other engineering-based requirements affecting manual control functions, wear and corrosion.

Colour is a more emotional area, and the choice of colour is determined largely by intuitive considerations as well as by the contribution colours can sometimes make to form, or to the ergonomic demands of efficient operation or safety. Colour is thus a difficult subject on which to give design guidance, but it is important that colours are appropriate to their purpose, and compatible with their environment.

Style too has an important part to play in engineering design, depending of course on the nature of the product. It can often be used with advantage to emphasise purpose and character, to soften appearance, or to suppress irrelevant details. Although style can never be a substitute for poor basic design, it should not be discarded as a design tool simply because it has sometimes been exaggerated or misapplied by being used inappropriately or wastefully [1.2].

Ergonomics is concerned with optimising the overall and detail design of the control, operation and maintenance interfaces between machines and men, in both normal and emergency working and in the light of the two-directional interaction between machines and the environment, all in terms of efficiency, reliability and safety and on a cost-effective basis. This involves a knowledge of human characteristics such as body dimensions, field of movement, field of view, speed of response, forces that can be exerted, and conditions that aid efficiency and comfort and avoid fatigue. It also involves physiological and psychological considerations such as perception, hearing, the mental processes used in accepting information and in decision making, and the causes of motion sickness, claustrophobia, fear or misuse. Ergonomics therefore aims to provide a reasoned approach to the solution of these problems in which the implementation of both engineering and industrial design technologies may be relevant.

Graphics as applied to engineering is the art of the organisation and presentation of visual information so as to ensure rapid and accurate appreciation by the operator. It can also add to the overall aesthetic balance and visual impression given by a product to a very remarkable extent. Graphic design embraces the selection and presentation of scripts, symbols, diagrams and words for instructions, nameplates and, most important, instrumentation.

With this background we can define some of the areas in which industrial design can help the engineer to achieve his objectives.

Figure 1.18   2750hp Rolls-Royce Type R aircraft engine, 1931. (*Science Museum*)

Performance. As mentioned earlier, this can be assessed either by measurable or by abstract criteria, often dependent on ergonomics. The performance of a power station is measured by thermal efficiency, and in service this will be affected by machine management, which will in turn be dependent on both the skill of the operators and on the ergonomic effectiveness of the control system in providing two-directional transmission of information and control decisions. The performance of a manually supervised machine tool is measured by accuracy and output and is influenced by the ergonomic effectiveness of its controls and operational functions. The performance of a household electric platten iron is measured by output, quality of ironing and operator fatigue and is therefore dependent on the ergonomics of its controls and presentation. The performance of the steering system of a car is dependent on a number of mechanical components which affect stability and feedback, but the response it generates in the driver can be assessed only in ergonomic terms. Ski bindings are designed to provide secure control of the skis under normal conditions combined with consistent release to limit the bending and twisting forces they can transmit under adverse conditions, and the quality of their performance, assessed by confidence generated and freedom from injury, is again dependent on close integration of mechanical engineering with ergonomics (Figure 1.21).

Reliability can be assessed in terms of failure rates, loss of use associated with repairs or preventive maintenance, and life expectancy. Reliability can therefore be improved by reducing risks of incorrect operation leading to damage or excessive wear, through attention to the control interface and operational instructions. It can also be improved by design for ease of maintenance, which involves ergonomic man/machine relationships affecting accessibility, the nature of adjustments, facilities for replacement, measurement and inspection, and by logical fault-finding procedures.

1.3 Industrial design techniques as applied in engineering

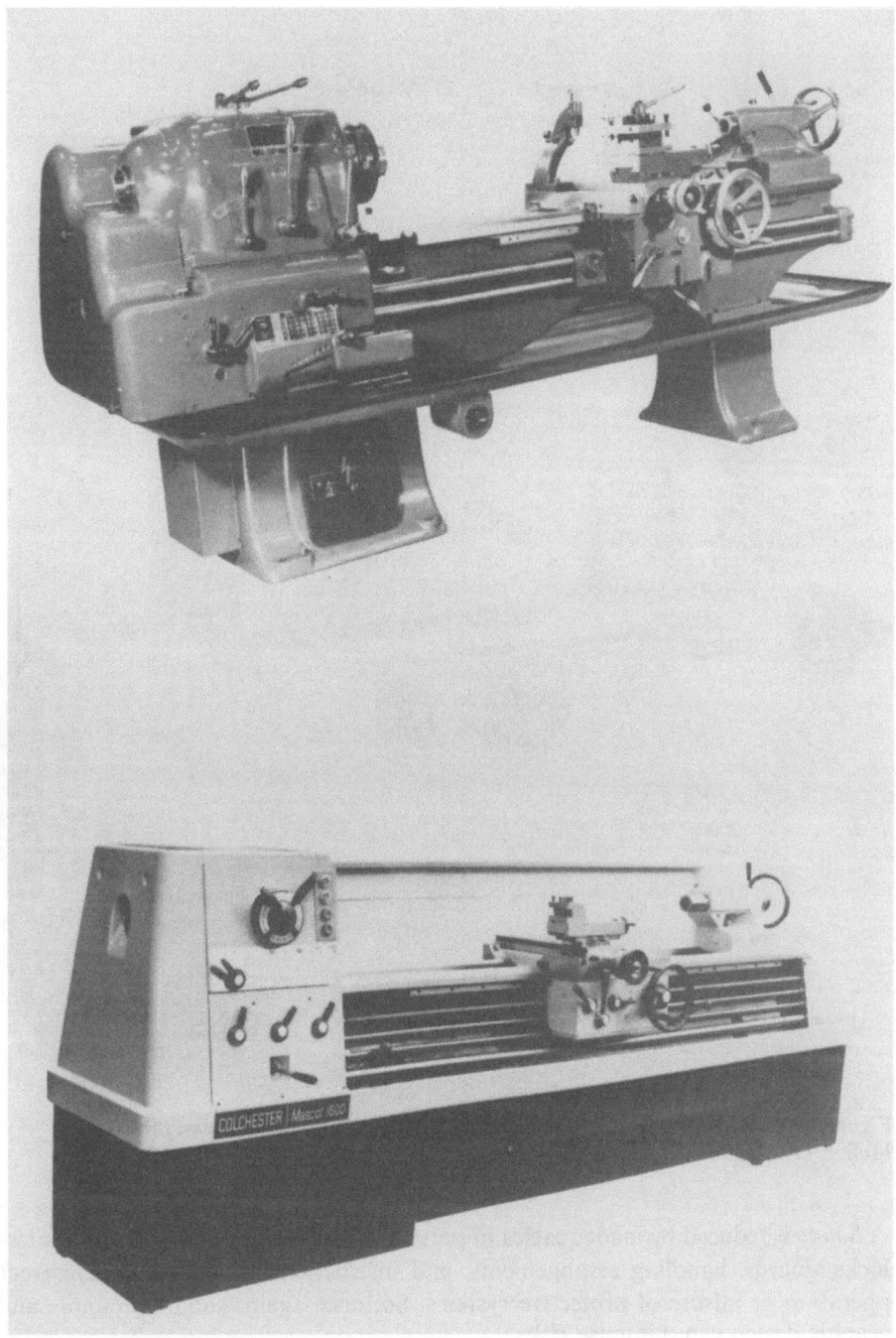

Figure 1.19 Colchester lathe, before and (below) after redesign.

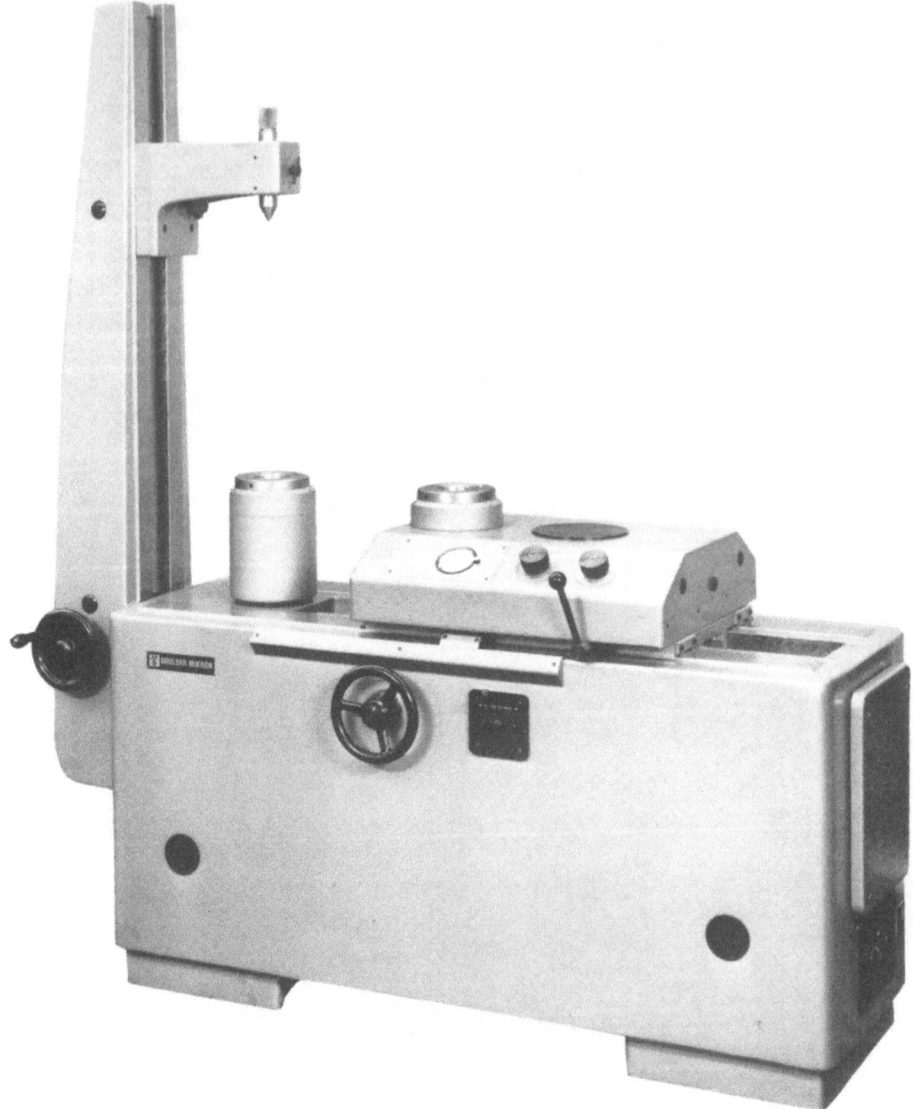

Figure 1.20 Goulder Mikron gear measuring machine, consultant designer David Mellor. (*J. M. Bray*)

Safety is reduced by inadequacies in presentation of information, control, interlocks, guards, handling arrangements, and instructions which permit incorrect operation or misuse of protective systems. So, once again, good ergonomic and graphic design can minimise risks.

The *overall appearance* of a product in relation to its environment is important in creating an impression of quality and functional efficiency, not only among potential buyers, but also among users, who will treat machines they respect with greater care. Environmental acceptability, including the influence the machine may have on the environment in terms of appearance and compatibility or interference with local amenity, and the effect environment can have on the machine and its operator in terms of working conditions, both affect the fitness for purpose of a machine under actual service conditions, as for example in hospitals, lifts or aircraft.

Cost effectiveness has two components, and the engineer can usually improve the latter, effectiveness, with the aid of industrial design techniques. Because of its basic philosophy of simplicity, and because of the specialised knowledge it provides in the interface area, industrial design combined with value analysis can also help the engineer to reduce actual costs, through the elimination of non-essential complication, the cost-effective selection of components, materials, finishes and processes, and by the standardisation of interface details. Industrial design techniques can help more especially with lifetime costs through improving facilities for maintenance.

A relevant factor in cost effectiveness is the extent and cost of the industrial design effort required in product development. One interesting estimate [1.3] suggests that, expressed as a percentage of the total design effort, the order of industrial design costs varies from around 80 per cent for simple tools such as garden shears, through 50 per cent for hand telephones, 10 per cent for instruments such as microscopes, 3 per cent for computers, 2.5 per cent for machine tools, 1.25 per cent for cars, 0.1 per cent for warships, 0.05 per cent for advanced aircraft, to zero for equipment such as submarine cables. The true contribution of industrial design techniques is, however, likely to be underestimated because of the ergonomic component of control, safety and maintenance design, which is often costed as 'engineering'.

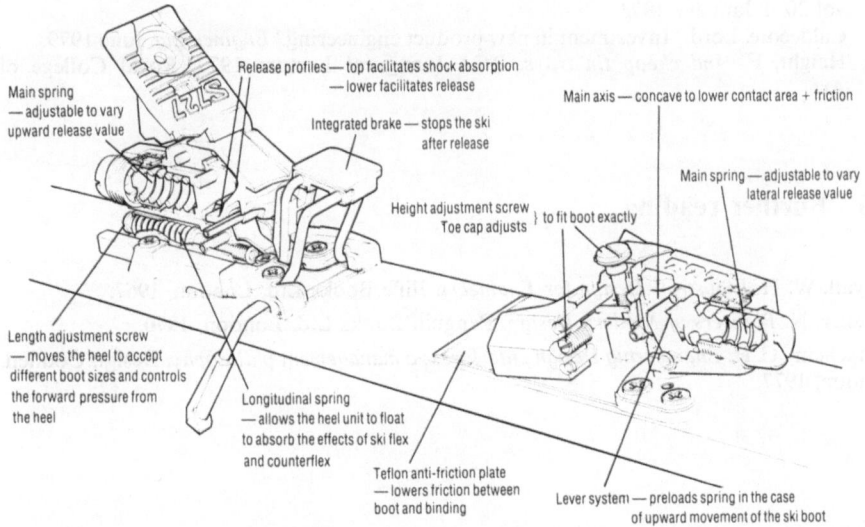

Figure 1.21 Ski binding showing ergonomic functions. (*S. A. Francois Saloman et Fils, Anneçy le Vieu*)

Beyond the design issues discussed here there are ethical factors that are now beginning to influence industry in its response to changes in the social environment. These influences, which require that machines shall be more acceptable to humans and not dominate their environment, are likely to increase in force with time and, since they are concerned largely with the man/machine interface, they fall naturally into the realm of industrial design [1.4].

For clarity of presentation, industrial design techniques are necessarily discussed here in separate chapters, but in practice they overlap both with one another and with engineering. The design of engineering products usually involves a number of different technologies which must be integrated in a process of optimisation, requiring the designer to select between the alternative solutions these technologies often provide, and to compromise between their individual demands. These important decisions necessarily often have to be made on the basis of incomplete information, and subsequent success, even in areas of high technology, is then dependent on the competence of the engineer in what can be called 'the art of design'. Because industrial design technologies are frequently less precise, and certainly less easily quantified, than those traditionally encountered by the engineer, they are particularly dependent on the designer's 'art', but they must nevertheless be given equal attention within the overall engineering design mix if the product is to be successful.

1.4 References

1.1 McCormick, E. J. *Human Factors Engineering*. McGraw-Hill Book Co, New York, 1976 (fourth edition).
1.2 Black, M. 'Engineering and industrial design.' *The Chartered Mechanical Engineer*, vol 20, 1 January 1973.
1.3 Caldecote, Lord. 'Investment in new product engineering.' *Engineering*, June 1979.
1.4 Height, F. *And cheap tin trays*. RCA Inaugural Lecture 1977. Royal College of Art.

1.5 Further reading

Mayall, W. H. *Industrial Design for Engineers*. Iliffe Books Ltd, London, 1967.

Pevsner, N. *Pioneers of Modern Design*. Penguin Books Ltd, London, 1970.

Flurscheim, C. H. *Engineering Design Interfaces: a management philosophy*. Design Council, London, 1977.

2 Introduction to ergonomics
John Wood

2.1 What is ergonomics?

Ergonomics, except as a word, is not new. As discussed in Chapter 1, it has always been necessary to design machines so that they can be used by men and women. However, high technology has put man in a position where errors can have very serious implications, as for example in the work of an air traffic controller, a surgeon, or an operator of a nuclear reactor or early warning radar system. A curious feature of the ergonomist's work is that though he may be dealing with what initially might appear to be very similar types of problem, context may make them very different. For example, technically there may be little difference between the tasks of checking a watch to see whether the pubs are open and checking an altimeter, but the implications of an erroneous reading are vastly different. Ergonomics became a recognisable discipline when the demands of operational and control systems approached the boundaries of human performance. In the last world war psychologists, physicians, anatomists and physiologists found themselves involved in the engineering design team to prescribe these limitations. The infusion of these biologically orientated disciplines into engineering was, and is, the essence of ergonomics; a short formal definition of ergonomics is 'the scientific study of human factors in relation to working environments and equipment design'.

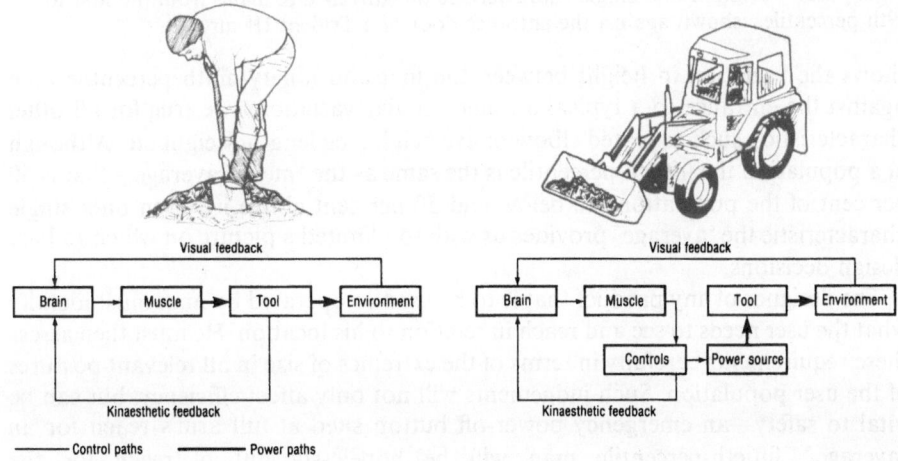

Figure 2.1 The changing nature of the human task.

The widespread introduction of automation is changing the nature of ergonomic problems as will the microprocessor revolution which is just at our doorstep. Today, machinery has largely replaced muscle power with engine power. Now even low-level control is beginning to be delegated to the machine, leaving only strategic planning and the handling of unusual occurrences to the human operator (Figure 2.1). Ergonomists, therefore, will in the future become more concerned with man's information-processing capabilities.

Safety must always be a prerequisite of any design task where an operator is involved. Ergonomics does not claim to have all the answers here, but clearly, by embracing its techniques and guidelines, the designer can reduce forseeable risks. Indeed, legislation in many industrial countries is increasing the designer's liability in this respect. The remainder of this chapter and those that follow on ergonomics aim to show how ergonomic techniques can be of assistance to the engineering designer [2.1, 2.2].

2.2 Body size

In considering body dimensions it is important to recognise that there are very wide variations between extremes in the human race, and that a person 'normal' or 'average' in every respect does not exist. The example given here (Figure 2.2)

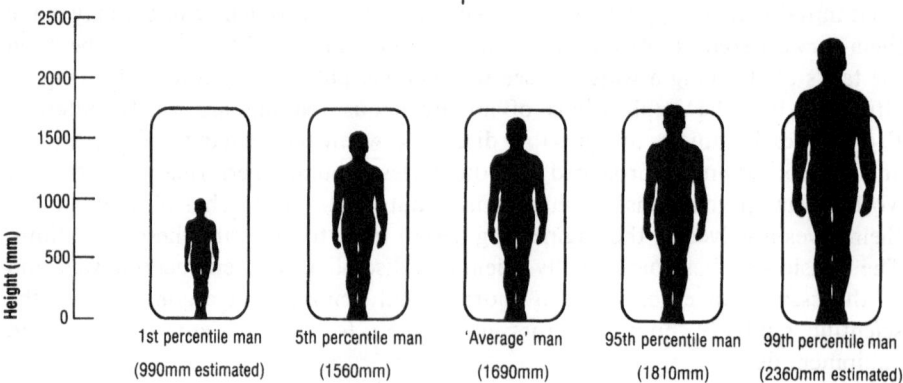

Figure 2.2 Variation of a single characteristic (height) in UK males from the first to 99th percentiles, shown against the entrance door of a Trident III aircraft.

shows the variation in height between the first and ninety-ninth percentile man against the entrance to a typical airliner; similar variations are true for all other characteristics, such as seated elbow or eye height, leg length, weight etc. Although in a population the fiftieth percentile is the same as the 'mean' average—that is 50 per cent of the population are below and 50 per cent above it—even on a single characteristic the 'average' provides us with too limited a picture on which to base design decisions.

The designer of any product that is to be used or operated by man must identify what the user needs to see and reach in relation to his location. He must then assess these requirements carefully in terms of the extremes of size in all relevant postures of the user population. Such judgements will not only affect efficiency, but can be vital to safety—an emergency power-off button sited at full arm's reach for an 'average', fiftieth-percentile man will be hopelessly out of reach for the fifth-percentile operator.

2.2 Body size

2.2.2 The available data

Measuring people accurately is a surprisingly difficult and tedious job, usually involving large numbers of people and the use of specialised equipment. Because of this, most of the data available is derived from specialist surveys, very largely military or university-based in origin, and tends to relate to relatively fit and young groups. The techniques used for measurement are being standardised, but again much of the data available in the literature can be many years old. The references at the end of the chapter list several basic ergonomics texts and military papers which give additional data to that illustrated in Figures 2.3 to 2.5.

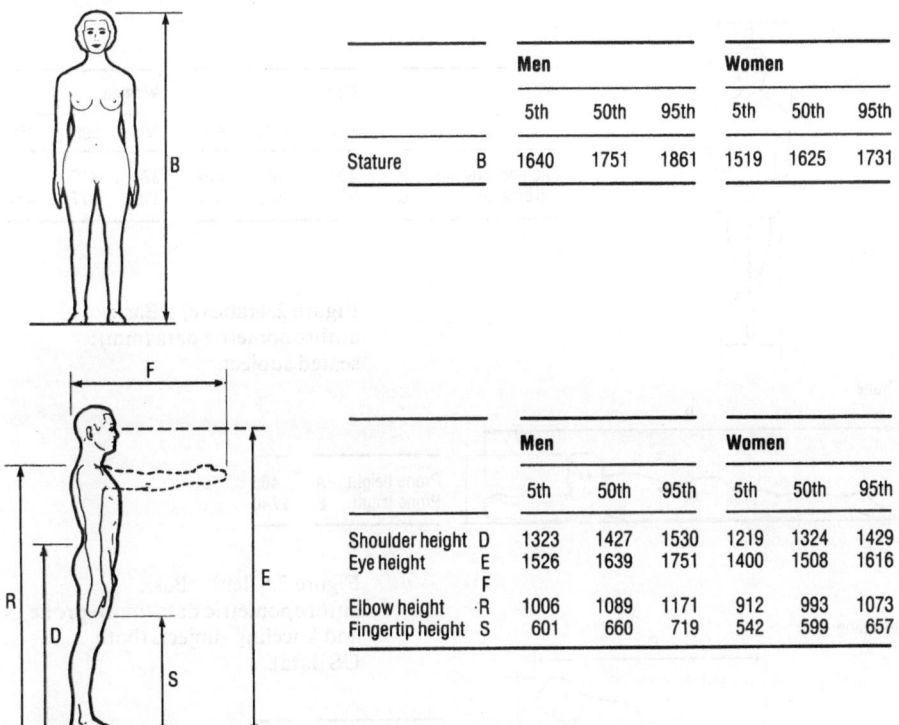

		Men			Women		
		5th	50th	95th	5th	50th	95th
Stature	B	1640	1751	1861	1519	1625	1731

		Men			Women		
		5th	50th	95th	5th	50th	95th
Shoulder height	D	1323	1427	1530	1219	1324	1429
Eye height	E	1526	1639	1751	1400	1508	1616
Elbow height	R	1006	1089	1171	912	993	1073
Fingertip height	S	601	660	719	542	599	657

Figure 2.3 Basic anthropometric data (mm): standing subject (US data).

Whenever anthropometric data is derived from a number of differing sources, care must be taken because of a number of possible limitations. These include measurement errors, variations in measurement techniques, sex of subjects, non-representative samples, and studies based upon the nude subject.

Two broad classes of anthropometric data are available, static or dynamic. The former, as the name would suggest, is derived typically from the measurement of static, naked subjects adopting rather stylised postures. Dynamic anthropometric data is more complex and three-dimensional, involving volumes of space swept by limbs and extremities. Not only are the static lengths of the body members considered, but also the restrictions imposed by joint type and clothing. An example is the difference between the freedom of movement of the 'ball and socket' hip joint and the restrictions of the 'hinge' elbow joint.

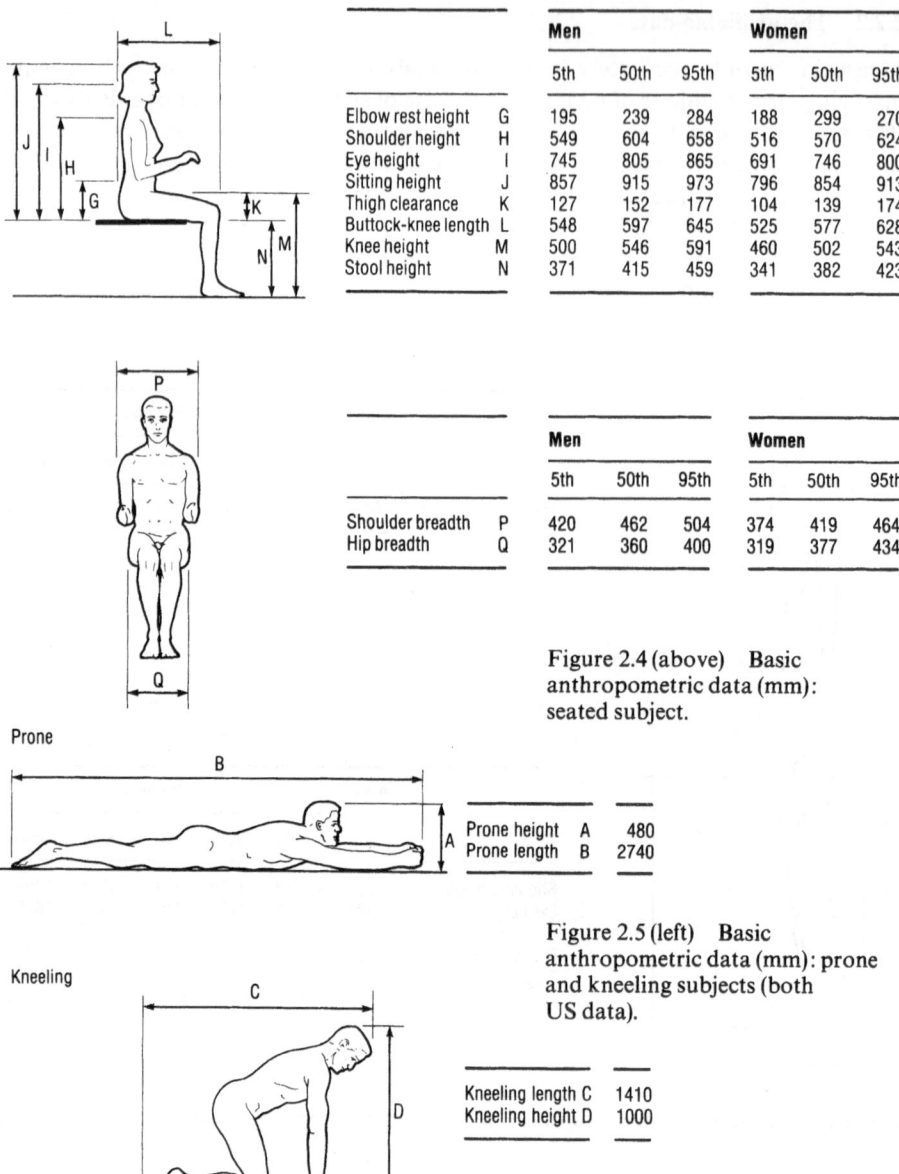

		Men			Women		
		5th	50th	95th	5th	50th	95th
Elbow rest height	G	195	239	284	188	299	270
Shoulder height	H	549	604	658	516	570	624
Eye height	I	745	805	865	691	746	800
Sitting height	J	857	915	973	796	854	913
Thigh clearance	K	127	152	177	104	139	174
Buttock-knee length	L	548	597	645	525	577	628
Knee height	M	500	546	591	460	502	543
Stool height	N	371	415	459	341	382	423

		Men			Women		
		5th	50th	95th	5th	50th	95th
Shoulder breadth	P	420	462	504	374	419	464
Hip breadth	Q	321	360	400	319	377	434

Figure 2.4 (above) Basic anthropometric data (mm): seated subject.

Prone height	A	480
Prone length	B	2740

Figure 2.5 (left) Basic anthropometric data (mm): prone and kneeling subjects (both US data).

Kneeling length	C	1410
Kneeling height	D	1000

2.2.3 Do's and don'ts of anthropometric data

The table gives illustrations of how to select the appropriate anthropometric value for a particular design (Figure 2.6). Ideally one should design to accommodate the entire population of users, though for economic reasons it is usual to account for 90 per cent of the population, hoping that the remaining ten per cent, comprised of both extremes, will be able to be accommodated.

Obviously, where safety is involved, accommodation of the entire population should be the aim, though economics may again preclude designing directly for the first and ninety-ninth percentiles. In such cases special consideration is clearly necessary, such as restricting potential users by selection.

2.2 Body size

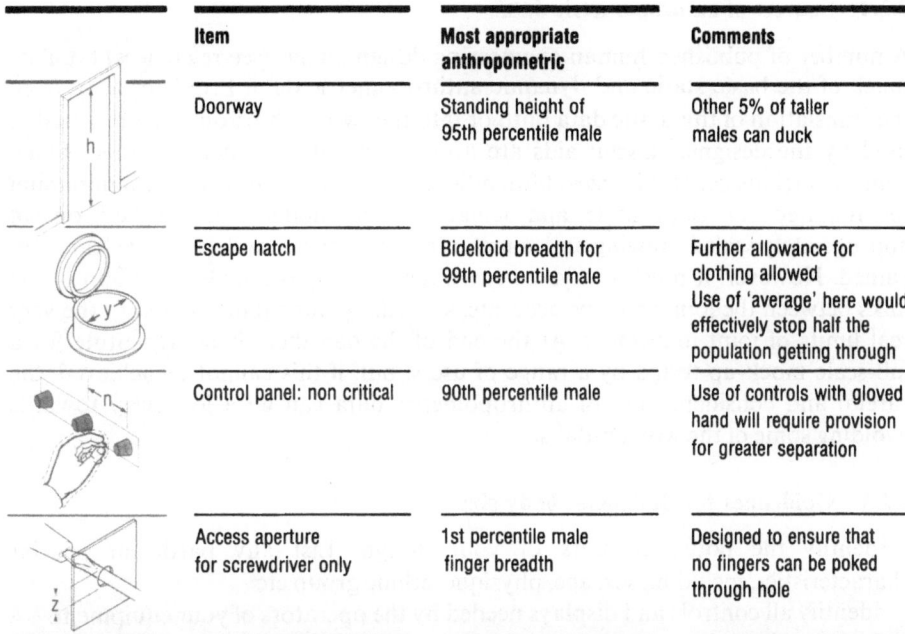

Figure 2.6 Critical anthropometric values for four design details.

Body dimension	Light summer wear	Winter over-coat	Light industrial clothing, boots and helmet
Stature	25-40	25-40	70
Seated eye height	3	10	3
Thigh clearance height	13	25	8
Foot length	30-40	40	40
Foot width	13-20	13-25	25
Heel height	25-40	25-40	35
Head length	n/a	n/a	100
Head width	n/a	n/a	105
Shoulder width	13	50-75	8
Hip width	13	50-75	8

Figure 2.7 Increase in body dimensions (mm) of male subjects as a result of wearing clothing (after Ashby).

Much anthropometric data is either based upon measurements taken using nude or lightly clothed subjects. The engineering designer must consider not only the type of clothes his users are likely to wear, but also any equipment they may be wearing or carrying, such as helmets, boots, and measuring and test equipment. The table at the end of the section (Figure 2.7) shows some of the common adjustments that need to be made when using basic anthropometric data. Emergency conditions must also be considered: a smoke-filled emergency passage carefully designed to accommodate 99 per cent of the population under normal conditions is not much help to rescue staff encumbered with oxygen cylinders, protective helmets, specialised clothing, torches, axes and stretchers.

2.2.4 Sources of anthropometric data

A number of published human engineering design guides (see references) tabulate much of the basic static and dynamic anthropometric data. Problems arise from the translation of this static data onto or into the two or three-dimensional models used by the designer. Useful aids are articulated anthropometric models which come in various percentile sizes; fifth, fiftieth and ninety-fifth would be a minimum set required for both male and female. These models can be laid on an appropriately scaled drawing and some estimate of reach envelopes and clearances gained. However, it must always be remembered that, although the length of the links between the joints may be accurate, swivelling pins cannot represent the very real limits of joint movement. At the end of the day there is no substitute for a full-scale mock-up tested by a range of users, but, if this cannot be achieved, the careful and considered use of anthropometric data can go a long way towards avoiding some of the worst pitfalls.

2.2.5 Guidelines for designers: body size

Identify the potential users of your design. List any particular special characteristics, including sex, age, physique, ethnic group etc.

Identify all controls and displays needed by the operators of your equipment.

In general aim to accommodate 90 to 95 per cent of the population of users, except where safety is concerned.

Consider the repercussions of being either very large or very small in relation to using your proposed design.

Consider the effects of clothing, especially specialised clothing, boots, helmets, gloves etc.

Gross dimensions (for passageways and access) must accommodate or allow passage of the body or parts of the body and should be based upon the ninety-fifth percentile values for the applicable body dimensions.

Where safety is involved, make sure that the really critical anthropometric dimensions in respect of usage and safety are identified. Aim to accommodate the first or ninety-ninth percentile as the case may be—that is, 99 per cent of the population. If this is not possible, restriction of users by selection or redesign may be necessary, but in all cases action needed to cover the remainder of users must be considered.

Reaching distance, control movement, displays, test points, hand rails etc which are limited by extensions of the body should be based upon the fifth percentile values for applicable body dimensions.

Adjustable dimensions (controls, seats, restraint systems, belts etc) should be adjustable to accommodate the fifth to the ninety-fifth percentile values of applicable body dimensions.

Postural problems are inextricably bound up with size problems, so that the two must be considered together.

The design should permit variation of posture, as restriction of movement usually increases fatigue.

The dimensions of a work space properly calculated for workers in one particular country may be quite unsuitable for workers elsewhere, since human dimensions vary considerably from one country to another.

Normal movement is important. For example, during walking, the top of the head moves vertically by up to 50 mm [2.3, 2.4, 2.5, 2.6, 2.7, 2.8, 2.9, 2.10, 2.11].

2.3 Body strength

2.3.1 Introduction

Although much of the physical component of work demanded from men has been replaced by machinery, some industries, such as agriculture and mining, still retain activities involving considerable energy expenditure. This can often be exacerbated by high temperature and uncomfortable postures which pose an additional physiological load. An ergonomist's task is to ensure that none of the activities involved in carrying out the task are likely to impose an unacceptable load on the operator. What constitutes an unacceptable load and what constitute limitations of human strength are the subject of this section.

2.3.2 General features of human muscular power

In man, power is transmitted by the interplay of muscles acting on the system of levers and joints that comprise the skeleton. Energy is released by a 'slow burn' process of nutrients within the muscles, fuelled by oxygen absorbed through the lungs into the blood and pumped to the muscle by the heart. Waste products are also removed by the circulatory system.

In this usual mode of working some of the effects of increasing output, such as higher breathing and heart rates, are understandable. Remarkably, however, the body has an alternative energy system which can function without external oxygen when extra power is required—a sort of physiological 'kick down'. This type of work cannot be carried out for long, and once stopped it is followed by a recovery period during which the 'oxygen debt' is repaid, a classic example takes place in the case of the sprinter.

It is evident, therefore, that we cannot sustain very high levels of work for long periods. Oxygen uptake rises with increasing work load and eventually reaches a maximum beyond which waste products cannot be removed fast enough and we grind to a halt to recover. At lower rates of work an equilibrium can be achieved and work sustained over long periods without fatigue (Figure 2.8). If continuous power is required over an extended period, the level must be considerably less than the peak effort that a healthy adult can deliver briefly for any given form of exercise. Physiological work does not necessarily imply movement; and when postures are maintained against gravity or forces exerted against controls which do not move, these too impose static work loads.

Similar limitations are to be found where smaller muscle groups are involved. With local muscular work, for example a hand gripping exercise, endurance for both static and dynamic work is related to the developed tension expressed as a percentage of the maximal tension that the muscle can develop. Thus for isometric muscular contraction (where work is done but there is no movement of joints) it is found that a 50 per cent load can be maintained for about one minute while a contraction less than 15 per cent of the maximum force of the muscle can be maintained almost indefinitely (Figure 2.9).

A similar relationship exists in the case of rhythmic contractions (Figure 2.10). If the load on a muscle corresponds to about 80 per cent of maximal strength, only ten contractions per minute can be maintained, but a reduction to about 60 per cent enables a rate of 30 contractions per minute to be achieved. These values appear to be independent of the size of the muscle group and there would seem to

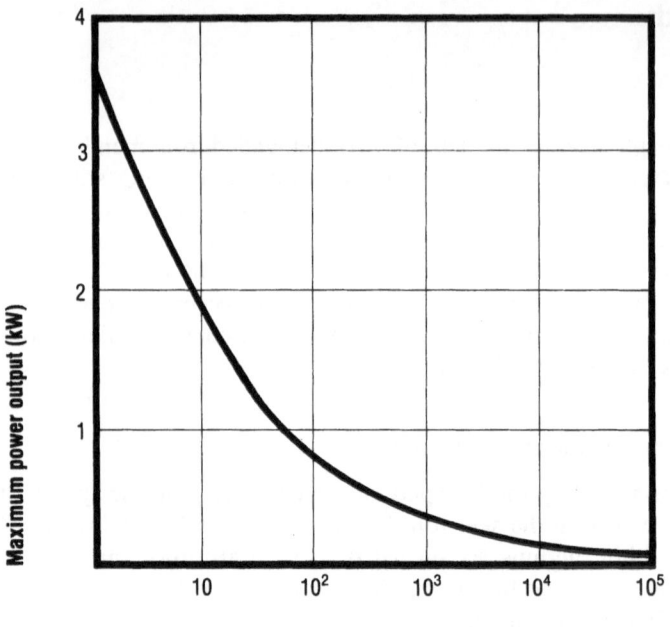

Figure 2.8 Maximum output of external mechanical power plotted against duration of exercise.

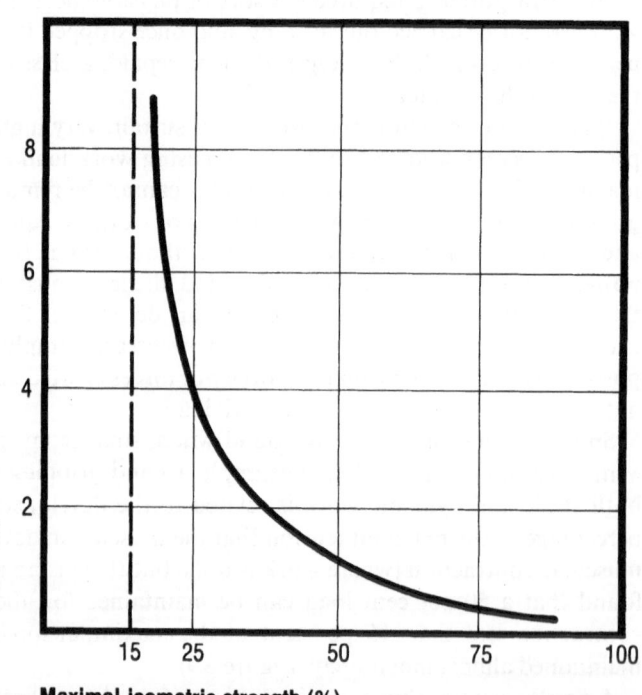

Figure 2.9 Maximum holding time plotted against the percentage of maximal isometric strength exerted (after Ashby).

2.3 Body strength

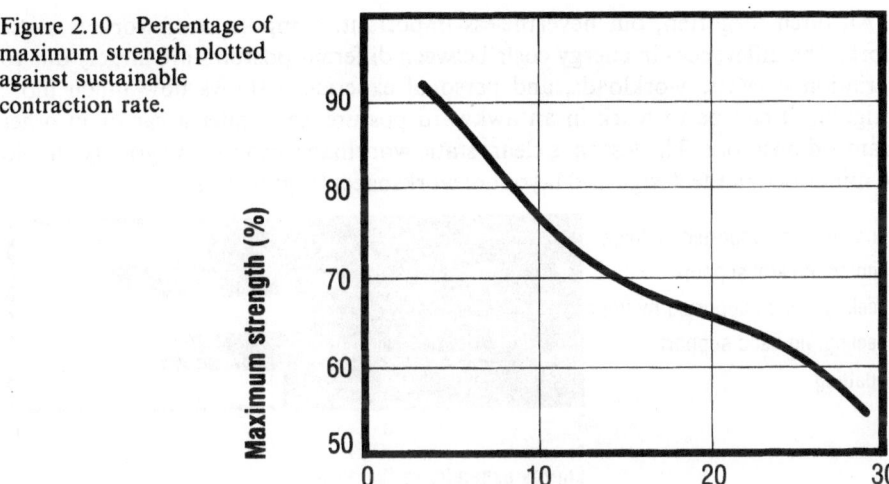

Figure 2.10 Percentage of maximum strength plotted against sustainable contraction rate.

be an optimum capacity when the ratio between work and rest periods is of the order of 1:2.

In order to ensure that excessive and unacceptable workloads are not imposed upon individuals, work physiologists have developed a number of techniques whereby the physical cost of a range of standard activities can be measured using a knowledge of the biomechanical breakdown of nutrients. By measuring the amount of oxygen an individual has consumed during a period of work and subtracting the amount that he would use anyway while resting, we can derive an energy cost for that work, which is usually expressed in kilojoules per minute (Figure 2.11). The available data does not cover all the various conditions outlined in the next section, but does at least provide the engineering designer with some indication of the likely loads imposed on the operator.

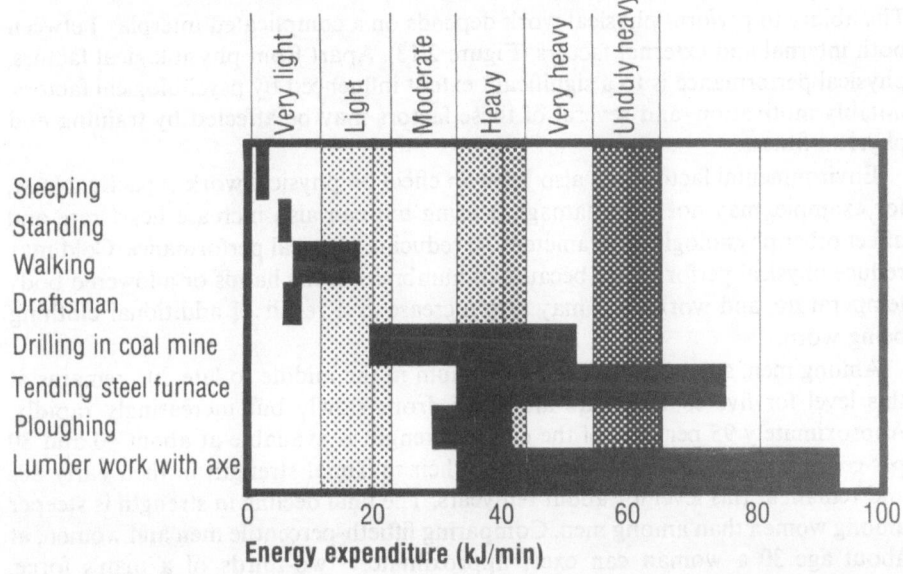

Figure 2.11 Energy expenditure for a range of activities (after Astrand, Rodahl and McCormick).

An often forgotten, but nevertheless important, component of work is static work. The differences in energy costs between different postures are largely due to variation in static workloads, and personal experience shows how much more fatiguing it can be to work in an awkward posture, say under a car or in other cramped positions. The lesson is clear: static workloads imposed by gravity should be minimised in the design and layout of workspaces (Figure 2.12).

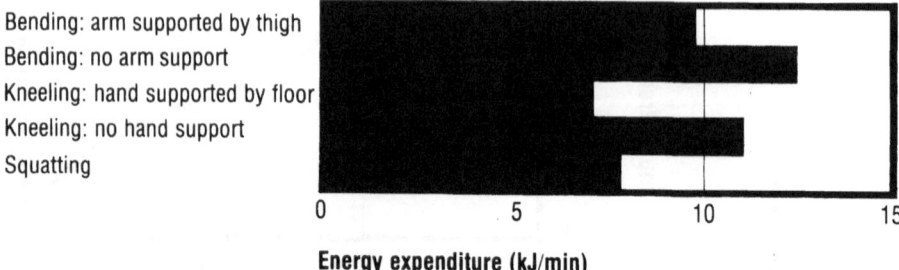

Figure 2.12 Energy expenditure for five postures used in picking up a light object from the ground (after McCormick).

The fact that the overall level of activity is low does not necessarily guarantee that small individual muscle groups will not be fatigued, as in the case of writer's cramp for example.

In summary, if there is a significant physical component in the proposed task check that the total load imposed upon the operator over time is not excessive. In doing this consider the energy cost of the main components as well as the factors described in the next section. Where the task demands local muscular work, whether static or repetitive, check that the rule of thumb limits outlined previously are not exceeded in respect of the weakest user.

2.3.3 Factors influencing body strength

The ability to perform physical work depends on a complicated interplay between both internal and external factors (Figure 2.13). Apart from physiological factors, physical performance is to a significant extent influenced by psychological factors, notably motivation, and several of these factors may be affected by training and physical fitness.

Environmental factors can also have an effect on physical work capacity. Noise, for example, may not only damage hearing but can also increase heart rate and affect other physiological parameters, so reducing physical performance. Cold may reduce physical performance because of numbness of the hands or a lowered body temperature, and workloads may also increase as a result of additional clothing being worn.

Among men, strength reaches a maximum in the middle to late 20s, remains at this level for five to ten years and then drops slowly but increasingly rapidly. Approximately 95 per cent of the earlier strength is available at about 40 and 80 per cent at age 50 or 60. Women reach their maximal strength in their early 20s and remain at this level for about ten years. The final decline in strength is steeper among women than among men. Comparing fiftieth-percentile men and women, at about age 30 a woman can exert approximately two-thirds of a man's force, whereas in their 50s they can exert only about half. However, the range of people's strength is considerable so some women are stronger than some men.

2.3 Body strength

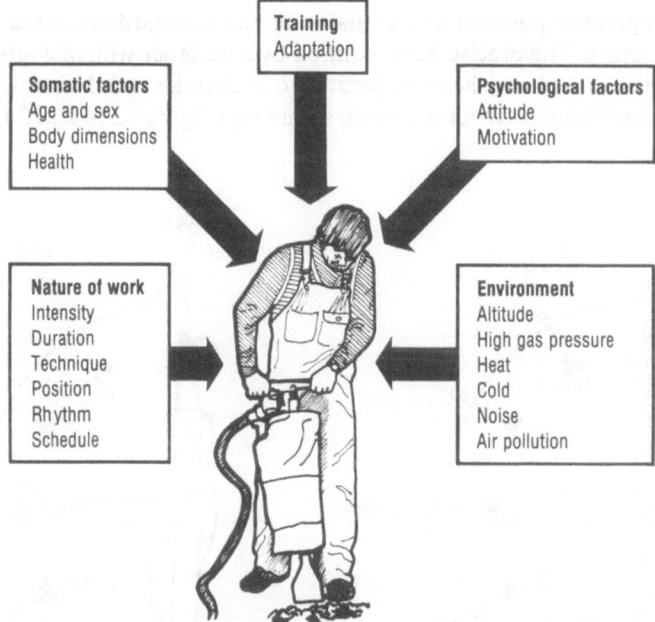

Figure 2.13 Factors affecting physical work capacity (after Astrand and Rodahl).

Working capacity similarly depends on age and sex. While ability for the short-term burst of energy declines distinctly after the 20s, ability for prolonged exertion is largely maintained into the 40s. On average the working capacity of women is two-thirds to four-fifths that of men.

Although the engineering designer may not be able to select individuals for the tasks he creates, he may be able to ameliorate heavy and medium workloads by his control over both the environment within which the work is conducted as well as in the design of the work.

2.3.4 Standard data on body strength

As with dynamic anthropometric data, detailed information on body strength is usually derived from experiments conducted on special-purpose rigs. Military interest has meant that the data tends to be biased towards fairly fit, youngish male subjects.

An interesting feature of human body strength, as compared with other physical characteristics, is its far greater range of variability. Whereas the range from the smallest to the tallest individual is of the order of thirty per cent and that between the lightest and the heaviest one hundred per cent, it is not uncommon to find a ten-fold difference in body strength from the weakest to the strongest. Clearly the engineering designer needs to be particularly cautious when applying published data on body strength, particularly to using 'average' data rather than ranges of strengths when applied to age or sex differences.

The maximum force that can be applied to a control will depend on its location with respect to the individual applying that force and thus, to a large extent, on the physical laws relating to levers, fulcrums and masses. By adjusting his posture and thus the efficiency with which he applies a force, an individual can sometimes compensate for particular weaknesses. Data for the various combinations of control and operator position are available in the standard reference books on human engineering. The precise data required by a designer will, of course, depend on the particular tasks he wishes his operator to undertake and the environment he will work in. The figures provide a general summary (Figures 2.14 and 2.15).

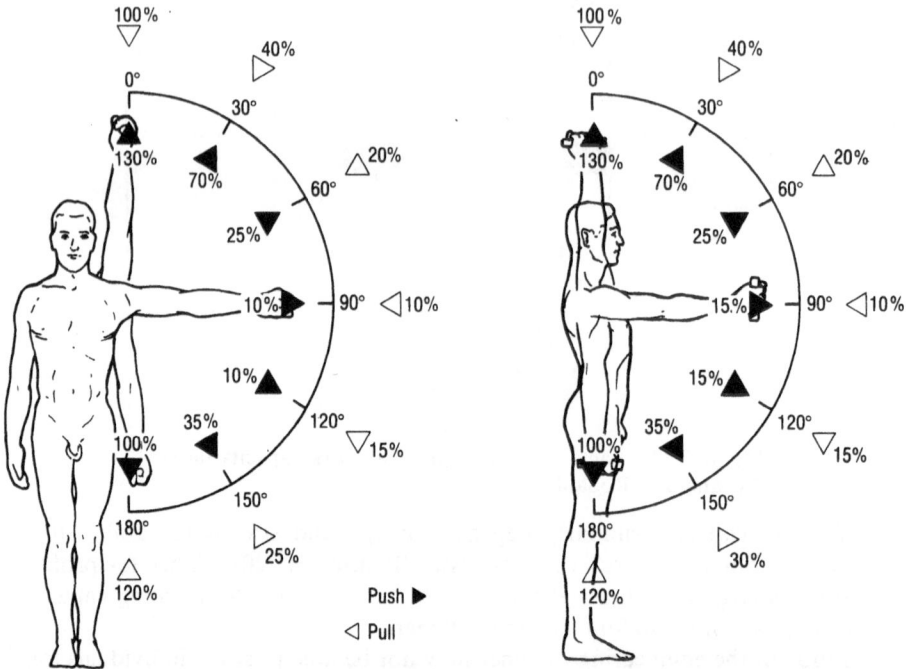

Figure 2.14 Approximate maximum possible forces exerted in standing position expressed as percentages of total body weight.

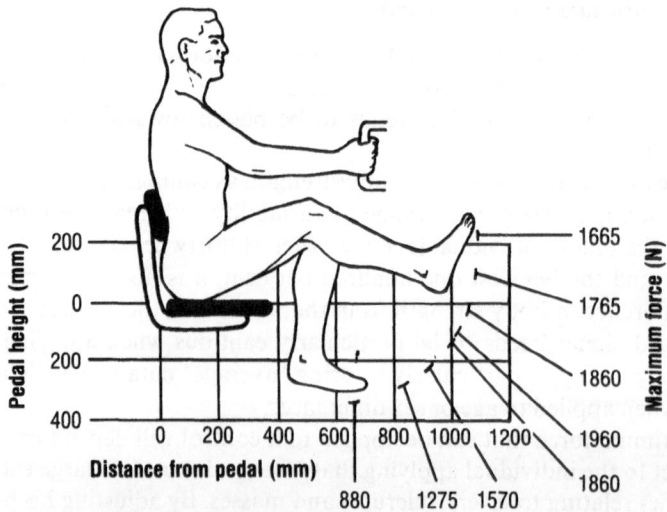

Figure 2.15 Variation in maximum possible static forces with changes in leg position.

2.3.5 Guidelines for designers: body strength

For extended period of whole-body or local muscular work, design the task so that it does not require more than 30 to 40 per cent of the predicted maximum power of the user.

Avoid static workloads. For example, the head is heavy, weighing four to five kg; work in which it is not more or less balanced on top of the spine will be more fatiguing than otherwise.

In using available data in the literature, check how relevant it is to your population of users—especially in relation to sex and age. If required use rule of thumb adjustments recommended in the text.

For maximum energy availability use joints that involve large muscle masses moving over the longest possible distance.

The application of large, steady forces depends on stability as well as on maximum muscular contractions [2.8, 2.10, 2.11, 2.12, 2.13, 2.14, 2.15].

2.4 Ageing and biorhythms

2.4.1 Ageing

The effect of ageing on body strength has already been mentioned, but age affects all our faculties (Figure 2.16) and it is salutary to remember that nearly half the working population in most developed countries is likely to be over 45 years of age. With some occupations minimum standards are set and checks are made periodically to ensure performance does not fall below them—airline pilots are a classic example. In other occupations a form of 'natural selection' occurs. For example, a study made in the engineering industry showed the more highly skilled jobs tended to be conducted by younger men, whereas labouring and store-keeping

	General effect	Specific example	Consequences
Sense organs	Decease in efficiency	From 20 years to 60 years performance drops by 60 per cent in relation to maximum	For vision better lighting needed For hearing no consequences of importance to capacity most work
Central Processes	Decreased information handling capacity		Loss of 'thread' in complex routines Decreased versatility
Physical work performance capacity	Some fall-off but not rapid until after 60 years	From 20 years to 60 years muscular strength drops by 25 per cent of maximum capacity	Increased importance of postural aspects of job design Heavy work possible but not at high speed
Compensatory Effects	Increased experience Narrowing interests		Greater range of situations already met Increased reliability

Figure 2.16 Effects of age on human performance (after Singleton).

were carried out by older men. The study suggested that the largest single determinant of age/work structure appeared to be the difficulty inherent in the job, and that by redesign of jobs in such a way that older personnel could continue in them, a substantial contribution to productivity could be made.

The working population tends to be much older than the samples from which laboratory data has been gained, so we need to build safety margins into our expectations of human performance if we are to avoid driving older workers beyond their limitations. While older men may be less able than younger men to cope with hard physical work and to work at speed, there is some evidence that they can compensate by developing 'knacks' to minimise the physical load of the task and to maintain a steady level of work with few errors. However, the lifetime of experience and skill development so important in a craftsman may be a positive disadvantage to a man expected to relearn new skills in a changing industry.

Chronic sickness and illness are also more likely to develop in the 50s and arthritis and heart conditions are the most common sources of disability in the population. A government survey in Great Britain found that over eight per cent of the population aged 50 to 64 had some functional impairment. In most developed countries firms have statutory obligations to employ people with disabilities and particularly to continue to employ employees who become disabled. The wider the range of jobs which can be adapted to their needs, the easier this will be.

2.4.2 Biorhythms

Many of us have experienced the disorientating effects of long-distance flights, where our 'internal clock' has to be nursed back to synchronise with the new local time. We have in fact underlying rhythms affecting sleep, appetite, alertness and defecation over which we appear to have no immediate control. These rhythms appear to be a fundamental feature of our biological make-up, and so far as work design and organisation are concerned they affect our physical and mental performance.

Measurements of such physiological variables as body temperature, urine volume and content and blood hormones present a common pattern when plotted over a 24-hour period (Figure 2.17).

There is a correlation between peaks and troughs in these physiological variables and levels of performance, and complications arise when we attempt to adjust our stabilised rhythms forcibly to fit somewhat arbitrarily dictated external work rest demands—such as in shift work. The body can and does adjust eventually to new work/rest schedules, but during the transition phase there are measurable physiological and psychological costs. Measurements on pilots have shown that full accommodation to major time displacement may take about six days, during which the various body cycles are seen to shift gradually so as to adjust to the new time zone.

Research into the effects of these so-called circadian rhythms on night work has suggested that production on a night shift is lower than that of the morning shift for about the first four weeks on continuous night shift; that absenteeism is higher on the morning shift immediately following periods on night shift; that process workers working on nights for a week at a time make about twice as many errors in logging at about 0.400 hours than during the morning and afternoon shifts; that the rate at which circadian rhythms adjust to a new activity pattern is likely to take place more quickly in an active job than in an inactive one; and finally that

2.5 A systems view

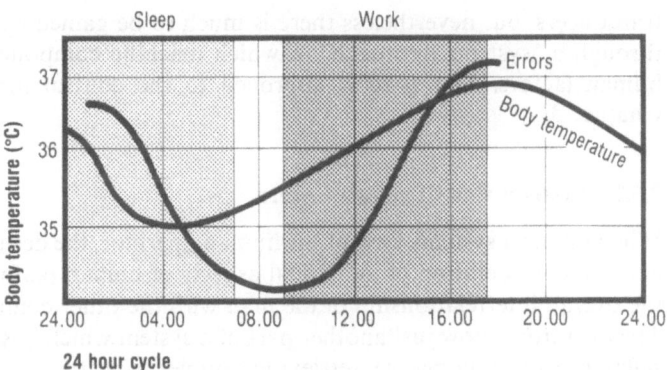

Figure 2.17 Variation in body temperature and error rate over a 24-hour cycle (after Singleton).

ageing progressively increases the strain of night work, and it is considered generally unwise to recruit workers aged 40 or 50 for night work.

2.4.3 Guidelines for designers: ageing and biorhythms

To reduce the effect of age on a particular task: reduce the required speed of work, particularly machine pacing; increase lighting levels; reduce short-term memory load; minimise strenuous movements or excessive postural changes.

Older people should not be expected to acquire habits that are in any sense a reversal of habits already established.

The speedier, more complex and difficult the motor task, the more performance is likely to be affected by age.

Older workers in particular will be adversely affected by technological change since they can no longer depend on their accumulated experience to compensate for the effect of ageing.

Experience enables older workers to maintain performance provided that they are working on a task with which they are familiar.

Job redesign may increase the performance differential between the older and younger worker due to the greater difficulty the older worker has in adapting to the new methods.

Jobs which could be done from adjustable seating with controls usable without fine finger grip dexterity and with well-designed large print labelling are more likely to be potentially available to disabled operators.

Workers aged over 40 or 50 should not be recruited for night work.

If the equipment is likely to be operated by men suffering from the effects of sleep disturbance, requirements/demands on operators should be fixed at a lower level [2.10, 2.16, 2.17, 2.18, 2.19, 2.20, 2.21, 2.22].

2.5 A systems view

2.5.1 Introduction

It was one of the pioneers of ergonomics who first drew up a table comparing the physical and mental performance of man with the machine equivalents, thereafter called a 'Fitts' list after its originator. It is possible to consider man as just another element in a system—a 'black box' with specific input and output characteristics. Of course man is much more complex than a bundle of relays, switches and transducers, but nevertheless there is much to be gained by the perspective given through a 'systems approach' in which machine components are matched with human factors. The systems approach to the control function is discussed in Chapter 3.

2.5.2 Systems view of human factors

When taking a systems view of the human operator, the designer steps back from a detailed consideration of individual aspects of human performance and examines the overall interrelationship of the man with the other components of the system. The operator is now just another part of a system which has defined goals, such as safely transporting people, retrieving samples from the moon, health care and so on. In some cases it may be clearly impracticable to use a man, for instance in the direct manipulation of radioactive samples; in other cases the system could operate either with a man or automatically—say in undersea mining operations or space travel—in which case the pros and cons of including the human operator need to be carefully weighed up.

Any man-machine system will only be effective if all its components are matched together and interact in appropriate ways. Machines designed without due regard to the mental and physical capabilities of those who are to use, control and monitor them are unlikely to be well designed. System design differs from engineering design in the importance it attaches to the human operator as an integral part of the system.

Systems can be broadly characterised by the degree of human versus machine control. *Manual systems* consist of hand tools and other aids that are coupled together by the human operator who controls the operation using his own physical energy. The operator transmits to and receives from his tools a great deal of information and his task is typically self-paced. *Mechanical systems*, such as powered machine tools, consist of well integrated physical parts designed to perform their task with little variation. The power is usually provided by the machine and the operator's function is essentially one of control. Individual mechanical systems can be linked together to form the much larger assembly line. In these systems the operator receives information about the state of affairs of the system via displays, performs an information-processing and decision function, and implements the decisions by the use of control devices.

With a *fully automated system* all operational functions, including sensing, information processing, decision making and action, are performed by the machine. Such a system needs to be fully programmed for all anticipated combinations of sensory input. In theory such systems do not need the intervention of the human operator, but in practice the human functions in such systems are monitoring, programming and maintenance.

An approach to identifying the role of the human operator in a system has been put forward by Singleton, who suggests that, to stimulate flexibility of innovation, systems need to be conceptualised in an abstract way with functional concepts separated from physical realisation (Figure 2.18).

Fitt's criteria for the allocation of tasks in military design were based on the relative abilities of men and machines. After 1950 the complexity of weapons systems increased to the point where cost became of critical importance and a new criterion of the cost/value function was added. When the concepts spread into the industrial field around 1960, two further criteria were added; the first was the need

2.5 A systems view

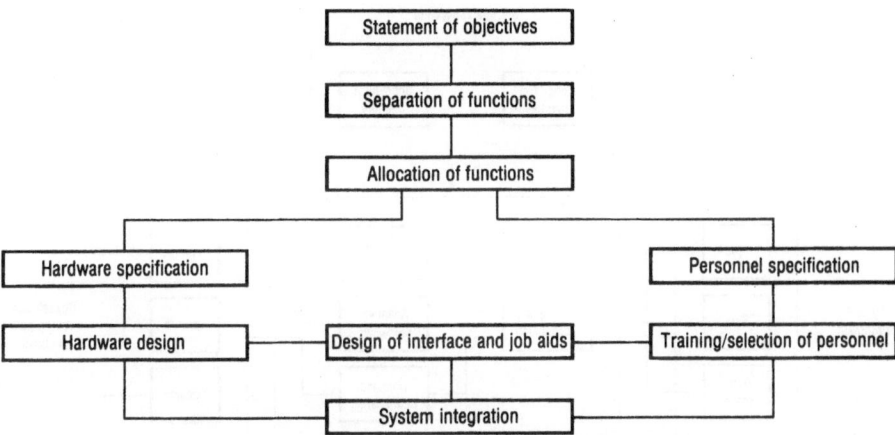

Figure 2.18 The systems design process (after Singleton).

for integrated tasks that adequately utilised the abilities of the human operator while at the same time making allowance for his limitations; the second was the need in large systems for graded tasks matched to the differing levels of ability and seniority to be found in every organisation.

Today it is increasingly man's psychological limitations that need to be borne in mind in allocating functions, together with his needs for an adequate social environment at his workplace. Boredom and lack of attention may be the new hazards for systems designers to deal with.

2.5.3 General model of man as an information processor

Although it may not appear so at first sight, the operators in an overhead crane, a railway cab and on a ship's bridge may all be considered as components of a closed-loop control system, with the human operator as an integral part of the loop (Figure 2.19).

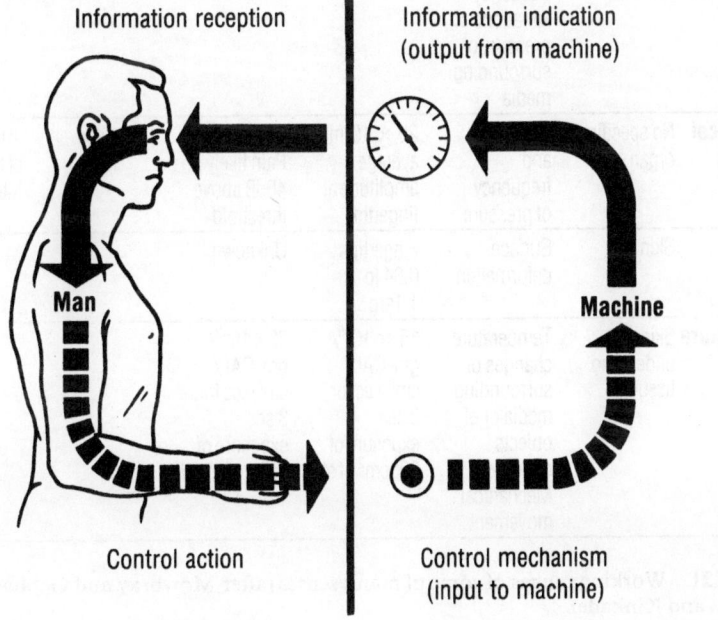

Figure 2.19 General relationships in a closed-loop control system (after Shackel).

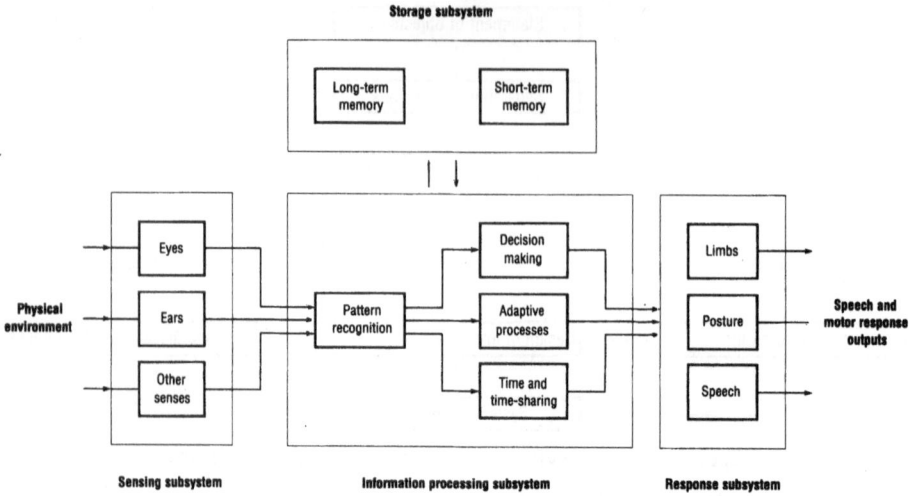

Figure 2.20 The human information-processing system (after Van Cott and Kinkade).

Sensation	Sense organ	Stimulation	Smallest detectable (threshold)	Largest tolerable or practical	Frequency – Sensitivity	
					lower limit	upper limit
Sight	Eye	Some electro-magnetic waves	10^{-6} ml	10^4 ml	300mμ	1500mμ
Hearing	Ear	Some amplitude and frequency variations of pressure of surrounding media	2×10^4 dynes/cm^2	$<10^3$ dynes/cm^2	20Hz	20,000Hz
Mechanical vibration	No specific organ	Amplitude and frequency of pressure	25×10^5 mm average amplitude at fingertip	Varies. Pain likely 40dB above threshold	Unlimited	10,000Hz at high intensities
Touch	Skin	Surface deformation	Fingertips 0.04 to 1.1erg	Unknown		
Temperature	Skin and underlying tissue	Temperature changes of surrounding media or of objects contacted. Mechanical movement	15×10^{-5}/ gm-CAL/ cm^2/sec for 3 sec exposure of 200cm^2 of skin	22×10^{-2}/ gm-CAL/ cm^2/sec for 3 sec exposure of 200cm^2 of skin		

Figure 2.21 Working ranges of some of man's senses (after Mowbray and Gebhard, and Van Cott and Kinkade).

2.5 A systems view

Developing the concept further, we can draw up a simple block diagram (Figure 2.20) in which the process which we might expect to occur *within* the operator while he performs his control task is presented. This model, and the three major sub-systems of sensing, information processing and response, are the main bases for structuring the remainder of this section.

2.5.4 Sensing sub-system

Our brains receive information about the body and its environment from a number of different types of nerve receptors. Individual receptors are sensitive to specific types and ranges of physical energy. The senses we use are not restricted to sight and sound, although these are particularly highly developed; they also include rotation, taste, smell, touch, pressure, temperature, pain, vibration and kinaesthetics, the last being an internally generated stimulus which enables us to be aware of the position and movement of our limbs. As with electrical transducers, stimulation beyond the working range leads to unreliable sensing, with missed signals at one end of the range and pain or permanent sensory damage at the other (Figure 2.21).

As the magnitude of a physical stimulus changes along a given dimension, the magnitude of the corresponding sensation also changes. The relationship is not linear, however, and equal increments in stimulus magnitude do not produce equal increments in sensation over the entire range of stimuli. For example, doubling sound energy does not double perceived loudness; a tenfold increase is required for a sound to seem twice as loud.

Also, while sensation along a given stimulus dimension varies primarily with that dimension, it is affected by changes along other dimensions of the same

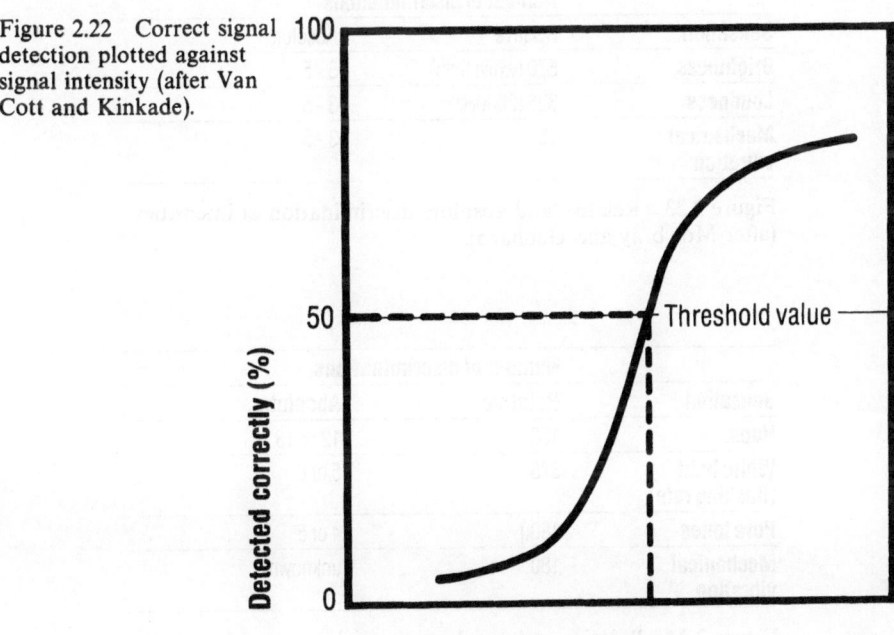

Figure 2.22 Correct signal detection plotted against signal intensity (after Van Cott and Kinkade).

stimulus. For example, loudness of a tone varies with its frequency as well as with its amplitude, and the apparent brightness of a light varies with the colour as well as with the radiant energy of the source.

Lower thresholds of sensitivity to stimuli are not really precise values, but rather the average, mean or median of a statistical distribution of energy values. Typically the threshold is defined as that stimulus value which, under ideal conditions, is detected approximately 50 per cent, or in some cases 75 per cent, of the time (Figure 2.22). In practical applications much higher values are required.

Man's ability to make absolute and relative judgements varies enormously. For example, a paint colour can be matched very much more accurately by comparison with a control sample than by relying on memory of the colour. The smallest change in one stimulus with respect to another that will result in a 'just-noticeable difference' can be established by careful measurement. Obviously the just-noticeable difference will vary in that we would need a much greater incremental change in stimulus level if we already have a strong light source than if we have a weak one. Fortunately it has generally been found that, over the useful range of some stimuli, the increment required bears a constant relationship to the magnitude of the reference stimulus. In other words, as the base or reference stimulus increases, larger increments in stimulus magnitude are required for detection.

In some dimensions the difference between relative and absolute discrimination is tremendous. In colour selection, for example, it has been estimated that about 10,000 times as many relative discriminations can be made compared with absolute ones. Figures 2.23 and 2.24 show discriminative ability for varying intensity and frequency.

	Number of discriminations	
Sensation	Relative	Absolute
Brightness	570 (white light)	3 - 5
Loudness	325 (2000Hz)	3 - 5
Mechanical vibration	15	3 - 5

Figure 2.23 Relative and absolute discrimination of intensities (after Mowbray and Gebhard).

	Number of discriminations	
Sensation	Relative	Absolute
Hues	128	12 or 13
White light (flashing rate)	375	5 or 6
Pure tones	1800	4 or 5
Mechanical vibration	180	unknown

Figure 2.24 Relative and absolute discrimination of frequencies (after Mowbray and Gebhard).

It must, of course, be remembered that there are certain stimuli for which man has no sense organs. Radiation and humidity are examples where, if information on levels are required, the physical stimulus needs to be transformed to produce a form of stimulation which man is sensitive to, for instance, by visual or auditory displays.

Before leaving this area brief comment needs to be made about commonly found sensory disabilities, illusions and fatigue. The major handicap imposed by a total loss of sight and hearing needs no emphasis, but the designer in specifying his displays needs to bear in mind that his operators are likely to exhibit such commonly found defects as short and long sight, various degrees of colour blindness, tunnel vision and impairment of hearing due to old age. Defective colour vision, for example, is exhibited in various forms by six per cent of the male population. Unless users are to be subject to rigorous selection procedures and regular testing of their sensory functions, as are airline pilots, it is advisable to use available data with caution and build in a safety margin wherever practicable. Environmental conditions and protective equipment can also reduce and distort the information received.

Unfortunately for systems designers, man can also quite easily be fooled into seeing things that are not there and not perceiving things that quite definitely are there. Most of us have come across the more common types of visual illusion, and know perfectly well that certain lines are of equal length and that what 'looks' like a bent ruler is in fact straight, but that is not what we actually see (Figures 2.25 and 2.26). Similarly, how often has one sat in a room in which there was a clock with a distinctly audible tick and after a while found oneself consciously making an effort to hear the ticking?

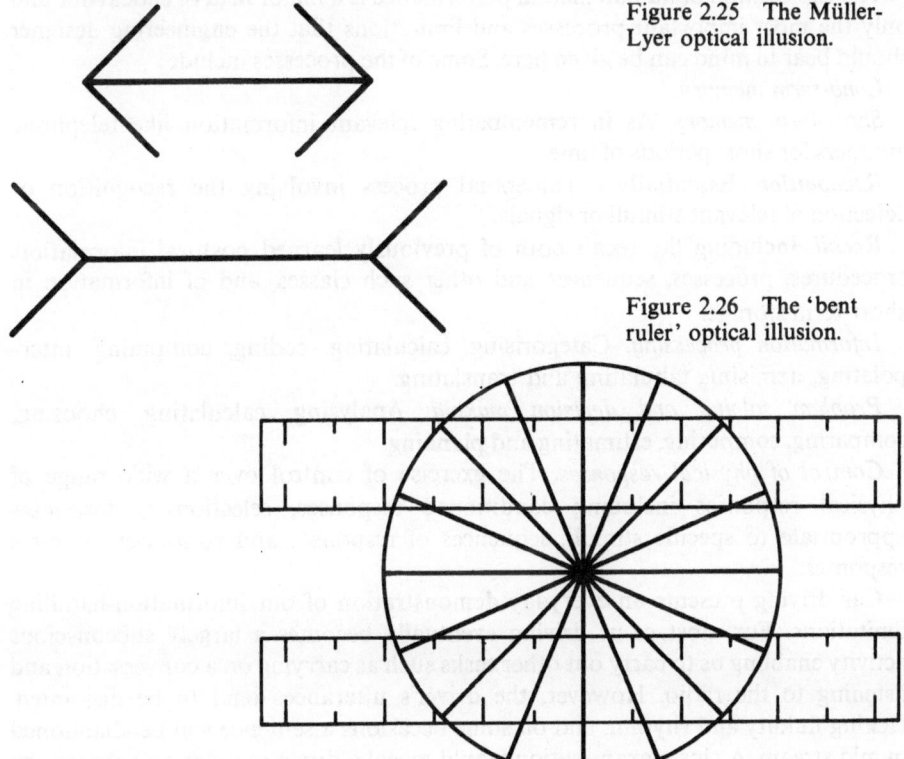

Figure 2.25 The Müller-Lyer optical illusion.

Figure 2.26 The 'bent ruler' optical illusion.

Extreme fatigue can also play havoc with our senses, producing missed signals or 'mirage' signals produced with no stimuli. Lack of stimuli also creates similar effects, as reported in extreme forms by prisoners in solitary confinement. Such effects may have significant implications for the vigilance of controllers of fully automated systems waiting only for very rare events such as major breakdowns, for example some airline pilots. Additional 'unnecessary' tasks may need to be included or a lower level of automation used to enable the critical job function to be safeguarded.

In summary, great inroads have been made in mapping and quantifying man's mechanisms for sampling his environment. However, his mechanisms do not obey the straightforward laws of physics associated with electrical transducers and care needs to be taken that, at the very least, stimulus ranges are selected that fall well within his bounds of sensitivity and that he is not expected to undertake tasks that are beyond his known capabilities.

2.5.5 Information processing sub-system

Between seeing a red stop light and taking the first externally recognisable action, a car driver will have rapidly gone through a number of processes—interpretation of the signal, checking against previous experience, deciding upon appropriate action and programming muscles to make the appropriate physical response. The various processes that have occurred in the period between the viewing of the red light and the start of the response are of immense importance to the human factors engineer, especially where he must match external system demands to performance limitations.

The exploration of human mental performance is a major field of endeavour and only the most important processes and limitations that the engineering designer should bear in mind can be given here. Some of the processes include:

Long-term memory.

Short-term memory. As in remembering relevant information like telephone numbers for short periods of time.

Recognition. Essentially a perceptual process involving the recognition or detection of relevant stimuli or signals.

Recall. Including the recall both of previously learned postural information, procedures, processes, sequences and other such classes, and of information in short-term storage.

Information processing. Categorising, calculating, coding, computing, interpolating, itemising, tabulating and translating.

Problem solving and decision making. Analysing, calculating, choosing, comparing, computing, estimating and planning.

Control of physical responses. The exercise of control over a wide range of physical responses including conditioned responses, selection of responses appropriate to specific stimuli, sequences of responses and continuous control responses.

Car driving presents an everyday demonstration of our information-handling limitations. For most of us driving eventually becomes a largely subconscious activity enabling us to carry out other tasks such as carrying on a conversation and listening to the radio. However, the driver's utterances tend to be disjointed, lacking fluidity and rhythm, and on some occasions a sentence will be abandoned in mid stream. A closer examination would reveal a direct correlation between the

2.5 A systems view

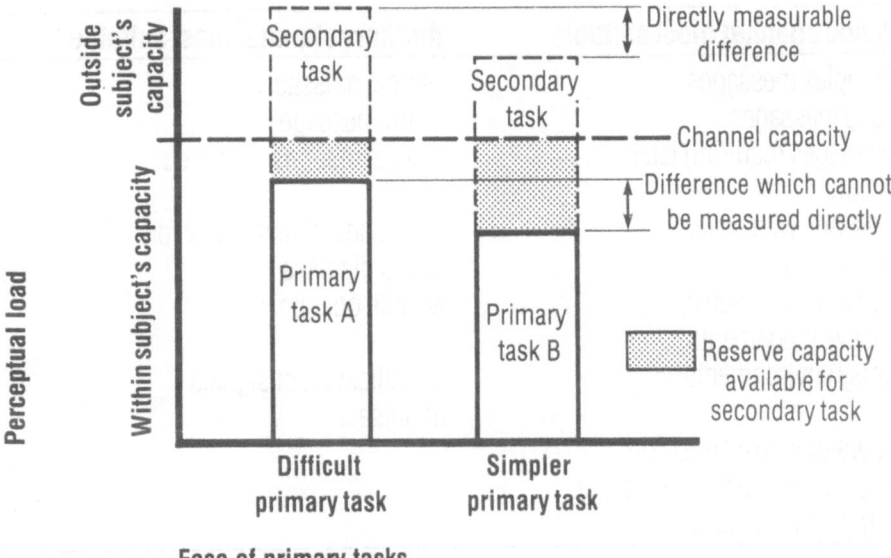

Figure 2.27 Use of a secondary task to occupy reserve mental capacity (after Rolfe).

complexity of the driving task for the driver and the quality of his speech; the easier the driving task, the more satisfactorily the secondary task of holding a conversation can be carried out. The measurement of performance on a simple secondary task has in fact been used by experimenters to gain some insight into the load imposed by the primary task (Figure 2.27).

As a gross simplification, one may view the human operator as a 'single channel' data-processing system with a central decision mechanism that must be allowed a finite time to process one stimulus/response before a second can be accepted. Although an operator cannot deal with more than one channel simultaneously, a limited amount of time-sharing can occur between sensory inputs. Of the two most important senses (vision and hearing), vision is the best sense to deal with complex stimulation since visual attention is under the direct control of the individual. Many things may be, and usually are, going on in the peripheral field of vision without intruding upon the central task upon which one is concentrating. Unless they are somehow important to the central task, these peripheral events will be ignored, and to all intents and purposes they do not exist. On the other hand, if peripherally occurring events are important to the central task, they may be taken in in quickly successive fixations. The auditory system is not similarly constituted and therefore offers some special problems in the realm of multiple inputs.

The choice of the communication channel between man and machine thus depends upon the type of information that is to be transmitted, the way it is to be used, the location of the man, the environment in which he operates, and the nature of the sense organ itself. For example, our ears, unlike our eyes, can receive information from all directions. Moreover, we cannot close our ears the way we close our eyes. These properties make the ears well suited to the reception of emergency warning signals. Other criteria can also be used to select between visual and auditory displays (Figure 2.28).

Visual channel most suitable	Auditory channel most suitable
Complex messages	Simple messages
Long messages	Short messages
Messages requiring later referral	Fastest response required
Message with spatial information	Environment unsuitable for visual displays
Information based upon relative judgements	Mobile operator
Noisy environments	Identification of signals in noise
Operator in fixed position	
Requirement to monitor range of different displays	

Figure 2.28 Comparison of suitability of visual and auditory channels (after Chapanis).

One of the questions most frequently asked of a human factors engineer is, if a single stimulus dimension, such as hue or brightness, is to be used to code and display information, into how many distinct steps or levels must it be divided for each level or step to convey a distinct meaning to an observer. A number of investigators have measured the amount of information transmitted by both unidimensional and multidimensional stimuli—as for example a combination of hue and brightness. The amount of information is frequently expressed in terms of 'bits'—the logarithm to the base two of the number of equally likely alternatives. The number of bits is equal to the number of two-choice discriminations required to specify a particular event from alternative ones.

The ability of people to make absolute discriminations among individual stimulus dimensions has already been shown to be not very large—typically seven, usually between four to nine. This can also be presented as a corresponding number of 'bits' from about 2.0 to 3.0 or 3.4; the greater the number of absolute judgements that can be made along a single dimension the more bits are transmitted. With combinations of diversions the information that can be transmitted is noticeably greater (Figure 2.29).

Several points of interest are revealed by a comparison of channel capacities of different senses. First, the channel capacity of vision is higher than that of other senses. Second, within a given sense, different stimulus dimensions are associated with different capacities for transmission. The greater the range between the upper and lower detection thresholds for a given sensory dimension, the greater the capacity of the sense organ to transmit information along that dimension. Third, the number of absolute judgements that can be made along any one dimension varies considerably from sense to sense and among stimulus dimensions. For example, while the eye can reliably identify only five different levels of brightness, it can identify nine different colours. In practical terms this means that colour codes can convey more information than brightness codes. Although the possibilities of multidimensional coding are attractive, if errors are to be minimised any codes used must be well within quoted maximum discrimination thresholds.

2.5 A systems view

Sense	Stimulus dimension	Discriminable categories	Channel capacity (bits)
Vision	Size of squares	5	2.2
	Dominant wavelengths (hue)	9	3.1
	Luminance	5	2.3
	Area	6	2.6
	Line length	7-8	2.6-3.0
	Direction of line inclination	7-11	2.8-3.3
	Dot position (in space)	10	3.2
	Size, brightness & hue (varied together)	18	4.1
	Position of dot in square	24	4.6
Audition	Intensity	5	2.3
	Pitch	7	2.5
	Loudness & pitch	9	
Vibration (on chest)	Intensity	4	2.0
	Duration	5	2.3
	Location	7	2.8
Electrical shock (skin)	Intensity	3	1.7
	Location	3	1.7

Figure 2.29 Amount of information in absolute judgements of various stimulus dimensions (after Van Cott and Kinkade).

It should also be remembered that such codes do not necessarily increase the speed or rate with which information is transmitted through the operator, and in many systems operator response time can be of critical importance. People cannot respond immediately to a triggering signal and, though the delay can be minimised under certain conditions, it is nevertheless an important parameter to bear in mind when considering human operator performance. Typically at least a quarter of a second may elapse before any physical response to a signal *starts*; a significant length of time when driving a train at 200 km/h. Some of the factors affecting human response time include the sense used, the characteristics of the input signal, signal rate, whether or not anticipatory information is provided, the response requirements of the task, individual differences and experience.

2.5.6 Response sub-system

In mechanical and automated systems, most motor responses involve the use of control devices such as levers, cranks, push-buttons and pedals. The use of speech, say for air traffic control, must also be considered.

It is interesting to examine the published records on maximum manual and electric typewriting speeds in this context (Figure 2.30). We would expect that the greater physical effort required to depress manual typewriter keys would have a limiting effect at high speed. When we examine some of the available data and compare the number of words typed over just a single minute, the records do show that the electric machine is considerably faster. Far more revealing than the results of an instantaneous burst of activity over one minute, however, is the more reliable

	Maximum rate over one minute (WPM)	Average rate over one hour (WPM)
Manual	170	147
Electric	216	149

Figure 2.30 Comparison of maximum typing speeds on manual and electric typewriters.

measure of words per minute averaged over an hour's typing. Here we see no significant difference, and can deduce that the real limiting condition to typewriting speed is the rate at which information can be transmitted by the human response system.

Information analysis techniques have also been applied to human motor activities. For most practical situations, the rate at which information can be transmitted by the response system is assumed to range from less than three bits per second to a maximum of nine bits per second, depending on the task, the organisation of the information and the readout elements involved. For verbal responses a maximum rate of approximately 7.9 bits per second has been suggested while for key pressing the maximum rate is about 2.8 bits per second.

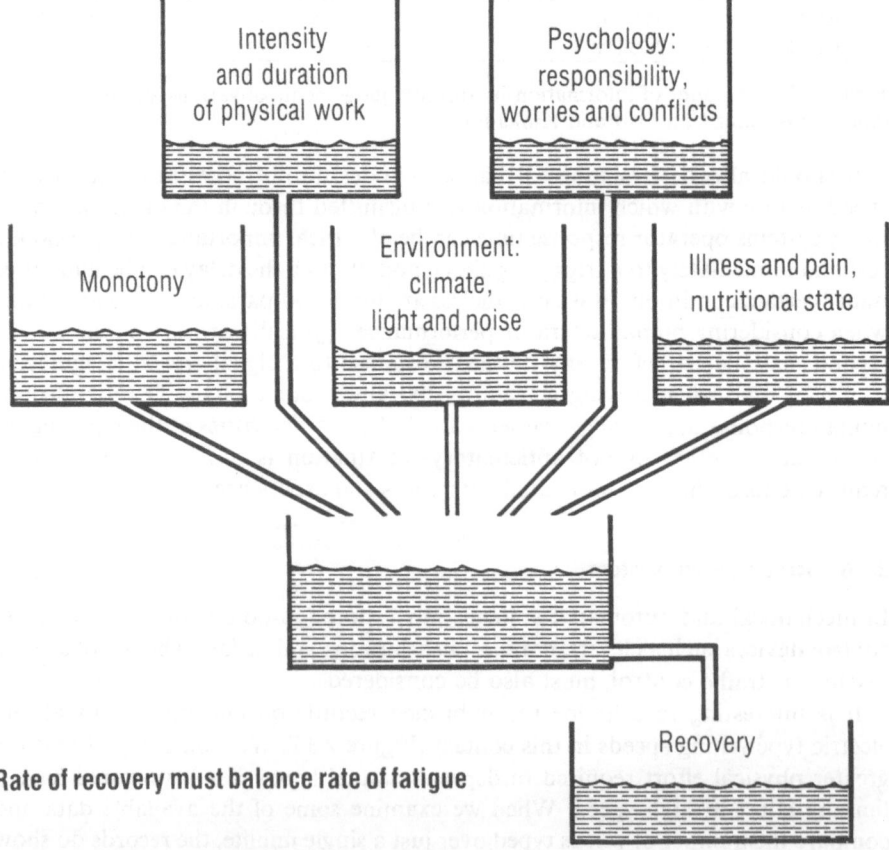

Rate of recovery must balance rate of fatigue

Figure 2.31 Model of fatigue (after Grandjean).

As with any other aspect of human performance, however, it is quite unrealistic to consider human motor output just in terms of mathematics and information transmission. These may give us some idea of the maxima that may be achieved but in practice performance at the workplace will be affected by a myriad of interacting factors, including fatigue (Figure 2.31).

In most instances we are quite unable to quantify the various parameters that affect motor output. This should not stop us, however, attempting to design tasks that can be reasonably expected to be within the grasp of the average operator.

2.5.7 Guidelines for designers: man as a systems component

Man's sensitivity ranges are limited 20 to 20 000 cycles per second in the sound spectrum and 4000 to 7000 angstrom units in the electromagnetic spectrum.

Man's input capacity is easily saturated and care should be taken to avoid messages that are overlapping, competing or visually and aurally incompatible.

Man needs a finite time to respond, typically greater than 250 ms.

All of man's outputs are motor responses and these are relatively slow and low powered.

Man has a good long-term memory. His access time is slow, compared with that of a computer, but he is able to recall generalised patterns of previous experience to solve immediate problems.

Man's performance may tend to deteriorate as a function of time on the job. This is seldom a result of physiological tiring per se but is likely to result from boredom, inattention and lack of motivation.

Care should be taken in designing the man's job so that he is not forced to operate near his maximum load limits for very long or at too low a level of activity [2.3, 2.8, 2.20, 2.23, 2.24, 2.25, 2.26, 2.27, 2.28, 2.29].

2.6 A general strategy for tackling ergonomic problems

2.6.1 Introduction

An ergonomic problem exists whenever there is a mismatch between the demands imposed upon the human operator by an item of equipment, the environment, or a combination of both, and the individual's ability to meet the requirements. The issue is made more complex by variations between individuals, and within an individual over time. As we reduce the demands of a task, so the proportion of people who will succeed will increase, although there may be some people for whom a task will always be insurmountable.

There are no standard problem-solving prescriptions for such ergonomic problems but this section outlines briefly a strategy for identifying and resolving them.

2.6.2 A general strategy

The flow chart (Figure 2.32) outlines the procedural stages suggested. Stage 1 is identification and checking, Stage 2 relates to specification of the ergonomic problems in the form of a brief, Stages 3 and 4 are procedures that would be appropriate for tackling the ergonomic problems specified.

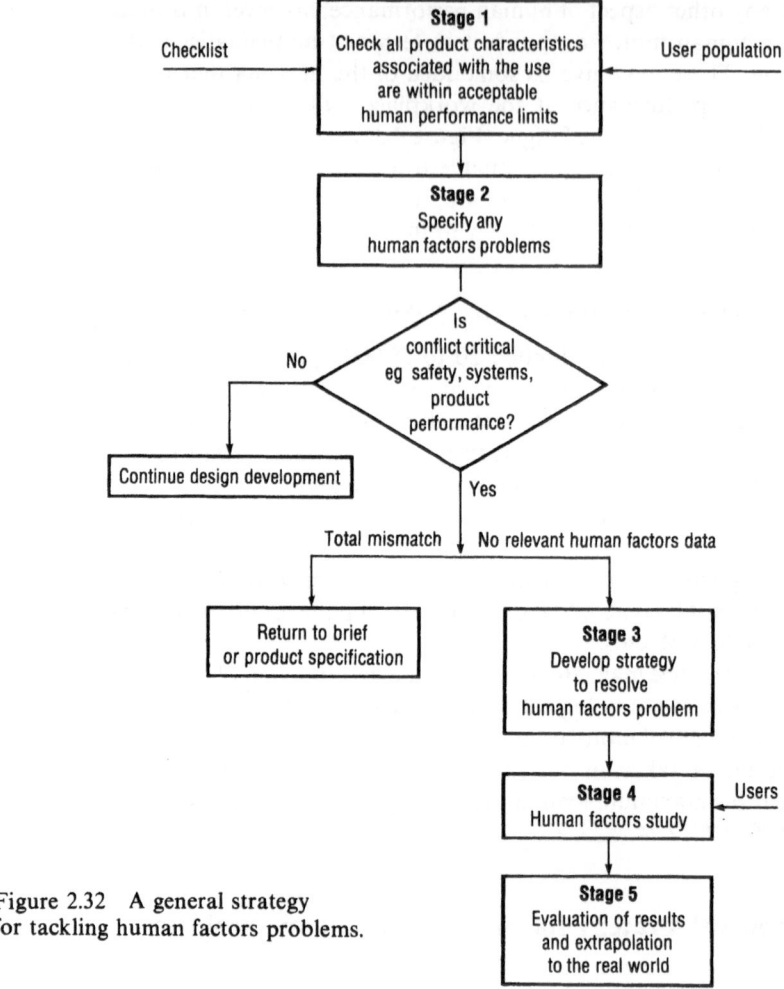

Figure 2.32 A general strategy for tackling human factors problems.

The first stage is then to identify human factors problems and establish whether the demands of the task fall within the range of human performance. If they do not, re-design is necessary to avoid the need for human intervention or to reduce the demands to ones that can be met.

The checklist (Section 2.6.5) offers a useful prompt for helping the designer look at features of his product in terms of human factors conflicts. The earlier problem identification can occur the better; potentially irreconcilable human factors problems are best identified at the briefing or at the outline design stage when flexibility still exists. The designer is recommended to run through the checklist and tick off all those items that are relevant to his design. Next these ticked-off items should be re-examined and those factors that are particularly critical to safety or operational efficiency sifted out. Either from his knowledge of the operation of the system or product, or by questioning and observing the potential users, the engineering designer will be able to identify the demands he will place upon the operator; a check with the human factors literature will reveal whether such demands fall within or outside typical operator performance. Should there be no data a problem obviously arises over whether the direct incorporation of a particular product characteristic or system feature will jeopardise system efficiency

2.6 A general strategy for ergonomic problems

or operator safety. At an early design stage it may be possible to 'design out' the particular demand placed upon the operator, thus avoiding the problem—for example, replacing an impossible visual monitoring task with an auditory alarm system. Where this cannot be done, answers need to be sought on human

Product/system demand			Human performance			Importance to system efficiency/safety		
Specific feature	Estimated Min	Max	Minimum acceptable	Comfort range	Maximum acceptable	Critical	Important	Not important
Response-time						√		
Environmental temp							√	
Weight							√	
Reach to control						√		

Figure 2.33 Interrelating product characteristics and human performance.

performance in relation to the particular product feature concerned. Figure 2.33 suggests one way in which a systematic examination of the implications of various product features can be explored and, prepared in this format, tables can act as a specification of the character of the ergonomic problems to be solved. The importance of conducting this type of analysis is to ensure that the man/machine interface will be adequately considered. Section 6 of the Health and Safety at Work Act obliges manufacturers 'to ensure so far as is reasonably practicable, that the article is so designed and constructed as to be safe and without risks to health when properly used' and 'to carry out or arrange for the carrying out of such testing and examination as may be necessary for the performance of the duty imposed upon him . . .'

It is relatively unusual for users of a product or system to be randomly drawn from the entire population, and even with widely used consumer products the actual range of users will usually be biased by sex of users, racial groups, intelligence or age. Provision to accommodate *all* users would be extravagant if only part of the population need be catered for. If, for example, an agricultural machine is to be solely marketed in a country where only men will ever operate it, to design a cab to accommodate from a fifth percentile woman to a ninety-fifth percentile man is clearly irrelevant. Conversely, the danger of designing a cab for men only when it may also be used by women is even more dangerous; it is clearly very important to define the user of the product or the system.

2.6.3 Resolving human factors conflicts

Assuming that the product feature for which there is no guiding human factors data cannot be 'designed out', and furthermore is critical to safety and/or performance, what then? Systems are nowadays far too complex, and the financial penalties too great, for a guess to suffice. Extrapolation from existing human factors data may be possible, but only by an expert and still with an element of risk. Should no other avenue be available, it may prove necessary to commission or conduct a limited experiment to resolve the issue.

Type of model	Source of data	Typical use
Activity flowchart 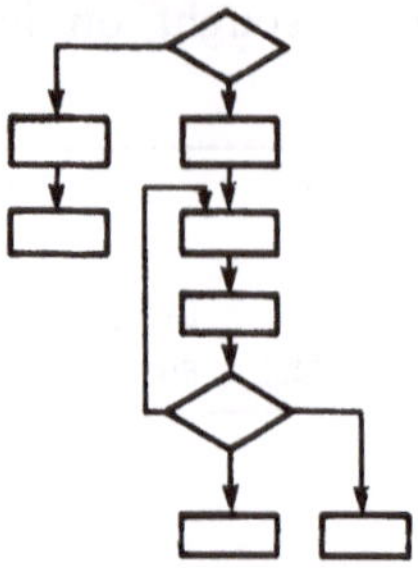	Photography Direct observation Interviews Simulation of system/ product use	Redesign of control panels Team usage of equipment or system
2-D mannekins 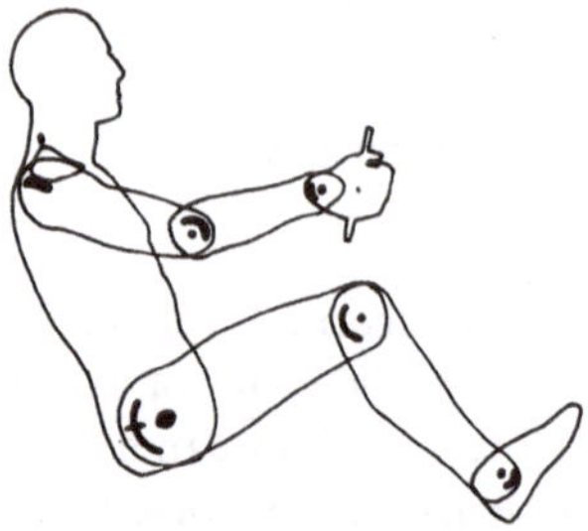	Anthropometric data in literature or bought off the shelf	Workplace layout
2-D mock-ups 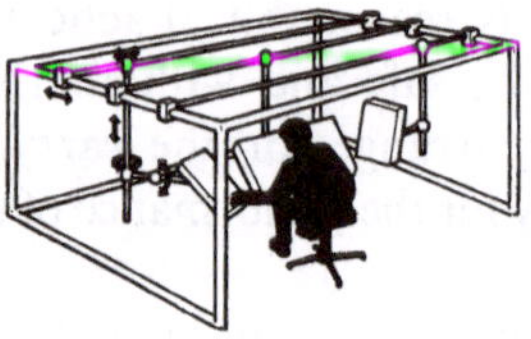	Equipment specifications	Workplace layout: anthropometric performance Workplace layout: physical workload Simulation of product usage
Matrices 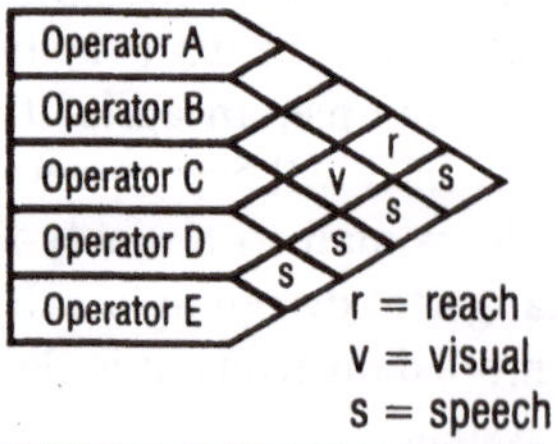 r = reach v = visual s = speech	Interviews Direct observation Analysis of predicted usage of system/product	Plotting communication links between man and machine or man and man Identification of conflicting tasks/overloads
Mathematical	Various	Models of physiological and psychological measures of performance

Figure 2.34 Modelling techniques relevant to human factors design.

Accuracy in human factors experimentation is expensive. It is, therefore, vital at the outset to define in the brief the quality of the result required; upon this will depend the experimental techniques selected, the numbers of subjects used and the scope of any trials. It is quite possible to employ a range of very different measuring techniques when examining human responses to a situation or product. For example, the comparison of different levels of physical stress imposed upon an operator by an item of equipment could be measured by subjective assessment based upon a questionnaire. On the other hand, physiological measurements might be made using measures of oxygen consumption, heart rate, muscle activity and an analysis of blood or urine samples. Each technique has its advantages and drawbacks from the point of view of relevance to the actual task, acceptability to subjects, ease of administration, ease of analysis for results, sensitivity, equipment and technical back-up required. Often a variety of techniques should be brought to bear upon the problem. Some common modelling techniques are shown in Figure 2.34 with their uses.

2.6.4 Guidelines for designers: tackling ergonomic problems

Make sure that you have correctly identified *all* critical product-system features and any variation they may exhibit.

Talk to potential users and wherever possible examine equivalent products and systems in use.

Prepare a specification of human factors problems, giving the order of critical factors identified in terms of their relative importance in respect of safety and system-product function; and stipulate whether high, low or medium-quality data will suffice for their resolution.

Check that the subjects and any trials are *realistically* representative of the range of potential users of the product or system.

Ensure that the measurements taken off the subjects are relevant and will result in useable guidance for setting the system-product parameters.

Through short *pilot trials* check that the trials techniques proposed are viable, measurements can be taken and results derived.

Design any trials in such a way that as little *redundant* data as possible is generated and so that that which is to be used can *easily* be analysed.

2.6.5 Checklist

This checklist is adapted from one published by the International Ergonomics Association. The questions are structured in the following manner:

Workspace: physical demands
Workspace: mental demands
Work method: physical demands
Work method: mental demands
Environmental load

Some of the factors discussed will be relevant to issues raised in later chapters concerned with ergonomics.

Workspace: physical demands

1. Is the workspace sufficiently spacious?
2. Is a favourable work posture promoted by the location of instruments, work pieces and controls?
3. Is sitting promoted by the location of instruments, work pieces and controls?
4. Is the height of the work surface adapted to the posture, and correct in regard to viewing distance?
5. Are the properties of the work surface correct in regard to:
   —hardness?
   —elasticity?
   —colour?
   —smoothness?
   etc
6. Is correct control by hand or feet promoted by the location of instruments, work pieces and controls?
7. Are the shape, size, surface and material of the hand controls correct in regard to the required forces?
8. Are chairs and supports available to obviate unnecessary standing?
9. Is a foot rest necessary?
10. Is a support for elbows, forearms, hands, back necessary?
11. Can the speed of the machine be adjusted in accordance with the skill of the performer?
12. Has the variability of operator performance time been taken into account?
13. Does the machine construction allow for good maintenance and repair (accessibility, risk of accidents, lighting, tracing of technical troubles)?
14. Is there any risk of burns?
15. Are any parts of the body exposed to undue constant or intermittent mechanical pressure?
16. Does the work require the use of personal protective devices:
    —clothing?
    —shoes?
    —gloves?
    —eye protection?
    —ear protection?
    —masks?
17. Does the machine cause significant vibration? If so, how will this affect operator performance?

Workspace: mental demands

18. Does the work imply high visual demands?
19. Does the work require a high illumination level?
20. Is general artificial illumination necessary?
21. Is local illumination necessary?
22. Does the work layout imply exposure to different illumination levels?
23. Are the visual data easy to discriminate in regard to variable daylight, reflection, etc?
24. Is there any glare from the workspace or surroundings?
25. Are there any special requirements in regard to colour perception?
26. Does the location of instruments, work pieces, controls etc aid good vision?

27 Are the controls located in front of the worker within optimal visual area and reach?
28 Are dials/displays used? Are they legible?
29 Are warning lights attention-getting and are they placed in the central part of the visual field?
30 Is the use of optical aids required?
31 Does the work imply high auditory demands?
32 Does the risk imply verbal communication?
33 Is normal verbal communication impeded by the noise level at the workspace?
34 Can auditory signals easily be distinguished from the normal noise of the workspace?
35 Does the task require a reduced noise level?
36 Can auditory signals with different meanings easily be distinguished from each other?
37 Can different parts, control knobs and tools easily be recognised by touch?
38 Can parts, control knobs and tools be recognised by their position?
39 Does the work involve position movements or precise force application?

Work method: physical demands

40 Does the work imply a large muscular load?
41 Is the work done sitting, standing upright or walking, or is there a combination?
42 Is the muscular load predominantly on:
    —arms?
    —legs?
    —neck?
    —trunk?
    —small muscles of hand, fingers?
43 Are small and/or large groups of muscles subject to static exertion induced by holding of the material or the tools?
44 Are large groups of muscles subject to static exertion induced by the work posture?
45 Is variation in work posture possible?
46 Does the work provide for a good alternative of work and rest and of static and dynamic elements in regard to muscular load?
47 Do secondary activities inherent in the work methods provide for changes in muscular loading?
48 Is the pattern of movement correct?

Work method: mental demands

49 Is the relation between the direction of movement of control and the effect compatible?
50 Is the use of tables (lists etc) required and is this efficient?
51 Can signals easily be confused?
52 Have signals always the same meaning?
53 Are the controls positioned in the sequence of task performance?
54 Can the controls easily be recognised by shape, size, labelling, colour, for

normal use and for emergency?
55 Are the controls located as near as possible to the corresponding sources of information?
56 Is the task adapted to the capabilities of older workers in regard to:
—pacing?
—visual demands?
—short-term (transitory) memory?
57 Is the rate of information likely to exceed the mental capacity of the operator and to overload him?
58 If any of the sensory channels is likely to be overloaded, can the load be more evenly spread?
59 Is the correct sense to be used in regard to the meaning of the signal (danger, alarm—ear; normal machine performance—eye; discrimination of controls etc—tactile sense)?
60 Is it possible for signals from different sources to occur simultaneously?
61 Have signals to which preference has to be given the highest attention-getting value?
62 Are different answers possible to one and the same signal whereas only one is suitable?
63 Is adequate time allowed in machine or process cycle for decisions and resulting action?
64 Can rapid feed back of the effects of adjustment to a system be given?

Environmental load

65 Are the climatic conditions within the comfort zone?
66 If climatic conditions are not within the comfort zone, is this primarily due to:
—air temperature?
—humidity?
—air movement?
—radiation?
67 Is there any risk of hearing loss due to noise?
[2.2, 2.30, 2.31, 2.32]

2.7 References

2.1 Singleton, W. T. 'Ergonomics in systems design.' *Ergonomics*, vol 10, no 5, September 1967.
2.2 Howells, R. and Barrett, B. *Health and Safety at Work Act 1974: a guide for managers.* Institute of Personnel Management, London, 1977.
2.3 McWirter, N. and Greenberg, S. *Guinness Book of Records.* Guinness Superlatives Ltd, London, 1978 (25th edition).
2.4 Warren Spring Laboratory. *Anthropometric Data for Designers: a critical evaluation.* Research Report no PR/ES/59. Department of Scientific and Industrial Research, May 1974.
2.5 Bolton, C. B., Kenward, M., Simpson, R. E. and Turner, G. M. *An Anthropometric Survey of 2000 Royal Airforce Aircrew 1970/71.* RAE Technical Report 73083.
2.6 Chapanis, A. (ed) *Ethnic Variables in Human Factors Engineering.* The Johns Hopkins University Press, Baltimore, 1975.
2.7 Meyer, R. 'Articulated anthropometric models in consumer product design.' *Symposium on Human Factors in Consumer Product Design.* Human Factors Society, Ohio, May 1977.

2.7 References

2.8   Van Cott, H. P. and Kinkade, R. G. (eds) *Human Engineering Guide to Equipment Design*. US Government Printing Office, Washington, 1972 (revised edition).
2.9   Diffrient, N., Tilley, A. R. and Bardaghjy, J. *Human Scale 1/2/3*. MIT Press, 1974.
2.10 Singleton, W. T. *Introduction to Ergonomics*. World Health Organisation, Geneva, 1972.
2.11 Ashby, P. *Ergonomics Handbook 1: Human Factors Design Data*. Design Institute Publication, South Africa, 1978.
2.12 Astrand, P. O. and Rodahl, K. *Textbook of Work Physiology*. McGraw-Hill Book Co, New York, 1977.
2.13 Wilkie, D. R. 'Man as a source of power.' *Ergonomics*, vol 3, 1960.
2.14 McCormick, E. J. *Human Factors in Engineering and Design*. McGraw-Hill Book Co, New York, 1976.
2.15 Royal Navy Personnel Research Committee of The Medical Research Council. *Human Factors for Designers of Naval Equipment*. MRC, 1971.
2.16 Murrell, K. F. H., Grien, S. and Tucker, W. A. 'Age structure in the engineering industry: a preliminary study.' *Occupational Psychology*, July 1957.
2.17 Carpenter, J. and Cazamian, P. *Night Work*. International Labour Office, Geneva, 1977.
2.18 Goldsmith, S. *Designing for the Disabled*. Royal Institute of British Architects Publications Ltd, London, 1976 (third edition).
2.19 Welford, A. T. *Ageing and Human Skill*. Oxford University Press for the Nuffield Foundation, 1973.
2.20 Shackel, B. (ed). *Applied Ergonomics Handbook*. IPC Science and Technology Press, 1976.
2.21 Wolcott, J. M. and McMeekin, R. R. ' Correlation of general aviation accidents with biorythm theory.' *Human Factors*, vol 19, no 3, June 1977.
2.22 Hunter, S., Schraer, R., Landers, D. M., Buskirk, E. R. and Harris, D. V. 'The effects of total oestrogen concentration and menstrual-cycle phase on reaction time performance.' *Ergonomics*, vol 22, no 3, 1979.
2.23 Mowbray, C. H. and Gebhard, J. W. 'Man's senses as information channels' in Sinaiko, H. W. (ed) *Selected Papers on Human Factors in the Design and Use of Control Systems*. Dover Publications, New York, 1961.
2.24 McCormick, E. J. *Human Factors Engineering*. McGraw-Hill Book Co, New York, 1976 (fourth edition).
2.25 Rolfe, J. M. 'The secondary task as a measure of mental load' in Singleton, W. T., Fox, J. G. and Whitfield, D. (eds) *Measurement of Man at Work*. Taylor and Francis Ltd, London, 1971.
2.26 Chapanis, A. 'Engineering psychology' in Dunnette, M. D. (ed) *Handbook of Industrial and Organisational Psychology*. Rand McNally College Publishing Co, Chicago, 1976.
2.27 Gregory, R. L. *The Intelligent Eye*. George Weidenfeld and Nicholson Ltd, London, 1970.
2.28 Grandjean, E. *Fitting the Task to the Man*. Taylor and Francis Ltd, London, 1969.
2.29 Meyer, R. P. 'Articulated anthropometric models.' *Consumer Products Technical Group Newsletter*, vol 4, no 2, 1979. (Available from Human Factors Society, PO Box 1369, Santa Monica, California 90406, USA.)
2.30 Nadler, G. *Work Design: a systems concept*. Richard D. Irwin Inc, Illinois, 1970.
2.31 Burger, C. E. *Ergonomic System Analysis Checklist*. International Ergonomics Association Conference Publication, 1964.
2.32 Wood, J. 'The application of ergonomics in the design of a computer console for emergency services: a case study' in *Proceedings of Sixth International Ergonomics Association, July 1976, Washington*. Human Factors Society, Santa Monica, 1976.

3 Ergonomics—display and control applications

David Davies

3.1 Introduction

Machines have been a part of human history for a very long time, as we have seen in passing in Chapter 1. However, up to the present century they were nearly all doing the same thing—extending man's muscle power. There were exceptions of course, such as the microscope and telescope, which enhanced the visual sense.

The Industrial Revolution brought many new problems as machines became more powerful and were coupled together, and the consequences of lack of control or error of judgement became more serious. Unforeseen difficulties arose when machines that would operate individually would not co-operate when assembled together, and this meant that such machines could no longer be designed in isolation, thus introducing the idea of what is known today as 'system design'.

Mechanisation and automation have further replaced muscle power with machine power, and there has been an increasing requirement for a systems approach to the design of equipment and plant. During the past 30 years, however, the information and communications revolution has brought a new range of problems.

The machines that are at the heart of this revolution are of course computers. These are not extensions of man's muscle power, but extensions of his capacities to recognise, process and transmit information. Control is again vitally important, and in this instance the control of information.

We have almost set the scene for this chapter with the introduction of these three vital factors—power, information and their control. However, there is one remaining factor to be introduced—the human factor—which requires that machines and systems should be created whose operation is within the capacities of people.

The anthropometrics and sense systems of humans have been discussed in Chapter 2. This chapter is devoted to ergonomic methodology and its essential role in design for control.

3.2 Separation and allocation of functions

A systems approach has already been used in Chapter 2 in relation to the capabilities of the human sensing system, and a similar approach will be followed here in relation to control. In this context an engineering system can be defined as a group of components—at least some of which are pieces of equipment—designed to work together towards a common goal. This definition allows for varying degrees of complexity and size.

Systems design is the design of the total entity, the system, according to a systematic procedure. It involves three major characteristics. The first of these is the separation of functional and physical problems, and the treatment of functional problems as a separate and definite phase at the very beginning of the design activity. This phase is dedicated to establishing the ends required by the system.

The second characteristic is the treatment of the human as an integral part of the system. Engineering principles cannot be applied to machines or equipment in isolation from the people who must operate and maintain them. When the functions are being defined and allocated, the capabilities and capacities of the human must be considered.

The third characteristic is definition and classification of the design decisions, with some attempt to put them into a serial order. This ensures a systematic framework for the designer, reducing the possibility of omissions, while proving useful to project planning and control [3.1, 3.2].

3.2.1 Definition of systems objectives

This phase should crystalise the objectives of the system with a statement of all the constraints under which it will be expected to operate. These should include all the outputs that will be required by the system, and their quality, the inputs the system will receive, permitted environmental effects, unit costs, availability of natural resources, and the relevant policies of the company in question.

Where systems are very large this procedure is followed for subsystems within the total system, and even for parts of the subsystem.

3.2.2 Separation and allocation of systems functions

Once the systems objectives have been clearly established, the next step is to determine what functions need to be performed in order to achieve these objectives. Consideration should not be given as to how the functions are to be performed, because this will almost certainly involve unconscious allocation of functions to people, machines or computers, rather than according to objective and systematically applied principles.

Once all the functions have been detailed, then decisions have to be made as to who will perform each function—men, machines or computers? In many instances certain obvious facts will dictate the answers, but there are many other functions which are within the scope of humans, machines and computers.

In the 20 years after the Second World War ergonomics provided guidelines based on general statements about the kinds of things humans could do better than machines, and vice versa. These statements can be generalised as follows—men are flexible but cannot be depended upon to perform in a consistent manner, whereas machines can be depended upon to perform consistently but have no flexibility. Such lists can still be consulted for general advice, and one example [3.1] can be summarised as follows

Humans are generally better in their abilities to:
Sense very low levels of certain kinds of stimuli.
Detect stimuli against a 'noisy' background.
Recognise varying related patterns of complex stimuli.
Sense unusual and unexpected events.

Remember strategies and principles over long periods of time.
Recall pertinent or related items of information, but reliability of recall is low.
Draw upon varied experience in making decisions and adapting them to situational requirements; act in emergencies.
Select alternative modes of operation, if certain modes fail.
Reason inductively, generalising from observations.
Apply principles to solutions of varied problems.
Make subjective estimates and evaluations.
Develop entirely new solutions.
Concentrate on most important activities, when overload conditions require.
Adapt physical response (within reason) to variations in operational requirements.

Machines are generally better in their abilities to:
Sense stimuli that are outside man's normal range of sensitivity, such as x-rays, radar wavelengths and ultrasonic vibrations.
Apply deductive reasoning, such as recognising specified stimuli as belonging to a general class.
Monitor for prespecified events, especially when infrequent.
Store coded information quickly and in substantial quantity.
Retrieve coded information quickly and accurately when specifically requested.
Process quantitative information following specified programs.
Make rapid and consistent responses to input signals.
Perform repetitive activities reliably.
Exert considerable physical force in a highly controlled manner.
Maintain performance over extended periods of time.
Count or measure physical quantities.
Perform several programmed activities simultaneously.
Maintain efficient operations under conditions of heavy load.
Maintain efficient operations under distractions.

3.2.3 An ergonomics strategy for allocation of functions

These lists of statements have obvious limitations; for example, they do not deal with the specific characteristics of each design, nor with its detail. The more up-to-date ergonomic approach has therefore been to follow one of two main strategies.

In the first strategy, where systems are complex and there is considerable automation and computerisation, prototype simulations of such systems are undertaken by ergonomists. This enables an investigation of alternative strategies of function allocation (as well as other more detailed aspects of the design). Performance indices and analysis of subjective feelings and responses of workpeople provide objective data for decisions on function allocation.

This process takes time and is expensive in terms of equipment, computers and people, but such a cost may only be a tiny fraction of the cost of large plants. Moreover, with the increasing use of computers and increasingly rigorous legislative demands on designers, it looks increasingly like an essential and not an optional cost, quite apart from the benefits such studies can bring in terms of improved plant efficiency and safety, and the health and well-being of operators.

In the second strategy, where systems are smaller and less complex, the functions are provisionally allocated according to generalised ergonomic principles. Then they are interactively refined as the design process proceeds. For example, the information analysis will establish what information and what information flows are required to execute the functions. This will often show up problems with function allocation. For example, it may be that the functions designated at an interface cannot possibly be fulfilled by one operator because there is too much simultaneous information to be handled. This may mean a new machine part to integrate a lot of this information, or it may mean a two-man interface, or two entirely separate one-man interfaces. The important point is that decisions on tasks and manpower are being dictated by human requirements to execute functions and to fulfil the system objectives.

Despite the fact that sections in this chapter appear consecutively and propose a systematic step-by-step approach, in practice a number of loops may well be required. Changes indicated by later stages require back-tracking through earlier stages to check for any consequential modifications.

So, by one strategy or the other, the functions are thus allocated to men, machines and computers. Provisional man/machine (including computer) interfaces are thus defined by a function or group of functions to be executed. In this provisional definition, consideration should not be given to details of physical constraint; rather the criteria for an interface should be ergonomic ones—the provision of meaningful and achievable control tasks which harmonise together and produce an efficient and safe total system (Figure 3.1).

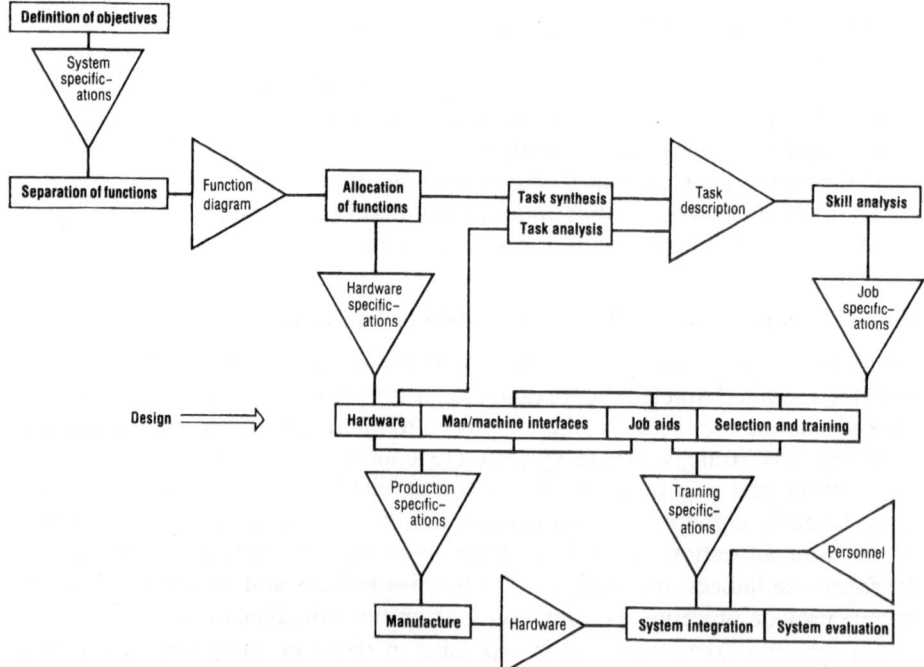

Figure 3.1 Graphical representation of systems design decision-making process (after McCormick).

3.3 Information analysis at the man/machine interface

3.3.1 The flows of information

Once functions have been allocated to humans and machines (including computers) there should be an examination of each man/machine interface that has been provisionally defined. At this stage it is the flows of information that we are concerned to determine and analyse and there are two key types of these, the first being a flow of information from the machine to the man, the second being from the man to the machine (see Figure 2.19).

These flows of information can be classified in different ways, for example as formal or informal. Informal flows occur when information is communicated without any special machine components or aids dedicated to the task. For example, the operator of a milling machine will be fed a lot of information by the sound being generated by the machine. A formal flow occurs when a means of communication has been designed to convey the information. For example, a machine may be fitted with a tachometer to provide the operator with precise information about its speed.

Informal flows of information can get overlooked, and yet they may be vital to the efficient and safe performance of tasks, particularly skilled ones. Because man is so adaptable he will often find some means of generating the information he needs if it is not available, and this adaptability can often mask serious information flow problems.

Similar classification of information flows is into direct and indirect [3.1]. Information which can be received by sensory and perceptual processes is sensed directly. Where information has to be transmitted in a reproduced or coded (symbolic) form, it is being indirectly sensed. These two classifications, of formal/informal and direct/indirect, are complementary, and so can be used together, providing a comprehensive method of analysing such flows.

In addition to the two key or primary flows of information there may be many more. In the case of a large production plant there may be many interfaces scattered over the shop-floor area and there may well need to be information flows between some of these. This might require machine connections between the interfaces, personnel might require to have direct sight of each other, or they might need an indirect means of verbal communication, such as a telephone. These flows must all be analysed and defined, along with the primary flows.

3.3.2 Analysing the flows of information

In analysis of these flows of information the objective is to establish what information is required for the allocated functions to be satisfactorily fulfilled, and how this information is being communicated.

Each function at every interface has to be examined in turn, and an information analysis carried out. If this is not done then it is very probable that the operator will find he is without certain information, or occasionally that he has too much information. Either way there will be impairment of performance and safety.

The information analysis will involve not only operational functions but also maintenance functions. Frequently the latter are not considered as an integral part of the design process, so that, for example, a fitter or engineer has to struggle to 'set up' a machine, having carried out routine or repair maintenance, adjusting controls at the operator's interface while finding certain key information he needs is not provided there. From this point in the text, therefore, when the word operator or man is used to describe the human at the interface or workplace, it should be understood to mean the operations man and the maintenance man in turn.

Five important facets have to be established for each item of information, as follows.

Precisely what item of information must flow or be transmitted? The word 'precisely' prefixes the question quite deliberately because precision is essential. For example, it is not enough to state that the temperature of a substance has to be transmitted to an operator. Does he need the exact temperature? −65.5°C— and if so, what are the required limits of accuracy, ±0.5°, ±0.1°C? Or, is it the fact that the temperature is within certain prescribed limits—say 40°C to 60°C?

What or who are the sender and receiver of the flow? For example, is the flow from the machine to the operator, or from the machine to the maintenance fitter? With two primary information flows this is usually clear. However, for the many secondary information flows it may be difficult to decide if, for example, the man-machine interfaces have not been decided and defined clearly.

Is this a formal or informal flow of information? If a person is involved in sensing the information is he doing it directly or indirectly? For example, can the operative obtain this information by listening to a sound, or by looking at a part of the machine? Or, will it be necessary to install an instrument to measure some machine activity and provide a display to transmit the information to the operator?

When is the information required? Often information is required at the time of generation. However, when information is needed some time after it has been generated, it must be stored, recalled and presented at the appropriate point. In some instances, too predictors of future states or variables can be invaluable in improving performance and warning of dangers ahead. The time basis of the information required must therefore be established.

For what purpose is the information to be used? As a warning, for a check, tracking, identification, or as an instruction? This question is important in ensuring that all the appropriate relationships between the different items of information are known and catered for.

3.3.3 Categorising the flows of information

An exhaustive list (Figure 3.2) is then drawn up of all the information requirements, categorised according to the principles enumerated in 3.3.2. This then provides the basic data for the next stage, which is the design of the means of conveying the required information.

3.4 Man/machine interface specification

The specification of the man-machine interface from an ergonomic viewpoint consists of three parts, first defining the sense or body part through which the

3.4 Man/machine interface specification

INFORMATION ANALYSIS SHEET No 23

INTERFACE: Main Control Room - Heating Control Station

FUNCTIONS: Control of furnace temperature within permissible limits

INFORMATION REQUIRED

FLOW 1   Furnace - Operator

1.1 Temperature at the six key furnace sectors. Information to be the exact temperature $\pm\ 0.5\,^\circ C$, available instantaneously and continuously (by formal and indirect means).

1.2 A historical record of the temperature in (1.1), which can be readily accessed at any time up to 8 hours after its generation (by formal and indirect means).

1.3 An indication of the speed of movement of material through the furnace. Information to be the exact speed $\pm\ 5$ mm/second, available instantaneously and continuously (by formal and indirect means).

1.4 A historical record of the speed of movement of material through the furnace (ie 1.3) which can be readily accessed at any time up to 8 hours after its generation (by formal and indirect means).

1.5 Information relating to exit door of the furnace, so as to allow the operator to check if it is open or closed (preferably by direct line of sight, otherwise by formal and indirect means).

1.6 Information relating to the conveyor bed onto which the steel slabs will pass from the furnace exit. Information as to whether or not the conveyor's rollers are running, and the disposition of the slab on the rollers (preferably by a direct line of sight, otherwise by formal and indirect means).

Figure 3.2 Categorising the flows of information, a typical example.

information will be conveyed; second, establishing the most appropriate mechanism for transmitting the information; and third, providing ergonomic data about the detailed design of this mechanism.

3.4.1 Displays—the choice of channel

The first question to be answered when deciding how to convey information to any human is: which sensory channel should be used to transmit this information? A comparison of the performance of different sensory systems has been given in Chapter 2.

There is often a choice between the senses, usually between sight and hearing. In certain cases the choice is easily made because of the requirements of the situation and the obvious advantages of one sense over another. Visual signals are able to convey the largest range and diversity of information, and are the most commonly used. Auditory signals are particularly appropriate as warning devices, or to attract attention, or to transmit information to a person who is frequently moving about or moving their visual attention. The auditory sense is thus most useful for transmitting relatively short, simple messages.

To supplement the general information in Chapter 2, the following checklist [3.1, 3.3] may be found useful.

1   Auditory stimuli are temporal; the information is extended through time. Visual stimuli are spatial; they have a location in space.
2   Auditory stimuli arrive sequentially in time; visual stimuli can be sequentially or simultaneously presented.
3   Auditory stimuli cannot be kept continuously before the hearer because of their sequential presentation, called poor 'referability' because the information cannot usually be stored in the display.
4   Auditory stimuli offer fewer dimensions for the coding of information than visual stimuli.
5   Speech, as a specialised form of auditory stimulus, offers greater flexibility, through variations in expression and nuances, than visual stimuli, which require advance coding.
6   Speech offers a time advantage. The relevant information is immediately available to the hearer. With visual stimuli a search may be necessary to find the appropriate information.
7   Speech only allows a rate of transmission equal to the speaking rate, whereas visual presentations can be faster.
8   Auditory stimuli are more effective at demanding and getting the hearer's attention. They are omnidirectional and break in on the hearer. Visual stimuli require the operator to be looking at or near them.
9   Hearing appears more resistant to the effects of fatigue than vision.

The sense of touch is the other sense which can be used as a means of communication. The most commonly known use is in Braille, which enables blind people to 'read', and mechanical and electrical stimulation can be used in touch displays. Research work has demonstrated that the sense of touch cannot resolve small intensities and qualitative differences as precisely as vision and hearing. On the other hand the tactile system can transmit a limited number of discrete stimuli, such as might be needed in a warning signal. When the visual and auditory channels are fully occupied or becoming overburdened, a tactile display could well be useful for getting warning or otherwise important messages through.

3.4.2   Visual displays

In designing visual displays there are four areas of ergonomic consideration.

The first of these is, what is the information to be transmitted? What is its purpose or function? And who or what is the sender/receiver? In previous discussion is has been mentioned that such information can, amongst other classifications, be divided into three broad categories.

This categorisation should lead us on to the second consideration, which is deciding what type of display is to be used. The display should be the most appropriate display to transmit the information concerned. For example, if the precise value of a certain temperature were required, then a digital display might be the most appropriate. On the other hand, if the temperature varied considerably and quite rapidly, then an ordinary digital display would not be appropriate, since the last digit or two could probably not be read because of their rate of change.

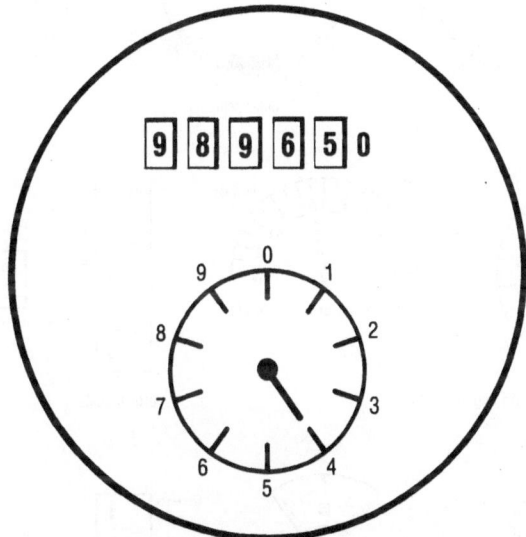

Figure 3.3   Typical hybrid display.

And if some indication of rate of change was necessary then a hybrid display (Figure 3.3) might be appropriate, where a digital display provided all but the last digit, which was displayed with a rotating pointer on a circular scale.

The third and fourth areas of consideration are very closely related, although they are separated here for the sake of clarity. The third consideration is the nature of the visual environment in which information is to be transmitted.

Important factors here are the intensity of the ambient illumination, location, intensity and direction of illumination of light sources, including daylit windows, roof windows etc, and reflectances of the surfaces in the work areas, including ceilings, walls, floors, consoles and desks. For example, if light levels in the immediate environment are very low then additional illumination of the displays will be needed. This can take the form of additional ambient illumination directed onto the displays, or internal illumination of each display individually. The former is obviously much cheaper and simpler to install and usually also to maintain. However, care has to be taken to avoid specular reflections on the face of the display.

On the other hand, if the light levels in the immediate environment are very high and the displays are, say, LEDs, then they will be 'washed out' and unreadable. It may then be essential to mask the displays so as to reduce sufficiently the ambient light falling upon them to permit acceptable loading performance.

The fourth consideration is the detailed design characteristics of the type of display chosen.

There is considerable ergonomic data available about the design of types of display. A summary of the major factors is given below, along with data on the comparative use of the different displays.

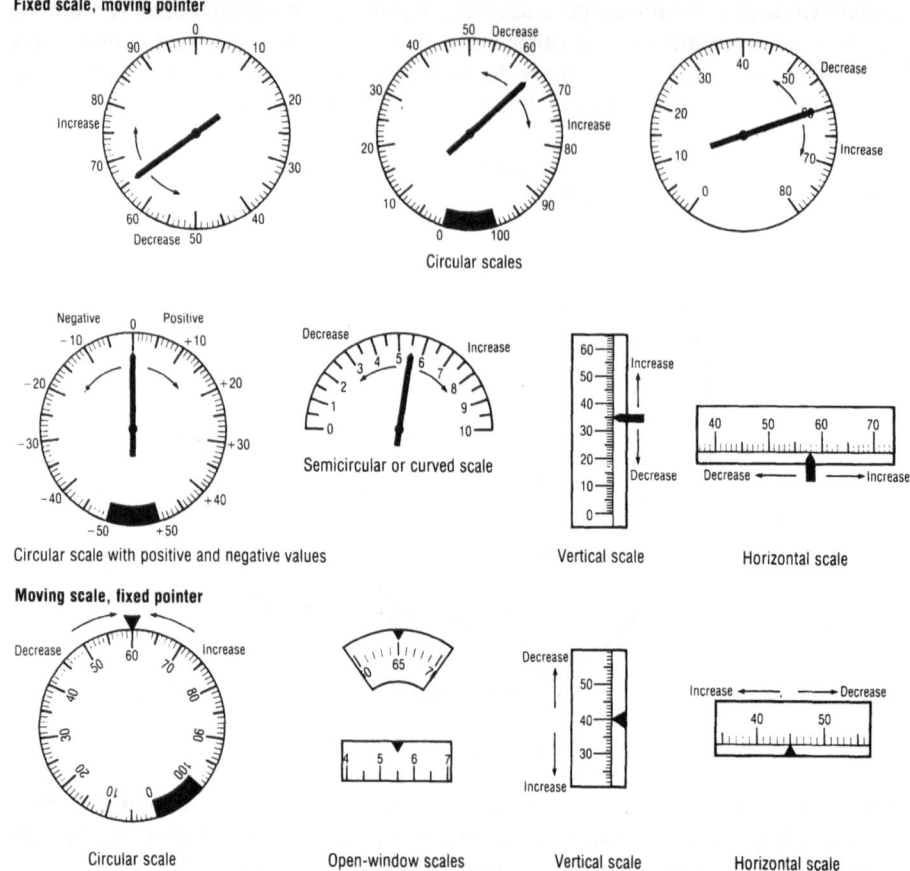

Figure 3.4 Displays used for the presentation of quantitative information (after McCormick).

Quantitative displays should be used when the receiver requires numerical information, and can be either analogue or digital. Figure 3.4 shows a series of analogue displays in which the position of a pointer on a scale is analogous to a particular value.

The digital display is superior for presenting individual quantitative values when they are not subject to continual or rapid change [3.4]. An example is the micrometer shown in Figure 3.5.

Analogue displays are best for check readings, and for showing rate and direction of change. Fixed scales with moving pointers are usually preferable to moving scales with fixed pointers. This is probably because the pointer position on a fixed scale adds a perceptual cue absent from the moving scale display. Where a quantitative display is also being used in a qualitative way—for example a display used to provide the actual value and deviation from a desired value, then use of fixed scale/moving pointer design facilitates the qualitative task.

Vertical and horizontal fixed scales with moving pointers can however be preferable where information has not only to be received but compared. For example, when controlling and comparing the performance of several related machines, direct comparisons can readily be made between the different performances using vertical displays in a bank (Figure 3.6).

Figure 3.5 A digital display on a Moore and Wright micrometer. (*Moore and Wright*)

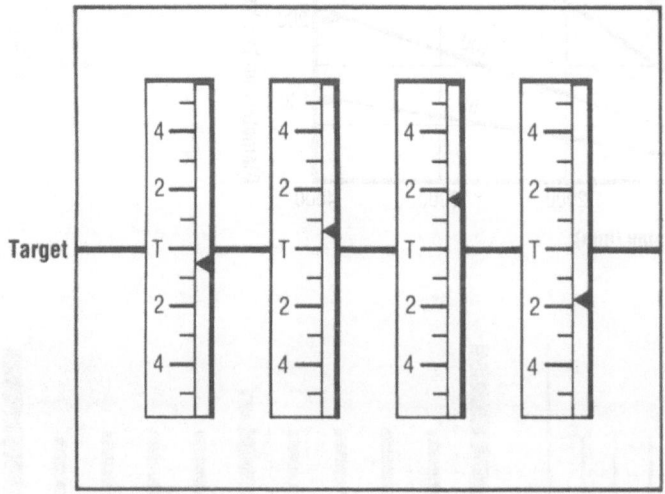

Figure 3.6 A display showing the performance of four interrelated machines.

More detailed design factors to be borne in mind are as follows [3.1].

Accuracy. Instrument accuracy is normally 2 per cent for commercial grade instruments, and the required reading accuracy should be twice as good in order not to degrade the overall accuracy.

Length of scale units. The basic scale dimension is the length of scale which represents the numerical value that is the smallest unit to which the scale must be read. Widely accepted standards in the USA stipulate a value of 1.3 mm in adequate illumination and 1.8 mm in low illumination, for a reading distance of 71 cm. Comparable UK standards are about half these values. When viewing distances are more or less than 71 cm the display features should be adjusted to maintain visual performance (Figure 3.7).

Scale markers. Where maximum accuracy is required, the scale should have a marker for every scale unit. If the scale becomes too crowded, then a marker for every other scale unit is preferable to every fifth or tenth scale unit. Three sizes of marker, and no more, should be used. A scale should generally have not less than five, and not more than 10 numbered divisions.

Scale progression. The number of unnumbered divisions between the scale markers (sometimes referred to as the 'called interval') should progress by 1s, 5s or 2s, in that order. Progression by 4s, 2.5s, 3s, 6s, 7s, 9s or 10 or more is generally troublesome and should not be used. The progression should be fixed to obtain the

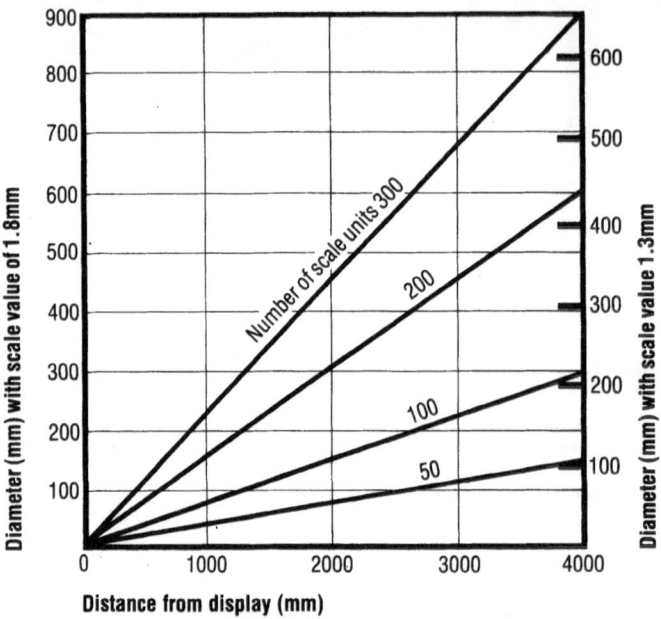

Figure 3.7 Diameters of circular quantitative scales required to provide constant visual angle of scale units.

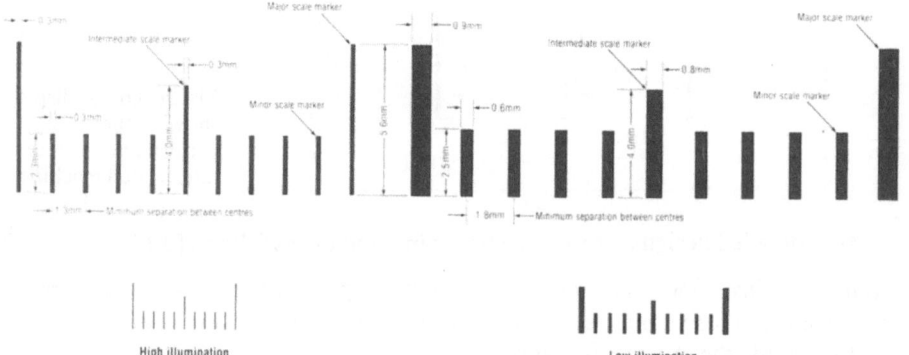

Figure 3.8 (above) Recommended formats for quantitative scales in high and low levels of illumination at a viewing distance of 71 cm.

Figure 3.9 (right) Representational displays in a railway signal box. (*British Railways Board*)

3.4 Man/machine interface specification

required reading accuracy, allowing for either interpolation by two or by five between the called intervals. Where large numerical values are used in the scale, relative readabilities will be the same if all values are multiplied by a factor of 10, 100, 1000 and so on. Decimals, however, may lead to confusion. The zero in front of the decimal point should be omitted if such scales have to be used.

Pointers. Pointers should be as wide as a minor scale marker (but not wider) and meet but not overlap the smallest marker on the scale. For circular scales the pointer length should be exposed for at least 80 per cent of the swept pointer radius. The tail of the pointer should be the same colour as the dial face and the whole pointer should be as close to the dial face as possible.

Recommended formats for quantitative displays for use in high and low levels of illumination at a viewing distance of 71 cm is shown in Figure 3.8 [3.1, 3.5].

Qualitative displays are appropriate when the receiver has to distinguish between a small number of different conditions. Sometimes quantitative displays are used for qualitative purposes, but it is preferable to provide a proper qualitative display, or attempt to incorporate a qualitative code on the quantitative display. The most important ergonomic requirement in the design of these displays is to ensure that the indicators or codings for each of the conditions are as distinctive as possible through differences in position, colour, shape and size. If possible it is best to employ more than one of these means.

Representational displays provide the receiver with some representational model of the system, machine or process being controlled. They are most suited for the control of large, complex and remote systems and are becoming increasingly widespread with the growth of automation and computerisation (see Figure 3.9). They enable the receiver to monitor the whole system and its constituent parts, and to detect errors or alarms and assess their significance. The most important design principle is to make the display as simple as possible, while not losing either essential details or important systems principles and relationships. It is difficult to provide much more detailed guidance because the systems represented vary so much. It is therefore desirable to construct models of a number of alternative displays, and to evaluate these in a simulated working situation. The final display can be chosen on grounds of optimum performance, safety and user acceptability.

There is an increasing trend towards the use of electronic displays, whch exist in two forms—visual display units (VDUs) and other flat screen devices.

The VDU is a general-purpose device using a cathode ray tube which permits a great variation of format, colour and display material. It can be programmed to provide qualitative, quantitative and representational displays. The ergonomic principles so far enumerated will apply to the design of displays on the VDU screen. Ergonomic aspects specific to VDUs include the selection of a phosphor whose persistence, colour and resolution are appropriate to the particular task concerned, and the use of legible alpha-numeric character designs. Since the user of the VDU is either sitting down to view it or standing in front of it, workplace design must also be considered [3.6].

Flat screen devices include special-purpose lamps and alpha-numeric displays. These display either numeric characters and/or alphabetic characters and/or two-state indication (usually on or off). There are two forms of display, active and passive, the former emitting their own light, for example light-emitting diodes (LEDs), and the latter reflecting or transmitting light from another source as in the liquid crystal display (LCD). However, either form may involve the use of a number of other technologies.

Flat screen device characters are commonly formed from a number of segments. Only seven segments are required for numbers, but sixteen segments are needed for all the alphabetic characters. However, many characters are now formed from a dot matrix and this may be more legible. Ergonomic factors relating to the characteristics of the displays and the ambient viewing conditions include luminance (or brightness), contrast, colour and character size; and viewing distance, viewing angle, and ambient illumination levels and sources, respectively. A summary of some general characteristics is given in Figure 3.10.

		Active displays					Passive displays			
		LED	PGD	VF	IF	EL	LCD	EC	EP	FD
		Light-emitting diode	Planar gas discharge	Vacuum fluorescent	Incandescent filament	Electroluminescent	Liquid crystal display	Electrochromic	Electrophoretic	Fluid dipole
PHYSICAL	Technology	Solid state	Gas sandwich	Phosphor tube	Wire in vacuum	Phosphor sandwich	Liquid sandwich	Liquid sandwich	Liquid sandwich	Liquid sandwich
	Life (hours) Ruggedness	>10^5 Good	>10^4 Good	2×10^4 Poor with vibration	>10^5 Poor with vibration	>10^4 Good	>10^4 Good	3×10^3 Not known	3×10^3 Poor with vibration	Not known Not known
	Temperature range (°C)					>100°C	Operational 0-60 °C store − 20 to + 80°C		−15 to +50°C	
ELECTRICAL	Current (ac or dc) Voltage	dc 2	ac/dc 170	dc 25	dc 5	ac/dc 200 - 600	ac/dc 5 - 10 or 15	dc <5	dc Low (10)	dc Low (1 - 10)
	Power (W) or current (A)	175mA	Low (mA)	2 - 20mA	10mA	Low	1μW/cm^2 current	High	Low	Low
	Inherent memory	No	Yes (ac)	No	No	No	No	Yes	Yes	No
ERGONOMIC	Brightness (ft L)	30– 300	50– 500	100	5000	dc 100 ac 1000 (at 300V)	Ambient illumination or auxiliary light			
	Contrast	Good (Using filter)	Good	Good	Good	Good (ac best)	Moderate	–	Good	–
	Viewing angle	Narrow to wide (20° to (140°)	Wide	Wide	Wide	Wide	Shadows at extremes	Very wide	Wide	Wide
	Response	Fast (10ns)	Fast (1 - 10 μs)	Fast (μs)	Slow (ms)	Fast	Slow (10-100ms)	Slow (20-200ms)	Slow (>10^2ms)	Slow (>10^2ms)
	Colours	R, A, G, Y	Red-orange	Blue-green	Any	R, G, Y, B,W	May use transmitted light and filters			
	Aesthetic quality	Fair	Good	Fair	Some good	Good	Fair	Good	Good	Not known

Figure 3.10 Display technologies and their main characteristics.

3.4.3 Auditory displays

When designing auditory displays three ergonomic factors should be considered.

The first of these is, what is the information we wish to transmit and what is its purpose or function? Knowing this, together with data relating to the relative merits and performance of different types of auditory display, we can ensure the right sort of display is chosen.

Secondly, what is the intensity and spectral composition of other environmental noise in which the display will be located? Since the display has to be detectable in its environment either an actual or estimated profile of the noise levels must be generated.

Thirdly there is the human factor [3.1]. The signals emitted by the display must be within the auditory capacities of the hearers. The efficiency of any given auditory display will depend upon the total situation in which the display is used. The signal dimensions and their encoding should exploit learned or natural relationships of the users, and where appropriate the signals should 'explain' the responses required, so as to maximise compatibility. Two-stage signals should be considered when complex information is to be displayed and a verbal signal is not feasible. An attention-demanding signal will attract attention, identifying a general category of information, and a second designation signal indicates the precise information within the general class indicated by the first. Auditory signals should be easily distinguishable from any other sounds present at that time, and they should not provide more information than is necessary. When more than one kind of information is to be presented to the hearer, the signals should be designed to prevent the hearer from listening to only one kind of information, or part of it. And finally, the same signal should designate the same information at all times.

Warning signals come into a class of their own, and the following general principles should apply.

Use high-intensity sudden onset sounds. Variable frequency sounds (such as sirens) are more alerting than steady state sounds.

The sounds should not distract or frighten hearers.

Alerting should take less than one second, so the sound should convey critical information as quickly as possible. Any subsequent signals should be to convey other information, and should be presented within the next two seconds. If different warning signals or signal components are used, each should be clearly discriminable from the others. Warning signals should be of such a nature that they will not mask other important signals or be masked by other signals. Avoid intensities that approach painful or damaging levels. If high intensities have to be used, avoid concentrating the signal in single or restricted frequencies, use lower rather than higher frequencies, and keep the signal short.

Finally, it may be that the information to be conveyed cannot be codified, in which case a direct verbal link may be required—usually either a telephone or some form of public address system.

Figure 3.11 provides a summary of intensities and frequency ranges for certain auditory signal devices.

For large areas: high-intensity coverage			Predominant audible frequency
	Average intensity level (dB)		
Device	At 3 m	At 1 m	
100 mm bell	65- 77	75- 83	1000
150 mm bell	74- 83	84- 94	600
255 mm bell	85- 90	95-100	300
Horn	90-100	100-110	5000
Siren	100-110	110-121	7000
For small areas: low-intensity coverage			
	Average intensity level (dB)		
Device	At 3 m	At 1 m	
Heavy-duty buzzer	50-60	70	200
Light-duty buzzer	60-70	70-80	400-1000
25 mm bell	60	70	1100
50 mm bell	62	72	1000
75 mm bell	63	73	650
Chime	69	78	500-1000

Figure 3.11 Intensity ranges and predominant frequencies of certain auditory signal devices.

3.4.4 Providing for secondary flows of information

This has to be achieved by two means. First, where they are the appropriate means, by the design of relevant displays and controls, using the principles outlined earlier in this section. Secondly, by laying out the interfaces and workplace so as to permit direct vision, direct auditory communications, and rapid and easy walking to and from interfaces, as described later in this chapter.

3.4.5 Controls—classifications

Just as displays are the means by which information is conveyed to the man from the machine, so controls are the means by which information is conveyed or transmitted to the machine from the man. Their importance is therefore obvious—if they are unsuitably chosen or badly designed then the possibility of a safe and accurately controlled machine must be reduced.

As in the case of displays, we can classify controls by their function, and relate them to the information they can best transmit.

Type of control	Type of transmitted information
Activation (usually on/off)	Dichotomous state
Discrete setting (at any discrete position)	Quantitative information Check information
Quantitative setting (individual settings of control at any position along quantitative continuum)	Quantitative information

Continuous control   Quantitative information
                              Qualitative information
                              Tracking information
Data entry (as in computer   Coded information
  keyboards)

3.4.6 Controls—their selection and design

The first question to be answered in control design is what information is to be transmitted by a particular control, and why? This will enable us to define the class of control required. The identification of controls is sometimes of major importance, and in this case the control has also to convey information to the operator about itself as well as enabling him to convey information to the machine. This identification is essentially a coding problem, and the primary coding methods used are shape, texture, size, location, operational method, colour and labelling [3.1].

The second question is what other factors about the environment have an effect on the control, and should be considered? Is the operator in a cold store at $-20°C$, in which case the controls must be operated by a heavily gloved hand? Or, is the environment very hot and humid, so the operator's hands will almost certainly be sweaty and greasy? Is there any vibration of the control panel, or the platform on which the operator stands? This could provoke unwanted and uncontrolled operation of certain types of hand-operated controls. All these environmental characteristics have a bearing on the selection and detailed design of the control, so a note should be made of them.

Third, a decision has to be made about the exact control to be provided. This, as in the case of a display, should be that control whose performance is best suited to conveying the information required, while surmounting any environmental difficulties. A comparison of control characteristics is given in Figure 3.12 [3.7].

Fourth, the detailed design of the control must be decided—size, shape, distances moved, forces required etc. This is intimately connected with the fifth factor, that of ensuring the operation of the control is within the capacity of the operator.

The range, speed, accuracy and force of the human movement imparted to a control depend on the parts of body involved. Some parts are better suited to certain purposes than others—for example the finger and hands are suited to fine and delicate manipulation, but not the exertion of large forces for long periods of time [3.7].

3.4.7 Controls—further design considerations

There are a number of other factors which relate to 'hardware' but have implications for human performance with the control.

Control/Display ratio. In continuous control tasks the ratio of the movement of the control to the movement of the display, or 'system response', is called the control display ratio (C/D ratio). A sensitive control has a low C/D ratio, since a small movement of the control produces a marked change in the display. When a person adjusts a control there are usually two motions taking place, though they may not be apparent. First there is a gross movement to approximately the desired position, and then there is a fine adjustment to the precise requirement. The deter-

3 Ergonomics—display and control applications

Characteristic	Type of control										
	Discrete adjustment					Continuous adjustment					
	Rotary selector switch	Thumb wheel	Hand push-button	Foot push-button	Toggle switch	Knob	Thumb wheel	Hand wheel	Crank	Pedal	Lever
Large forces can be developed	n/a	n/a	n/a	n/a	n/a	No	No	Yes	Yes	Yes	Yes
Time required to make control setting	Medium to quick	n/a	Very quick	Quick	Very quick	n/a	n/a	n/a	n/a	n/a	n/a
Recommended number of control positions (settings)	3-24	3-24	2	2	2-3	n/a	n/a	n/a	n/a	n/a	n/a
Space requirements for location and operation of control	Medium	Small	Small	Large	Small	Small to medium	Small	Large	Medium to large	Large	Medium to large
Likelihood of accidental activation	Low	Low	Medium	High	Medium	Medium	High	High	Medium	Medium	High
Desirable limits to control movement	270°	n/a	3mm × 38 mm	12mm × 100 mm	120°	Unlimited	180°	±60°	Unlimited	Small	±45°
Effectiveness of coding	Good	Poor	Fair to good	Poor	Fair	Good	Poor	Fair	Fair	Poor	Good
Effectiveness of visually identifying control position	Fair to good	Good	Poor†	Poor	Fair to good	Fair‡ to good	Poor	Poor to fair	Poor§	Poor	Fair to good
Effectiveness of non-visually identifying control position	Fair to good	Poor	Fair	Poor	Good	Poor to good	Poor	Poor to fair	Poor§	Poor to fair	Poor to fair
Effectiveness of check-reading to determine control position when part of a group of like controls	Good	Good	Poor†	Poor	Good	Good†	Poor	Poor	Poor§	Poor	Good
Effectiveness of operating control simultaneously with like controls in an array	Poor	Good	Good	Poor	Good	Poor	Good	Poor	Poor	Poor	Good
Effectiveness as part of a combined control	Fair	Fair	Good	Poor	Good	Good¶	Good	Good	Poor	Poor	Good

Adapted from AFSCM 80-3
*Except for rotary pedals which have unlimited range
†Exception: when control is back-lighted and light comes on when control is activated
‡Applicable only when control makes less than one rotation. Round knobs must also have a pointer attached
§Assumes control makes more than one rotation
¶Effective primarily when mounted concentrically on one axis with other knobs.

Figure 3.12 Comparison of common control characteristics.

3.4 Man/machine interface specification

mination of an optimum C/D ratio needs to take into account these characteristics of human performance. Gross movement time drops off sharply with increasing C/D ratios and then levels off, and the reverse pattern is shown by adjustment time (Figure 3.13). The optimum is therefore somewhere around the point of inter-

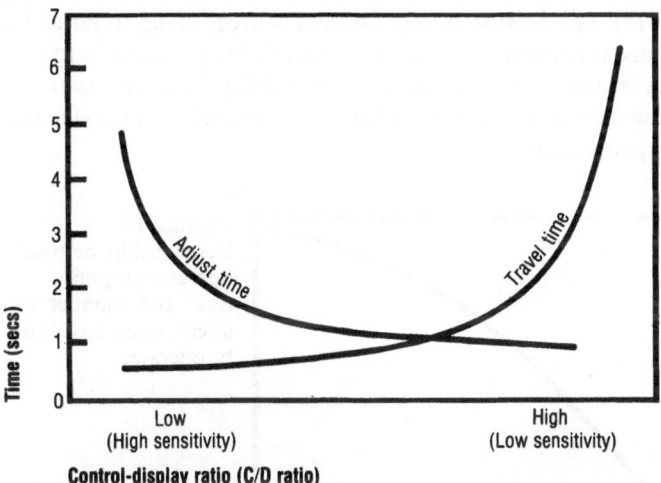

Figure 3.13 Relationship between C/D ratio and movement time (after McCormick).

section. Since C/D ratios vary according to the physical nature and size of controls and displays, the optimum ratio should be determined experimentally for the equipment being considered [3.8].

Resistance in controls. All mechanical devices have some resistance to their operation, and in some controls resistance can be built in by means of servo-mechanisms, hydraulic power, or mechanical linkages. Static and coulomb friction are the resistance to initial movement and sliding respectively. The former is greatest at the initiation of movement but then reduces rapidly. The latter is a continuous resistance to movement once initiated, and is independent of velocity or displacement. Both generally tend to cause degradation in human performance, because there is no systematic relationship between the resistance and any aspect of the control movement so there is no meaningful feedback to the user. Generally speaking increased resistance causes increased travel time, though not necessarily increased adjustment time. Elastic resistance varies with the displacement of a control device. The greater the displacement the greater the resistance, and the relationship can be linear or non-linear. Such resistance serves as a valuable source of feedback information to the operator. A designer can utilise this fact to build in additional cues, such as having distinct gradients in resistance at critical positions.

Viscous damping is caused by a force operating opposite to that of the control output, proportional to the output velocity. This generally has the effect of helping to execute smooth control, especially in maintaining a set rate of movement, and minimises accidental operation. However, the feedback to the operator is not readily interpreted.

Inertia is the resistance to movement caused by the mass of the mechanism involved. It varies in relation to acceleration. It aids in smooth control and minimises the possibility of accidental operation. Because of the forces required to overcome inertia, it does complicate the action of changing directions, and its use is probably only warranted when considerable friction is involved in a control.

Response lag. Lag is inherent in any man-machine system. It is the delay in the response to a changed input, and basically has two components—first the human reaction time, and second the lag in the system itself. Although human reaction time varies considerably between people, what is often overlooked is the fact that the control configuration can itself cause the human reaction time to vary considerably (Figure 3.14). The most important system sources of lag so far as the operator is concerned are transmission lag and exponential lag. Transmission lag refers to a situation in which there is a constant time delay between input and output. Exponential lag is where the output follows essentially an exponential function following a step function.

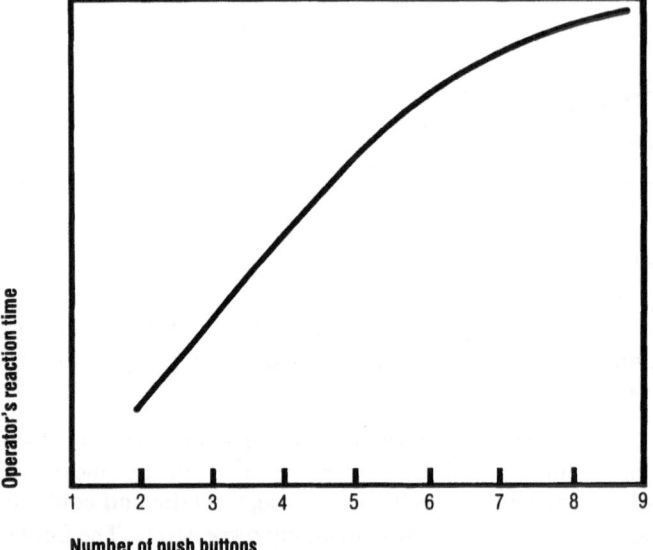

Figure 3.14
Relationship between operator response time and number of alternative controls to be operated.

Studies have indicated that appreciable lag brings about degradation of performance. However, in some situations, the effect of lag has been shown to be very dependent on the C/D ratio.

In some circumstances, too, the operator can anticipate input changes, and there is evidence that operators can compensate fairly well for lag if they can anticipate future inputs [3.9].

Backlash in a control system is a tendency for the system response to be reversed when a control movement is stopped. Usually operators find it difficult to cope with, especially with high gains. If a high display gain is indicated (for instance, in a high-speed aircraft) backlash must be minimised for optimum performance. Alternatively, if it is not possible to minimise backlash, then the display gain should be as low as possible, for optimum performance [3.10].

Deadspace is the amount of control movement that results in no response of the device being controlled. Deadspace of any consequence usually degrades performance, though the degradation is related to the sensitivity of the system [3.11].

Control-display compatibility. Not only should there be a compatible relationship between controls and displays in terms of the C/D ratio, but the relationship between direction of control movement and the response of the display should follow population motion stereotypes. In the population at large there are generally expected responses for certain specified control movements, such as turning a car steering-wheel clockwise and expecting the car to respond by turning to the right. If relationships are incompatible there are at least three penalties. First, training time is increased; second, control by the operator is degraded; and third, in an emergency the operator may move the control so as to produce the stereotype response, which will obviously produce a totally inappropriate result [3.1, 3.12].

3.5 Interface and workplace layout

There are three main areas for consideration. First is the interface itself—that is, the layout of the displays and controls required at that interface for that operator. Second, there is the more general layout of this operator's workplace—his seating and his position relative to the production line, which he may need to see. And then, third, there is the layout and relationship of this interface and workplace to the larger system within which it functions. There are many common principles in each case, and it is difficult to discuss one aspect without impinging on the others. This section concentrates on the first two aspects but includes an outline discussion on the third.

3.5.1 Principles of layout of displays and controls

There are a number of ergonomic guidelines, or considerations. First, the displays and controls must be grouped at the interface, and there are four methods of doing this—by importance, function, frequency of use, or sequence of use. Using the importance criterion the most important controls and displays to the task occupy the prime positions on the interface.

Using the functional principle, displays and controls related to particular functions are grouped together and segregated. There may also be sub-grouping within the functions, so that a power function might have displays and controls sub-grouped into AC and DC functions. Applying the frequency of use principle, the most frequently used displays and their associated controls should be located nearest to the operator, and those used only occasionally will therefore appear on the periphery of the interface. The exception to this would be a control for an emergency stop, which must be located so that the operator can rapidly and successfully operate this control when required.

Finally, displays and controls can be grouped according to sequences or patterns of machine operation that occur frequently. As a general rule, grouping should provide operator movements from left to right and top to bottom. Sequential grouping can be carried out for the whole interface panel, or within a functional grouping. This sequential type of grouping is particularly helpful in reducing operator omissions.

A further word about functional grouping. It may be that a considerable number of displays and controls are required exclusively for the maintenance and setting up of the machine or plant. In this case it is as well to have the maintenance equipment clearly isolated, but at the interface, so that maintenance staff have access to all the operational displays and controls. As much care and attention should be given to the design of displays and controls for maintenance as for operation.

3.5.2 Other work activities

Consideration must be given to the needs of all the other activities that will go on around and about the interface. The operator may need to make reference to operations manuals or maintenance handbooks, or write upon production documents, and all these activities will make demands upon space.

Often training activities take place at interfaces. Again, this will make demands upon space and even possibly the whole layout of the console. If two men will generally be operating the interface because training will be going on most of the time, then it may be best to design a two-man interface, allowing one and two-man operation. There may also be a need to accommodate other visitors at the interface from time to time.

Maintenance considerations are an integral part of the operation of modern equipment, but there may well be special maintenance considerations. For example, test equipment may have to be brought to the interface and connected to parts of it, and there must be sufficient space and access for such equipment and the human activities involved with it [3.7].

3.5.3 Detailed physical layout of the workplace

Two basic human factors should dictate the detailed physical layout of the interface. These are the reaching and seeing requirements of the operator. The two factors are intimately related, and will vary depending on whether the operator will sit, stand or do both at the interface, and whether he has to remain at the interface to obtain all his information.

The earlier information analysis should have defined all the machine parts and other persons in the work area who have to be directly viewed from the interface by the operator. Seeing requirements can therefore be established by means of lines of sight from the interface to all these points, for seated and standing operators, based on anthropometric data.

A decision will thus have to be made at this stage, if not before, as to whether the operator will sit, stand or do both. A standing operator is more mobile, but it is more tiring to work standing than seated, and the operator is less stable. When seated the operator can maintain a stable posture, relax those muscles not needed for his task, and generally exercise more accurate control.

The reaching capacities of the operator are based on anthropometric and biomechanic considerations. Use should be made of data of the type referred to in Chapter 2.

It is important to ensure that there is adequate space between the various controls to ensure efficient operation without inconvenience or injury to the operator, or inadvertent operation of a control. Consideration must therefore be given to any needs for simultaneous or sequential use of controls, the part of the body being used, control size and amount of movement, effects on the system performance of inadvertently using wrong controls, any personal equipment or clothing that might hinder control manipulation, and requirements for 'blind'

3.5 Interface and workplace layout

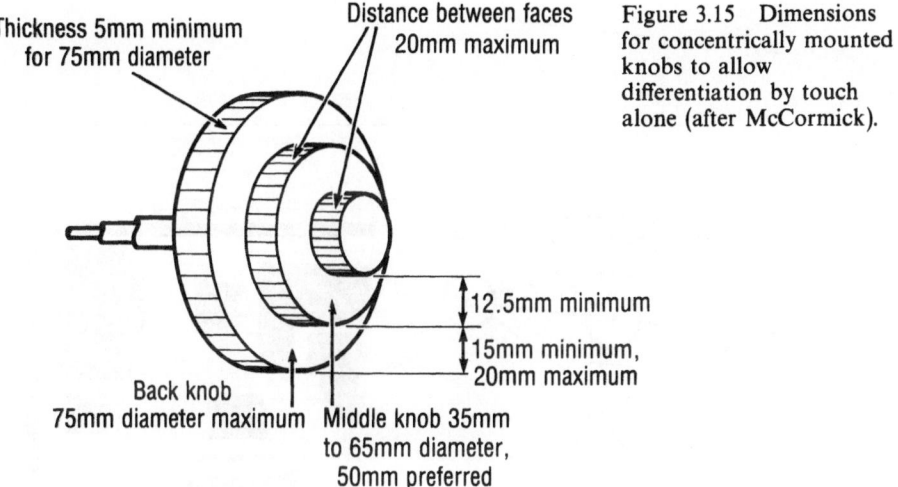

Figure 3.15 Dimensions for concentrically mounted knobs to allow differentiation by touch alone (after McCormick).

reaching for controls. As a general guide, controls located forward of the operator should be at least 150 mm apart, and 300 mm for areas behind or above his shoulders or on either side [3.7, 3.13].

It may be that, having tried to take these matters into consideration, there is found to be too little room! In this case a reduction in the number of controls must be attempted. For example, two or three knobs may be mounted on concentric shafts, though if their relative diameters are not chosen carefully the probability of accidental actuation increases (Figure 3.15).

3.5.4 Environmental considerations

Environmental conditions will have a significant impact on the interface if they depart far from the normal. In these cases often the simplest solution is to house the whole interface within its own building or substructure, isolating it from undesirable or dangerous environmental extremes around it, which may include high noise levels or high air temperature and humidity. Checks should be made that such a structure does not impair the operator's performance by, for example, blocking the view of a machine part, and that it allows sufficient space for access for maintenance, or for training purposes.

Chapter 4 provides more detailed information about environmental problems and their solutions.

3.5.5 Consideration of the wider system

We have been concentrating on the man-machine interface with a single operator, or possibly two. However, in the case of large industrial plants there may be many such interfaces scattered about the shop-floor area and it is vital that these interfaces are not considered in isolation, but as an integrated whole, fulfilling an overall system or plant objective. Link analysis provides a suitable tool for ensuring an optimum compromise layout of interfaces and workplaces.

The term 'link' in link analysis refers to any connection between man and a machine, or between one man and another. If the man must see another this represents a link. Other links include walking, talking, seeing, and movement of

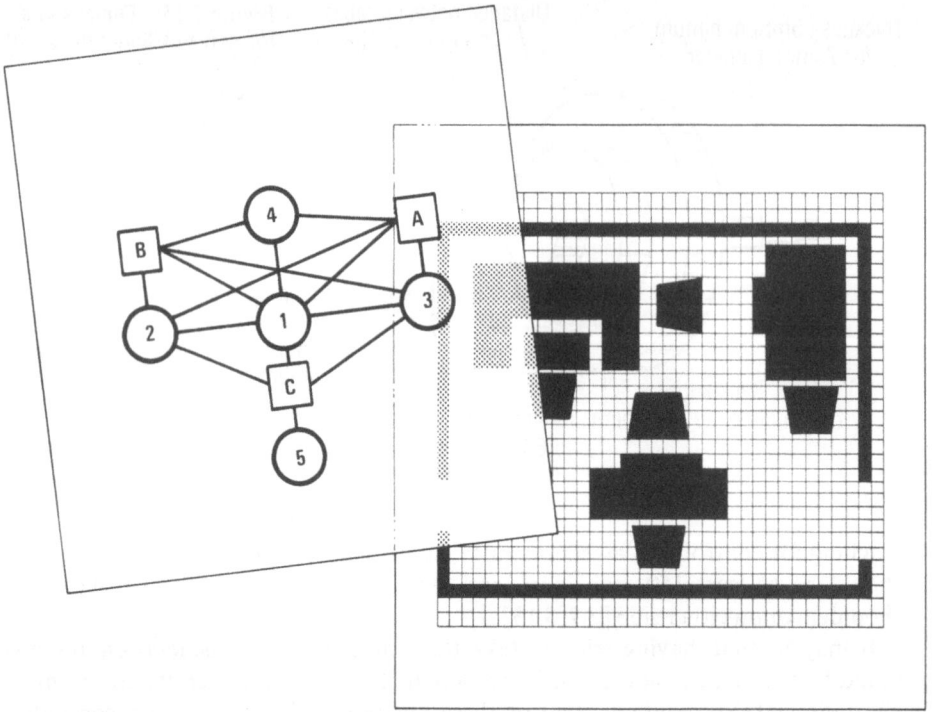

Figure 3.16 A link diagram and its superimposition (right) on a scale drawing of a shopfloor area.

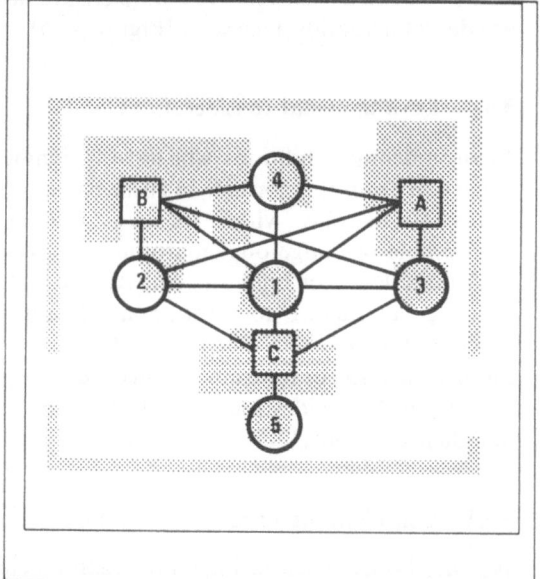

information (and material). The importance of each link and its frequency of use must be established, and relative values of each link are formed from some suitable composite of these two factors.

A link diagram is then drawn with circles representing people and squares representing equipment or parts of them, in which lines between any of these represent the links. The links are drawn with higher values having shorter links and so that there is a minimum of crossing links. The diagram should be confirmed by, at the very least, a scale drawing of the shopfloor area with the locations drawn on it as indicated by the link diagram (Figure 3.16). For very large systems computer programs are available which will carry out such exercises rapidly and efficiently.

3.5.6 Use of mock-ups and models

When detailed drawings have been prepared showing the layouts of each interface and their workplaces, it is highly desirable to carry out a last check to allow for final refinement. This can usually be done in one of two ways.

First, if it has not been possible to obtain anthropometric data from a similar population to the one using the equipment, and generalised data have been used, then a mock-up may be used for fitting trials. Ideally, however, this should be done earlier in the design in order to provide the initial data for the layout.

The second, and preferable, method is to use a full-size model of the interface. This enables potential operators to simulate use of the interface and identify any aspects they find undesirable. It also allows ergonomists and engineers an opportunity to run through all the operational and maintenance routines and to be reassured that all known eventualities have been catered for. See also Chapter 14.

3.6 References

3.1 McCormick, E. J. *Human Factors Engineering.* McGraw-Hill Book Co, New York, 1976 (fourth edition).
3.2 Singleton, W. T. 'The systems prototype and its design problems.' *Ergonomics*, March 1967.
3.3 Henneman, R. H. 'Vision and audition as sensory channels for communication.' *Quarterly Journal of Speech*, vol 38, 1952.
3.4 Zeff, C. 'Comparison of conventional and digital time displays.' *Ergonomics*, May 1965.
3.5 Morgan, C. T., Cook, J. S., Chapanis, A. and Lund, M. W. (eds) *Human Engineering Guide to Equipment Design.* McGraw-Hill Book Co, New York, 1963.
3.6 Cakir, A., Hart, D. J. and Stewart, T. F. M. *The VDT Manual.* John Wiley and Sons, New York, 1980.
3.7 Van Cott, H. P. and Kinkade, R. G. (eds) *Human Engineering Guide to Equipment Design.* US Government Printing Office, Washington, 1972 (revised edition).
3.8 Jenkins, W. L. and Gunov, M. B. 'Some design factors in making settings on a linear scale.' *Journal of Applied Psychology*, vol 33, 1949.
3.9 Poulton, E. C. 'Perceptual anticipation in tracking with two-pointer and one-pointer displays.' *British Journal of Psychology*, vol 43, 1952.
3.10 Rockway, M. R. and Franks, P. E. 'Effects of variations in control backlash and gain on tracking performance.' *USAF, WADC*, TR 57–326, January 1959.
3.11 Rockway, M. R. 'Effects of variations in control deadspace and gain on tracking performance.' *USAF, WADC*, TR 57–326, September 1957.
3.12 Murrell, K. F. H. *Ergonomics.* Chapman and Hall, London, 1969.
3.13 Fitts, P. M. and Crannell, C. W. 'Studies in location discrimination' in Hunt, D. P. *The Coding of Aircraft Controls. USAF, WADC*, TR 53–221, August 1953.

4 Ergonomics—environmental factors
Rod J. Graves

4.1 Introduction

The objective of this chapter is to make the design engineer aware of the importance of environmental effects on systems design. It can only be an overview, but it is hoped that enough detail is included to make the reader aware, not only of the limitations that there may be in human performance under certain environmental conditions, but also of such flexibility that exists as a result of human adaptation.

However, it is important when designing man-machine systems to be aware of the possible longer-term effects on both health and efficiency, and to utilise a total system specification approach by taking human limitations into account.

The role that ergonomics plays in the environmental aspects of man/machine interface design is essentially threefold. First, in identifying the effects that the environment has on man's physiological and psychological processes. Second, in ensuring that work patterns, equipment and machine interfaces are designed to minimise or take account of any reduction in performance and/or individual variations in performance. Third, in ensuring that any necessary protective systems are designed to take account of the physiological and psychological variations found in man.

Man has a sophisticated physiological system that performs optimally under conditions that do not disturb his internal balance (homeostasis). Becoming too hot or too cold results in both physiological and behavioural responses designed to re-establish homeostasis, and these responses may well affect his performance within the man-machine system. Consequently, it is important to be able to detail any localised climatic effects on man.

Man's perceptual processes, designed to gain information about his immediate environment, are also subject to limitations, and so his performance varies within certain boundaries. Sight and hearing are both important information gatherers in man. However, low light levels or noise may limit the amount of information these senses provide. Other perceptual processes, such as touch, taste, smell and body position information, play roles in this information-gathering process. Motion and air pressure can affect performance in many ways and should also be considered. Finally, certain types of airborne pollution and radiation may have to be guarded against, possibly by personal protection, and this protection may itself influence performance at the man/machine interface.

4.2 The climatic environment

The effects of climate on man are complicated by his work rate, his 'acclimatisation' or adaptation to the climate, his fitness, and the amount and type of clothing worn, as well as the mainly physical effects of the environment such as air temperature, humidity, radiant heat of the surroundings and air speed. The complexities of the interrelationship between physiological and climatic factors can be clarified by a brief review of the physiological processes involved in maintaining homeostasis.

4.2.1 Metabolic rate and bodily heat exchange

The body must balance the heat produced within itself against heat gains or losses to the environment, and so maintain its internal or core temperature at the optimum level of about 37°C in order to maintain homeostasis. Metabolic processes produce energy by oxidation of food, using oxygen in the air breathed in. Heat is a by-product of these processes, and even when the human body is asleep or lying still, about 60 watts of metabolic heat is produced. During physical work the muscles convert about 20 per cent of available chemical energy into mechanical power, while the remaining 80 per cent is converted into heat. Figure 4.1 shows the relationship between metabolic heat production (M) and heat loss or gain through radiation (R), conduction (Co) and convection (C).

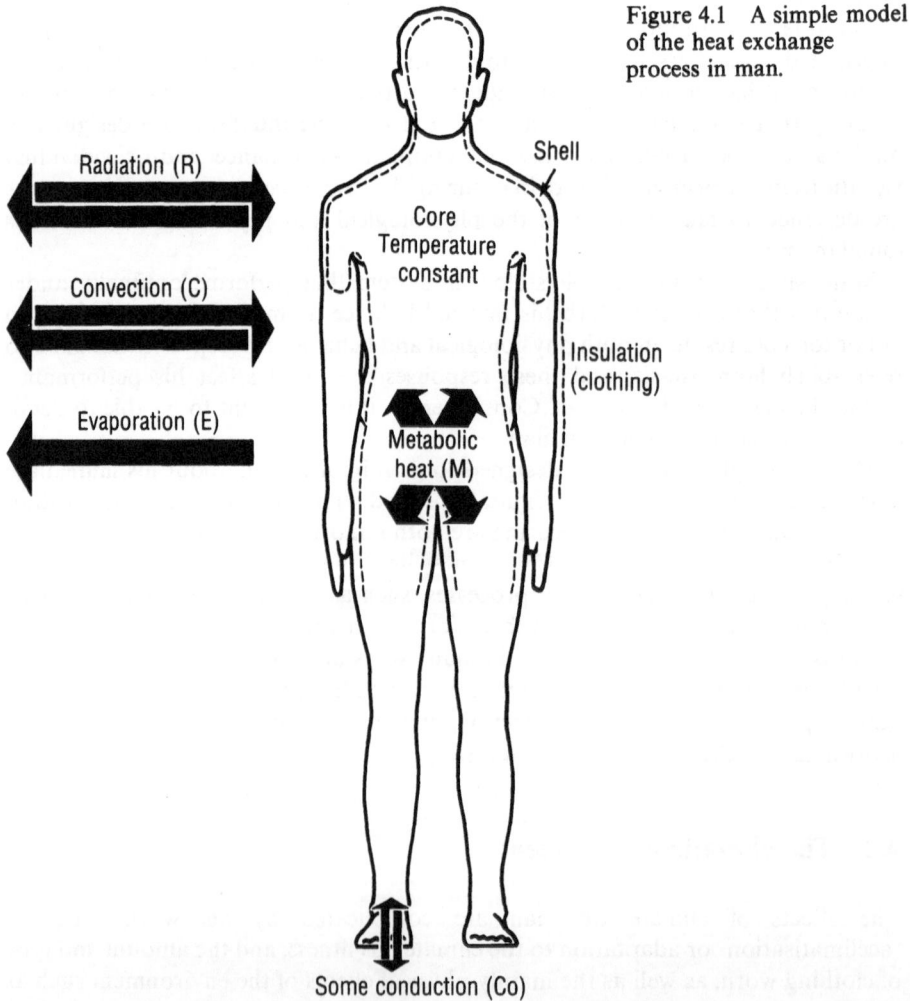

Figure 4.1 A simple model of the heat exchange process in man.

The heat balance equation: $M \pm Co \pm C \pm R - E = \pm S$

If the heat produced is likely to raise the internal body temperature above the normal core temperature, the blood flow in the skin is increased. Likewise, if the heat losses are greater than the metabolic input can compensate for, the blood flow to the skin is reduced. When the temperature difference between the skin and the environment is small, or the body is producing too much heat by physical exertion, then not enough heat can be lost by the skin to the air by simple convection, and body temperature begins to rise. Sweat secretion is then stimulated, and the evaporation of the sweat requires energy which is taken from the skin in the form of heat. The heat exchange system depends upon this sweat evaporation/skin heat loss/blood flow cycle. The body must carry out extra work as a result of the heart pumping blood around the body and to the skin to ensure adequate heat loss. This extra work load may create fatigue which will affect general performance, especially when coupled with physical work.

Under cool conditions, cold receptors in the skin induce vasoconstriction (closing of the blood vessels) which reduces the amount of blood flowing in the skin tissues. This lowers skin temperature and so limits heat loss by convection, radiation and, of course, by conduction. Pilo-erection ('goose pimples') improves the insulation properties of the skin and shivering, which generates localised heat by muscle action, together with a generally higher metabolic rate, may also take place.

Acclimatisation occurs when the body is repeatedly exposed to hot or cold conditions. In hot conditions there is an increased capacity to sweat, an increase in the amount of blood pumped by the heart through the skin, and a faster response of these mechanisms. In cold conditions there is evidence that man also acclimatises and that behavioural changes occur. For example, high metabolic rates and lower skin temperatures have been noted and sleep periods are shorter.

4.2.2 Factors affecting climatic comfort

Convection, conduction, radiation and evaporative cooling are thus important heat transfer mechanisms. The degree to which these mechanisms work, and whether a man feels hot, cold or comfortable—and by implication how his efficiency is affected—are determined by four important factors. These are air temperature, radiant temperature, air humidity and rate of air movement. Their measurement has been adequately described elsewhere [4.1], but it is important to note the following. Air temperature influences heat exchange mainly through convection. Radiant temperature can be important in industrial situations where radiant heat sources may contribute to a high degree of heat gain. Air humidity is important because as it increases so the ability of sweat to evaporate decreases. Air movement exposes the body surface to more air than under still conditions and so affects its rate of cooling or heating.

4.2.3 Climatic descriptors and thermal indices

It is important to take more than one climatic factor into account when determining a worker's comfort. Attempts have been made to produce values that integrate the four factors given above so as to describe the effects of climate in terms of sensation, comfort and physiological response. Three main approaches have been used to develop thermal indices: for hot conditions the use of instruments to simulate heat exchange; the development of empirical scales relating physiological and subjective responses to climatic variables; and the development of indices based on the mathematical analysis of heat exchange. They are discussed in more detail elsewhere [4.2]. Effective Temperature (ET), Wet Bulb Globe Temperature (WBGT), and the Index of Thermal Strain (ITS) are examples of thermal indices.

ET is widely used in the British mining industry, while WBGT has been adopted by the American Conference of Government Industrial Hygienists to define the threshold limit values for heat stress [4.3]. ET combines air temperature, humidity and air speed, but does not include work rate and has been shown to underestimate strain at high levels of heat stress. WBGT combines air temperature, humidity and radiant temperature, but no direct account is taken of air speed. Both scales are based on the subjective responses of seated men or those carrying out light work, and are commonly used in industry.

The ITS index value is the quantity of sweat that must be evaporated in order to maintain body heat balance. The method provides a way of calculating the theoretical amount of heat that can be actually lost, depending on body size, weight, environmental temperature, humidity and air velocity.

The difference between the theoretical heat production and heat loss is the amount of heat that must be removed to maintain equilibrium. This can be achieved by one of several of the following: lowering environmental heat gain; reducing work load by lowering its intensity or reorganising it; and using appropriate clothing (see Section 4.2.7). This approach provides a means of identifying the primary sources of heat gain and hence the factors that must be taken into account in redesigning the working environment.

The effects of cold conditions are not readily dealt with by the index approach, but one measure—the Wind Chill Index—provides a method of describing the physical effects of temperature and wind speed, and a coarse relationship to sensation at lower values. A nomogram containing temperature and air velocity gives a scale ranging from hot (80) to beyond bitterly cold (1200). High air velocities combined with low air temperatures have a greater cooling effect than low air speed and very low air temperatures. Some measure of thermal insulation is useful in the control of heat loss and gain, and the CLO unit is a measure of the thermal insulation necessary to maintain the comfort of a sitting, resting person in a normally ventilated room at 21°C and 50 per cent relative humidity.

4.2.4 Human response to hot and dry, and hot and humid environments

In some industrial settings men have to enter hot areas to inspect and maintain equipment, or to carry out repairs. Where the climate is hot and dry and there is no air movement it is possible for operators to survive for short periods in very hot environments—for example, for 20 minutes at 130°C—but it is recommended that men should not enter areas with surface temperatures above 54°C because of the danger of burns. Under such hot, dry conditions acclimatised workers will sweat profusely and their heart rates will go up. Dehydration and fatigue are possible results.

Sweat cannot evaporate effectively under hot and humid conditions, so the core temperature will start to rise at relatively much lower temperatures. For example, if the wet bulb temperature exceeds 32°C, even fit young men will not be able to work hard for long. Figure 4.2 shows the upper limits of tolerance for heat loss by evaporation for various humidities and dry bulb temperatures. Figure 4.3 shows schematically the effects of heat stress on the human body.

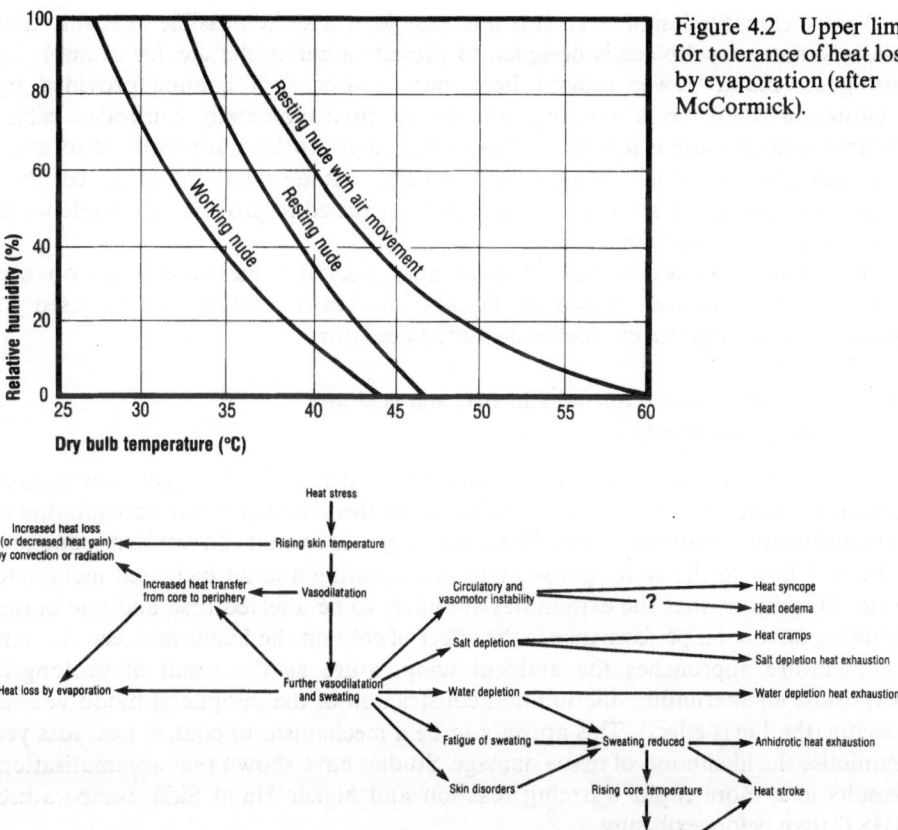

Figure 4.2 Upper limits for tolerance of heat loss by evaporation (after McCormick).

Figure 4.3 Effects of heat stress on the human body (after Leithead and Lind).

Psychological performance is affected as well as physiological performance. Studies have shown that, as the temperature increases beyond 30°C ET, there is reduced performance in manual tracking, morse code use and signal detection. However, there is wide variability in response, and tasks involving response time or time estimation are usually carried out better in hot conditions. Attention-dependent tasks are usually impaired above about 27°C ET, while other tasks are usually unaffected up to 29.5°C ET, but are impaired above this value.

4.2.5 Techniques used to reduce the effects of hot environments

Amelioration of the effects of heat is achieved by one or several of the following: climatic design, workplace protection, climatic limits, work reorganisation and microclimate protection.

Climatic design, where heat loads are reduced at source, is limited to indoor climates, and it is often not feasible to modify these in an industrial setting. Examples are seen in the cotton industry, where warm, moist environments assist the weaving process, or in coal mines, where ventilation, dust suppression and geological conditions limit climatic control. In other industrial processes, for example in coke ovens and blast furnaces, workers can be exposed to high radiant heat sources but still have to wear clothing to protect themselves from the weather.

Where climatic design as such is not feasible, it may be possible to ensure that the operator's workplace is designed to provide a better climate, for example by using screens to lower radiant heat gain, and/or spot cooling provided by ventilation where he is working, or with an environmentally controlled cabin. Where none of these is feasible, for example where workers are required to enter furnaces that are still cooling down to carry out essential maintenance, then climatic limits can be set and work can be reorganised to provide acceptable work and rest periods and refreshment.

Where the heat stress is high, but economic, safety or military considerations are paramount, ventilated, cooled or liquid-conditioned clothing can be used to provide an appropriate microclimate for the operator.

4.2.6 Human response to cold and wet, and cold and dry environments

When the body is cooled below normal core temperature (37°C), consciousness is lost at a core temperature of about 31°C. Below this level body thermoregulation is lost and death eventually occurs. The initial response to cold can result in reduction of blood flow to the skin, 'goose pimples', shivering and an increased metabolic rate. This means that the extremities are likely to be affected first, and one of the limiting factors on performance is the effect of cold on the hands and feet. As skin temperature approaches the ambient temperature as the result of prolonged exposure, an alternating dilation and constriction of the peripheral blood vessels occurs (the Levis effect). This appears to be a mechanism to control heat loss yet minimise the likelihood of tissue damage. Studies have shown that acclimatisation results in a more rapid warming reaction and higher Hand Skin Temperature (HST) than before exposure.

In general, lowered HST has two effects: a numbing of skin sensitivity and a loss of manual dexterity. These seem to occur at 8°C HST for sensitivity and between 12 and 16°C HST for manual dexterity. How quickly a critical HST is reached appears to depend entirely on localised cooling, and maintaining the body at a high temperature does not seem to overcome the localised effects. This implies that the HST must be maintained by localised protection or warming.

There is also evidence that exposure to cold produces psychological stress. Errors increase during exposure to cold, and such factors as wind noise, severe cooling of the extremities and the perceived threat of physical injury from the cold appear to distract the attention. However, once again acclimatisation results in better tolerance to cold exposure and improvements in metabolic performance.

4.2.7 Techniques to reduce the effects of cold environments

As with hot conditions, climatic design, workplace protection, climate limits, work organisation and the design of an appropriate microclimate can lessen the effects of cold conditions.

Climatic design would be the ideal solution. However, when men are required to work in artificial climates such as refrigeration rooms, or in natural climates as on North Sea oil rigs or in the Arctic, alternative methods must be found. Thermal insulation and the design of clothing play an important integral part of all solutions.

Workplace protection may be possible using localised heating and/or protection from high air speeds. Alternatively, a warmed refuge may be made available to minimise exposure time. Where these are impracticable, carefully designed protective clothing plus exposure limits may be adequate. Care must be taken to prevent localised freezing of exposed surfaces such as the face or hands, and workers should not work alone. Work rate can be an important factor because if sweating occurs in over-protected areas of the body, this can result in wetting of the clothing which will increase heat loss tremendously once the activity is reduced.

Heated clothing has been designed in which skin temperature is maintained by the circulation of heated liquid, or by electrical elements. Where protection for both hands and feet becomes too cumbersome it is possible to supplement the CLO values by the use of such techniques.

4.2.8 Machine system design and climate

Machines themselves can be made more acceptable to work with in both hot and cold climates with attention to some aspects of their design. Controls and tools should have low-conductivity surfaces to reduce heat flow. Precise control movements should not be required, especially in extreme cold. Panels and other man/machine interface surfaces should be of low reflectivity, with matt or non-metallic finishes. Insulating floors, ceilings and walls with suitable materials, and double glazing, will also help to reduce heat flow inwards or outwards. Tinted or semi-reflective windows in control cabins will reduce the amount of solar gain or radiant heat gain from industrial processes. In cold climates external glass windows and those over instruments should be protected against frosting and condensation by internal ventilation or by integral heating systems, such as electric elements.

Design for maintenance should aim to reduce to a minimum the extent of accurate work required on site for both very hot and, in particular, very cold climates. Throw-away assemblies, or modular replacement components which can be serviced in workshops, are to be preferred. The use of colour schemes that suggest high or low temperatures, such as red and blue, has been found not to affect a person's response to climatic conditions.

4.3 The visual environment

4.3.1 The basis of visual perception

The process of visual perception provides man with most of his information from the environment (Figure 4.4). The external input to this system is light, defined as radiations that the eye is sensitive to. For present purposes it is important to note that the quantity and type of light reflected to the eye, as well as the physiological and psychological mechanisms involved in its detection, are of paramount importance to visual perception.

The retina, that part of the eye in which light is converted into nerve impulses by chemical action, contains two types of light receptors—cones and rods. Cones respond to the various wavelengths that make up light under normal levels to give the sensation of colour, and they are concentrated mainly in the centre of the retina. Under low light conditions the rods come increasingly into action to give vision that is extremely sensitive, but only in black and white. Time is required to

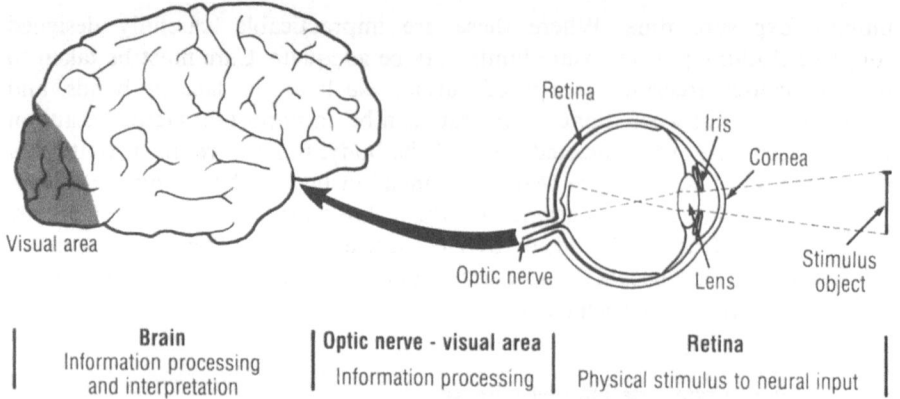

Figure 4.4 Simplified scheme of the arrangement of the visual perceptual system.

adapt from cone to rod vision and the period needed is related to the amount of light available.

4.3.2 Measurement of the visual environment

Light reaches our eyes from two sources. The first of these is from luminous objects, for example a light bulb or a candle. The second, and by far the most common source, is surface reflection. Both of these are measured in luminance units. Although it is useful to know how much light is being reflected from an object and its background to enable contrast to be determined, normally it is the amount of light falling on to the visual attention area that is measured in illuminance units. Figure 4.5 shows the relationship between luminance and illuminance units.

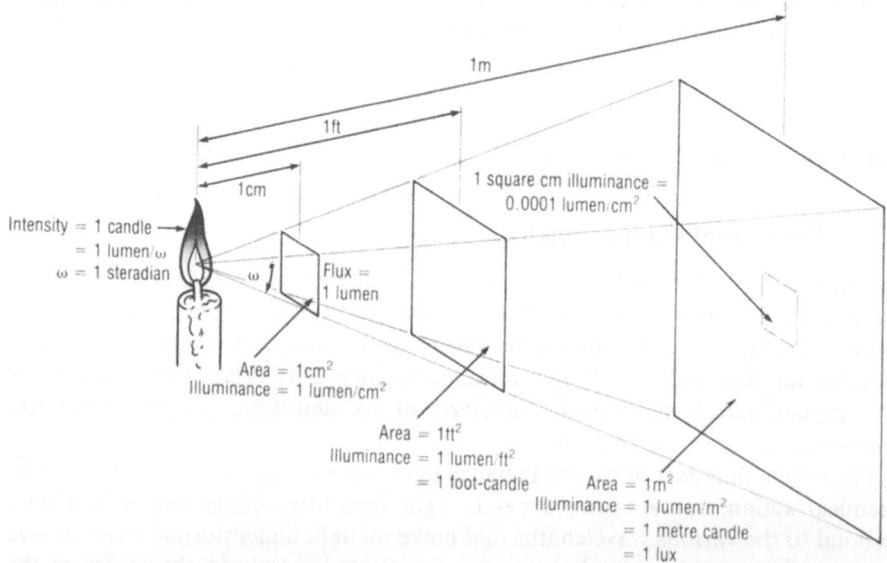

Figure 4.5 Relationships between some intensity units of source and illuminance units on surfaces at various distances.

4.3 The visual environment

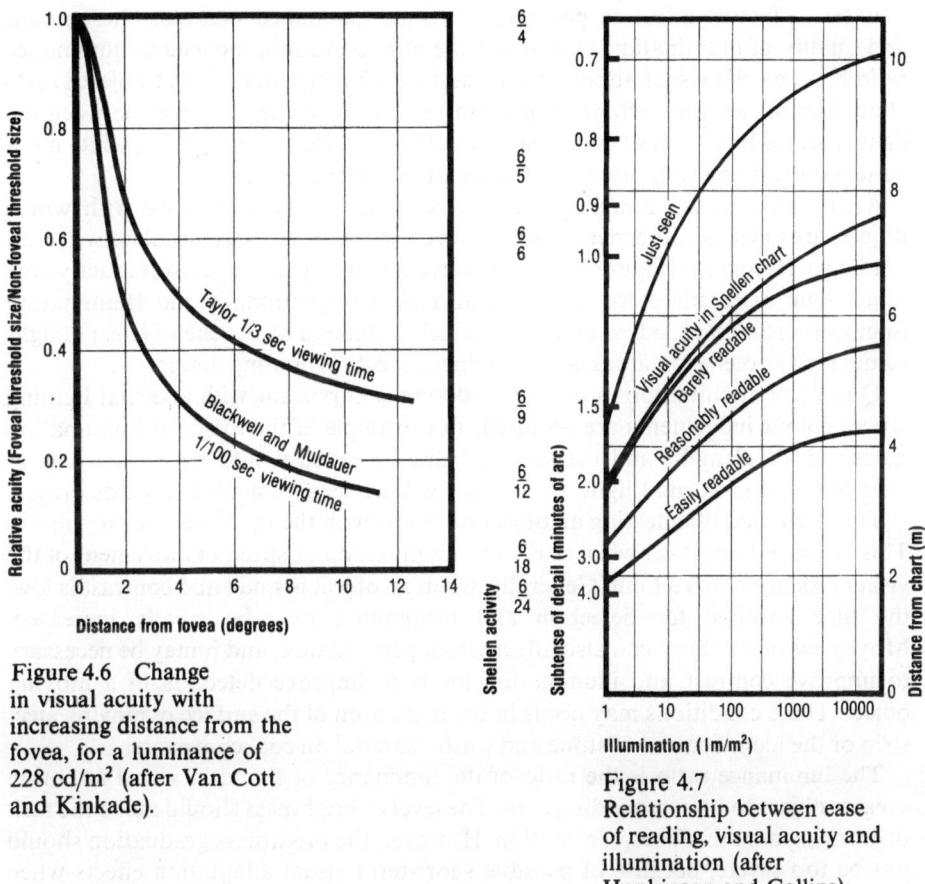

Figure 4.6 Change in visual acuity with increasing distance from the fovea, for a luminance of 228 cd/m² (after Van Cott and Kinkade).

Figure 4.7 Relationship between ease of reading, visual acuity and illumination (after Hopkinson and Collins).

4.3.3 Factors affecting visual performance

Visual performance depends upon the interaction of both external and internal factors. Internal factors are mainly physiological, while external ones depend on the environment.

Normal performance depends upon the eye/brain system working at its optimum efficiency and being able to provide 'normal vision'. Visual acuity is the ability to resolve black-and-white detail. Figure 4.6 shows how acuity decreases with distance from the eye. Convergence, in which both eyes are directed on to an object so that the images fuse together, provides an impression of distance or depth. If either of these functions is impaired, vision will be defective. The normal solution is to correct vision by the provision of glasses or contact lenses, but with ageing a higher proportion of people need to wear glasses.

Colour blindness is also found in a surprisingly high proportion of the population. Almost ten per cent of men have some form of colour blindness and are unable to differentiate between certain colours if they are of equivalent brightness. This has important implications for the selection of employees in industries where colour identification is important, either functionally in engineering or aesthetically, as in dyeing.

Visual adaptation to light levels is also an important physiological mechanism. Temporary blindness can result from a sudden change in light levels, for example, on entering a cinema auditorium.

External factors affecting perception include luminance, contrast, the amount and quality of illumination available, time of observation, movement, luminance ratio between the visual attention area and field background, size of object(s) and glare. Luminance contrast, or simply contrast, refers to the difference between the light reflected by an object and its surroundings. For example, black on white has a better contrast and so is more easily detectable than black on grey.

As the amount of illumination available varies, so does the ease with which objects are detected. In general, as light levels increase, so does visual activity and reading performance (Figure 4.7). However, controversy still rages over exactly how much light is required for a particular task. Conventionally the Illuminating Engineering Society Code is used [4.4] which includes a comprehensive list of light values for various work areas, and examines aspects of lighting design.

Quality of illumination or colour rendering is important with artificial lighting where colour judgements are required. For example, if the artificial illumination lacks red wavelengths and the surface being observed would have been seen as shades of red in normal light, the surface will be seen as shaded towards grey.

The likelihood of detecting an object increases with the time available to view it. This becomes important when size, time, illumination or speed of movement of the visual task approach a limit. Generally, when an object is small and contrast is low, the time required for detection and recognition may be greatly increased. Movement of an object can also affect visual performance, and it may be necessary to improve contrast and illumination levels to improve detection of a moving object. These conditions may occur in the inspection of the surface of moving steel strip or the identification of stone and waste material on coal conveyors.

The luminance ratio is the ratio of the luminance of the task, visual attention area or object to its surrounding area. The level of brightness should make the task or working area the focus of attention. However, the brightness graduation should not be too severe, because of possible short-term visual adaptation effects when looking from the task to the background, and vice versa.

Glare occurs when an object in the field of vision, or the field of vision itself, is brighter than the luminance to which the eyes have adapted, producing either discomfort or disability glare. Discomfort glare may not affect visual performance directly, but it is not conducive to comfortable viewing. Disability glare affects the ability to see. Both types of glare may cause overall distraction by drawing the eyes to the glare source and away from the task.

4.3.4 Principles involved in improving visual performance

An operator will gain most of his information about a system through his eyes via artificial or real visual displays. Individual performance may vary because of poor eyesight, visual correction requirements, adaptation and the possibility of colour blindness, and these factors should be borne in mind during the design of the visual environment. These internal factors interact with the external factors, such as contrast, light levels and types of illumination.

It is difficult to suggest solutions in the abstract, but some general advice is possible. Since the eye takes time to adjust, design should avoid the need to identify visual information at widely different distances or in widely different levels of illumination. It is also helpful to minimise the extent of eye movement required to resolve visual information, by organisation and arrangement of displays rather than by reducing size of detail.

Taking a specific case, the increasingly common use of cathode ray tube (CRT) type visual display untis (VDUs) has created two distinct problems. To maximuse screen-generated contrast, background illumination needs to be low. However, a large number of tasks involve reading a har-copy input such as typescript which must be copied on to the VDU screen. The typescript needs reasonably high illumination to be read, and therefore the whole task has two incompatible requirements. One solution is to shield the VDU screen from background illumination and provide a shielded spot source to enable the typescript to be read. Care must be taken, however, to ensure that the brightness of the screen is comparable with that of the hard copy, so that substantial adaptation changes are not necessary. This example shows that normally problems in the visual environment are of a task-specific nature, so that factors mentioned above should be kept in mind when designing a specific system.

4.4 The auditory environment

Auditory information can be as essential as visual information at the man/machine interface. Communication may be necessary to and from the system, as well as between operators. Evidence also exists that certain auditory environments can affect individual performance, and there may be both long and short-term effects on perceptual performance. Where the former is in evidence, system interface design may well have to include individual protection and so take account of auditory perceptual limitations.

4.4.1 The basis of hearing and sound measurement

Auditory perception, like vision, depends upon the transformation of a physical stimulus to a physiological stimulus by physiological mechanisms (Figure 4.8). Essentially, external air pressure waves are transformed into mechanical impulses which produce fluid waves that eventually stimulate nerve endings in the cochlea of the ear [4.5, 4.6].

The two primary attributes of sound are frequency and intensity, and these correlate with the psychological sensations of pitch and loudness. Normally, sound is measured in terms of intensity in decibels (dB) by using a sound-level meter. However, once a noise problem has been identified, it will be necessary to look at how the intensity and frequency combine, because the auditory system varies in its sensitivity to different frequencies.

This analysis can be done 'on line' by passing the noise directly through a real-time analyser which gives a visual representation of the intensity at different frequencies. Usually, however, the sound is recorded on tape so that it can be analysed as required and a visual record of frequency (in Hz) and intensity (in dB) produced.

4.4.2 Human response to noise

Noise has been defined as unwanted sound, and it has been shown to have both short and long-term effects on human performance. These effects may be internal and physiological in nature, resulting in the auditory system being unable to perceive sound. One type is called nerve deafness, in which nerve cells of the inner ear have reduced sensitivity to certain frequencies, usually the higher ones. Ageing and long-term exposure to high noise levels can result in this type of deafness. Deafness can also result from the physical inability of the ear to conduct sound. This conduction deafness is incomplete, because sound can still be transmitted via the bones of the skull.

It is important to know what types of operator population will be using the system being designed. An ageing population, or one that has been exposed to high noise levels, will have a different ability to detect sounds at particular frequencies than will a 'normal' population. Hearing loss can be measured objectively by presenting pure tones at different frequencies, varying their intensity, and plotting an audiogram which maps out the intensity of tone at which a person just hears that frequency (Figure 4.9). Hearing limits for a particular operator population can be determined, and when it is difficult to select workers for jobs, or when union resistance might prevent them being moved, these limits can be used as a basis for signal design. For example, only frequencies that can be heard by the great majority of a particular working population should be used as the basis of warning signals.

There is evidence that the amount of hearing loss is related to the level of noise to which the operator is exposed. Also, hearing loss is greater in the higher frequency range (over 4 kHz) than in the lower frequency ranges (1 kHz and 2 kHz). Studies also indicate that hearing loss is related to exposure time for higher-frequency intensities, but that it is not related at all, or to a lower extent, to the lower frequency intensities.

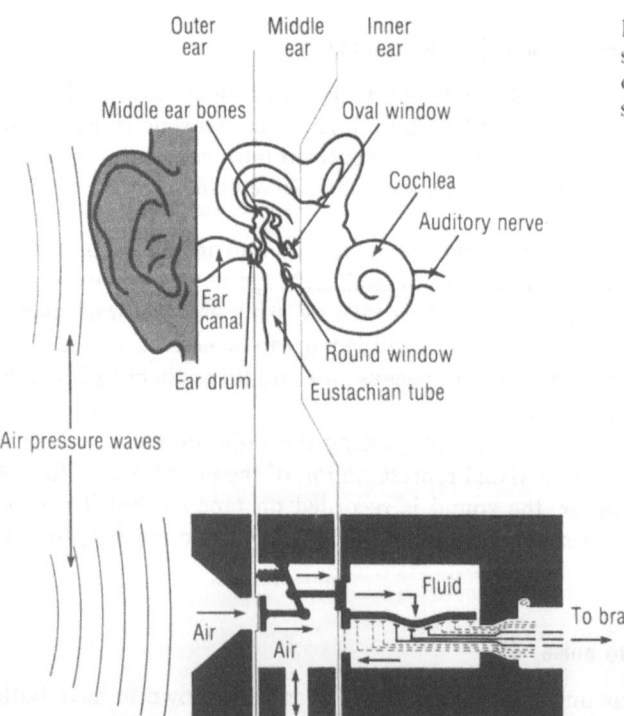

Figure 4.8 Simplified scheme of the arrangement of the auditory perceptual system.

4.4 The auditory environment

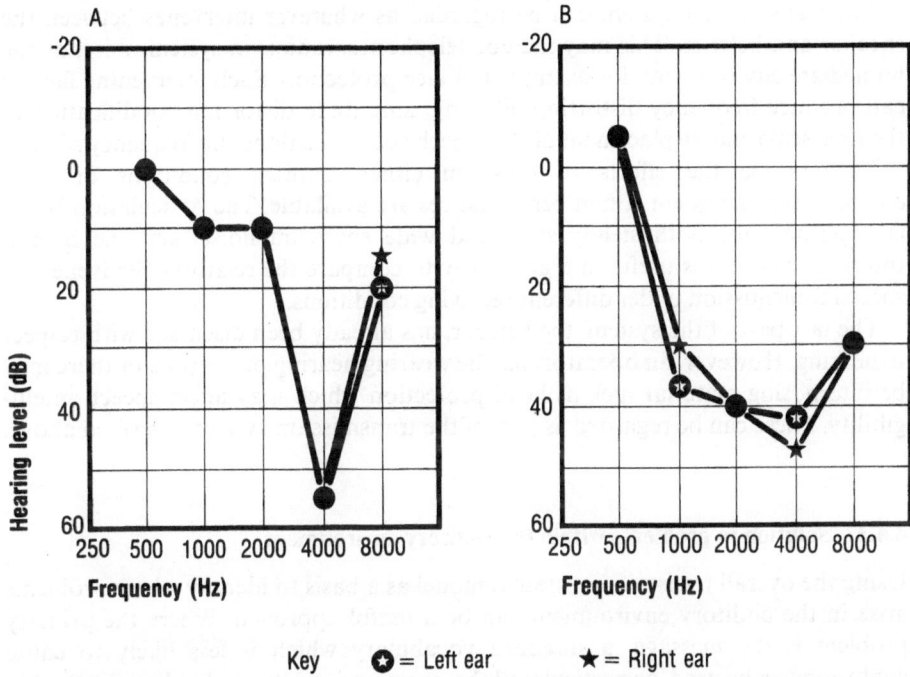

Figure 4.9 Sample audiograms demonstrating hearing loss with age and noise exposure.

Studies have also shown that noise also affects both physiological and psychological processes [4.6, 4.7]. Sudden loud noises which are entirely unexpected can produce an increase in blood pressure, sweating, heart rate, respiration rate and muscular contraction, and repeated exposure may affect a person's digestion. Also tasks involving the monitoring of information over long periods, complex mental operations, skill and speed, and a high level of perceptual activity, are all affected by noise. However, evidence suggests that sound levels of above 90 dB are required to achieve this.

Evidence suggests that operators acclimatise to continuous noise levels after about five minutes, and once they become accustomed to the noise are unlikely to think that it will have an effect on their performance. Also, certain individuals may become irritated as the result of exposure to certain types of noise. Higher pitched noises appear to have less effect than lower pitched ones, but both frequencies below 200 Hz and above 1200 Hz are more annoying than those within that range. Noise that varies in either loudness or pitch also appears to be more annoying, with that varying in loudness having a greater effect. Intermittency, unexpectedness, inappropriateness and reverberation are other noise attributes that can contribute to annoyance which in turn may affect performance.

Speech is liable to be affected at various points in a communication system chain in terms of the message, the talker, the transmission system itself (such as the telephone), the other sounds present, and the listener [4.5, 4.6].

Under certain poor conditions some messages are less likely to be intelligible than others. This relates to the vocabulary used, the context, and the phonetic aspects of the message. As for talker characteristics, speech intelligibility is dependent upon intensity, enunciation, dialect and so on.

The transmission system can be regarded as whatever intervenes between the speaker and listener. This may include telephones, radios, in-system noise, or the immediate environment, for example full-face protection. Such intervening factors can produce frequency distortion, filtering, amplitude distortion, modification of the time scale and displacement of the speech spectrum along the frequency scale.

To describe the effects of noise in either ambient conditions or in a communication system, a number of indices are available. The Articulation Index is applicable for both steady state and wide spectrum noise, and the Speech Interference Level is useful in a gross way to compare the relative effectiveness of speech transmission under different receiving conditions.

The last part of the system, the listener, has already been discussed with respect to hearing. However, an operator may be wearing hearing protection, or there may be interposing material such as head protection which may affect speech intelligibility. These can be regarded as part of the transmission system described above.

4.4.3 Solution to problems within the auditory environment

Using the overall transmission system model as a basis to identify where problems arise in the auditory environment can be a useful approach. Where the primary problem is the message, a standard vocabulary which is less likely to cause confusion can be used, or particular alpha or numeric parts can be eliminated, as in air traffic control transmission.

Where the problem is the talker, speech training may be necessary to correct those aspects that are at fault. In a large number of industrial situations, however, such a remedy is unrealistic. The transmission system should also be designed to maximise intelligibility. Where a system is designed to filter speech and it filters out frequencies above 4 kHz or below 600 Hz there would be little effect on intelligibility. However if those between 1 and 3 kHz were filtered out then the transmission would be unintelligible. Where power constraints determine system characteristics amplitude distortion can be utilised because the peaks of the sound intensity waves can be 'clipped' without losing too much intelligibility, but when the centres are removed the result is practically unintelligible.

Where noise is present and has been identified as a problem, it should be controlled in the following ways. Ideally it should be reduced at its source where it is feasible to do so. This may involve the redesign of an engine combustion chamber to eliminate troublesome frequencies and intensities, or it may involve the redesign of an exhaust system. Alternatively the noise source may be the result of inadequate maintenance rather than design, in which case the remedy is straightforward. Where the noise is generated mechanically at a secondary level this can be prevented by the insulation of the machine from the secondary source, for instance by the use of rubber mountings.

Noise can be isolated by the use of enclosures, rooms and other barriers, and air transmission can be reduced by ensuring airtightness. For example, closing the windows in a home can typically achieve a 10 dB reduction in noise levels. Where system design requires the operator to see operations outside an enclosure, noise transmission into the enclosure through glazed areas can be reduced by the use of double and triple glazing. The use of baffles, sound absorbers, and the acoustical treatment of walls, ceilings and floors can also reduce sound reflection.

Where none of these measures reduces the sound to an acceptable level, auditory protection is necessary and the present Department of Employment Code of Practice on Noise [4.8] recommends a limiting exposure of 90 dB(A) for eight hours or, where the sound level fluctuates, an equivalent continuous sound level (Leq) of 90 dB(A). Auditory protection must be provided to reduce operator exposure to these levels. Finally the attributes that the listener must have in order to be most successful in understanding speech in noise are as follows. He needs to have normal hearing, to be trained in the type of communication used, to be able to understand the stresses of the situation and to be able to concentrate on the required stimulus.

4.5 Motion effect on the human body

This section is concerned with the effect of vibration, acceleration, deceleration and weightlessness on the body. Essentially this involves the effect of whole-body mechanical motion or lack of it. Again there is inadequate space to look at these aspects in detail but the essentials are discussed below to give an introduction to the subject.

4.5.1 Vibration

The field of vibration research does not provide such clear-cut results as that of noise. There do not appear to be clearly demonstrated 'vibration-induced symptoms which are both general and permanent', symptoms, nor occupational disease equivalent to noise-induced hearing loss. However, there is evidence of vibration injury to the hands under certain conditions, and those who have travelled by train and tried to write or drink while crossing points are only too aware of certain aspects of lateral vibration.

Usually vibrations of the air are detected as sound, but air vibrations below approximately 20 Hz are not heard but can be felt. Movement below 1 Hz is termed low-frequency vibration and is met in ships at sea, aeroplanes and cars. It can be responsible for motion sickness. The range 1 to 100 Hz contains high-frequency vibrations. Vibration, like sound, is described in terms of frequency (in Hz) and the amplitude or intensity of the individual waves is normally expressed as acceleration (in ms^2).

Acceleration can be measured in three dimensions based upon an orthogonal co-ordinate system having its origin at the heart (Figure 4.10). The x, y and z axis accelerations are commonly referred to as sway, lateral and heave vibrations [4.9]. The accelerations are measured using transducers (accelerometers) that pick up the vibrations, which are recorded on a tape recorder for later analysis. Vibration is important in industrial settings because, in the same way as sound, there is some evidence that it can affect performance and cause physical damage.

The field is normally divided into the examination of whole-body and localised (for example hand-arm) vibration. When the whole body vibrates, the various parts of it do not vibrate in unison nor at the same frequency. This can result in alternating displacements of body parts and organs creating general discomfort and anxiety. Also the vibrations can be amplified or attenuated by the body posture, the type of seating and the vibration frequency. For example, if 5 Hz were being transmitted up the body of a pilot in a cockpit, the result could be a doubling

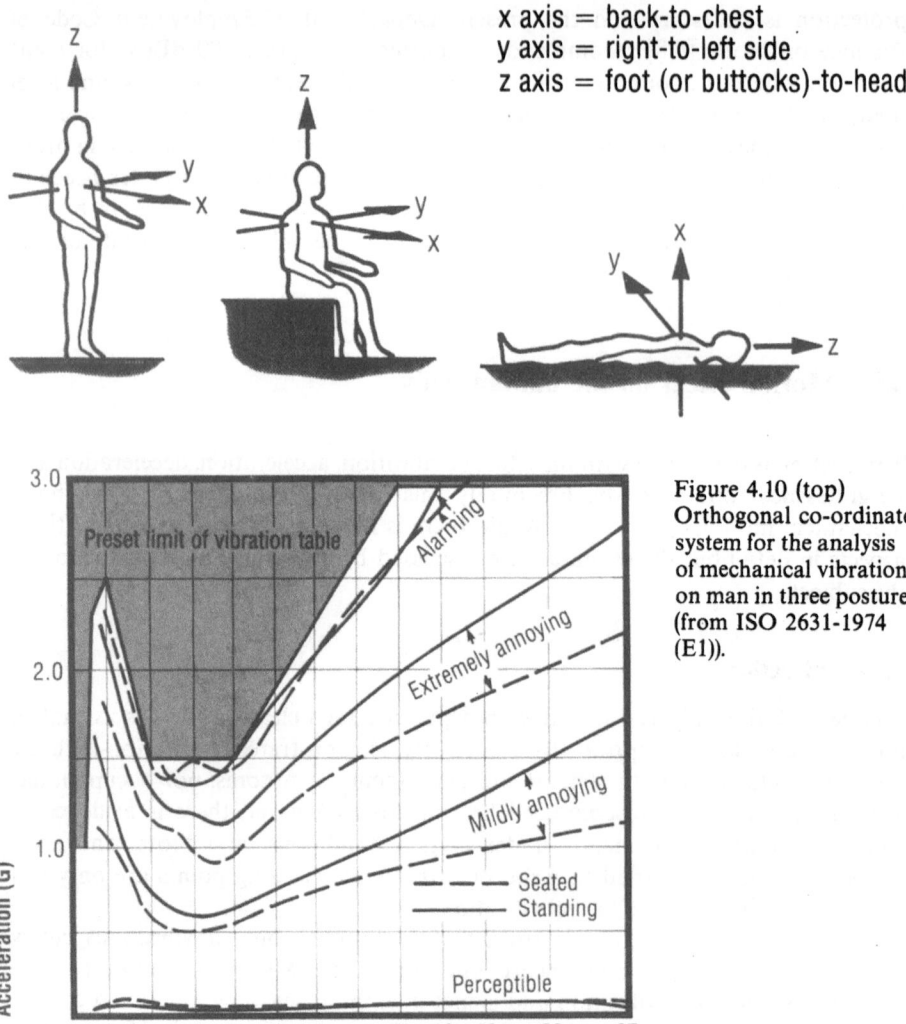

Figure 4.10 (top) Orthogonal co-ordinate system for the analysis of mechanical vibration on man in three postures (from ISO 2631-1974 (E1)).

Figure 4.11 Subjective reactions of seated and standing subjects to whole-body vibration (after McCormick).

of the frequency applied at the buttocks compared with that at the head. The instrument panel would be vibrating at a lower frequency than this, so instrument reading and control button selection could be difficult. At this frequency (5 Hz input) handwriting also deteriorates and the size of letters is considerably increased.

Localised vibration normally refers to the situation when a part of the body, such as the hands, is subject to vibration. The localisation will depend upon the source, so that in the case of a chain saw the vibrations will be almost totally restricted to the hands and arms, but with a pneumatic drill they are normally transmitted to the upper part of the trunk and the head. One well documented effect of vibration occurs when the hand or hands are subjected to vibration over a period of time as the result of holding a vibrating tool or workpiece. Vibration-

4.5 Motion effect on the human body

induced white finger (VWF) can develop as a result, which is a disorder of the finger blood supply that becomes evident in cold conditions where finger numbness occurs and there is pain on recovery. This condition has been noted in users of pneumatic tools, hand-held and pedestal grinders and chain saws.

Both subjective and performance effects of vibration have been reported. Figure 4.11 shows subjective reactions to vibration of various frequencies. Research has also indicated that vibration can affect performance on certain types of visual task, motor tasks that require precision and co-ordination, and some complex psycho-motor tasks, such as target tracking. Ideally vibration should be minimised at source. Normally protection from residual vibration is achieved by reducing the forces transmitted, by converting vibration energy to thermal energy using mechanical or hydraulic dampers (these being mounted as close as possible to the source of vibration) or by altering the body position and body support. Additional protection can be incorporated in seat suspension systems by introducing flexibility provided by springs, and again by addition of damping. A man's legs are a most effective attenuator of vibration, but standing postures have ergonomic disadvantages which discourage their adoption, unless otherwise essential.

4.5.2 Acceleration and deceleration

When the whole body is forced to change its velocity this results in a degree of movement of the body tissue, organs and blood, but acceleration below about 1 G is not significant. The acceleration that arises in collisions, in the catapult take-off of aircraft, or in the launching of space vehicles, can be very much higher. Individuals vary in the effect acceleration has on them, and training can increase their capability of resisting it. Figure 4.12 shows voluntary tolerance levels for linear accelerations.

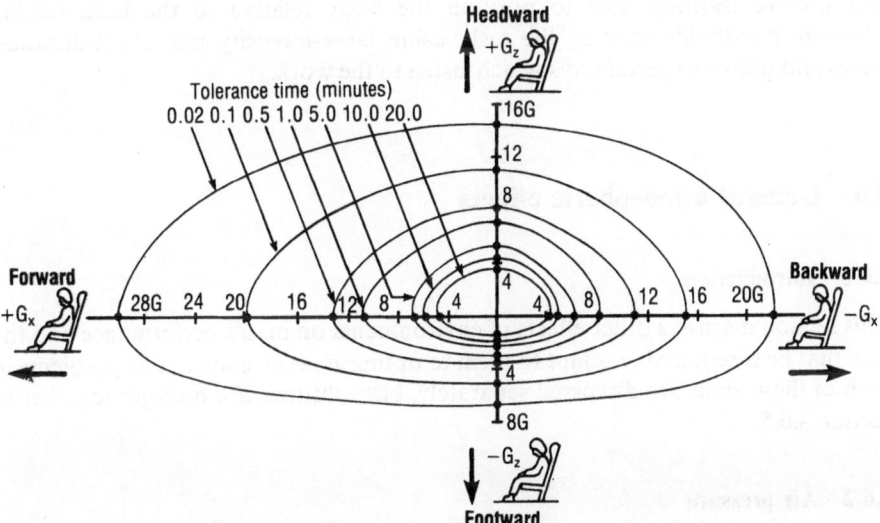

Figure 4.12 Average linear acceleration levels for different directions voluntarily tolerated for periods of 0.02 to 20 minutes (after McCormick).

Two other types of acceleration other than linear can be distinguished. Radial or centrifugal acceleration is produced when the body rotates 'head over heels'. Angular acceleration occurs when the body rotates about a feet to head axis, and 'twirls' round laterally.

For all these accelerations, too little or too much blood in the head has major effects. Visual dimming ('greyout') and blackout, and the loss of consciousness are two effects that occur when the blood pools in the lower part of the body, while confusion, pain and haemorrhage ('red out') occur when blood and fluid are forced into the upper parts of the body. Visual and motor performance can be expected to be affected adversely at forces of 2 to 4 G.

Protection against acceleration and deceleration take the forms of body restraint, changing body posture by means of seat or couch design, and by using compensating pressure suits and helmets. Obviously these solutions all have a marked effect on the man-machine interface design.

4.5.3 Weightlessness

This condition is the result of the absence of gravity (zero g) and is of greatest importance in manned space flight. Short-term physiological effects include nausea, vestibular disturbances, blood pooling, blood pressure drop and various effects on heart rate. Longer-term effects so far have shown weight, muscle and strength loss as well as height increase. The last applies specifically to the trunk and neck, which may have implications for the design of pressure suits, clothing, work stations and control stations where eye level is critical to visibility.

The body also assumes a 'weightless posture' and as a result effort is needed to change this posture, leading to discomfort, fatigue and inefficiency. Muscle and strength effects can be minimised by onboard training.

Performance in weightlessness varies. Research has shown that a free-floating worker may have problems of orientation, inadvertent tumbling, locomotion, stabilisation and material handling. Manual tasks require increased time to use push-buttons, toggle switches and rotary switches. There are, however, solutions that involve learning how to position the body relative to the task, having adequate handholds near to the task, using large-intensity and short-duration forces and utilising special tools which fasten to the work.

4.6 General atmospheric effects

4.6.1 Introduction

This section discusses other effects of environments on man's performance and the way that he is required to adapt to them to optimise performance. The problems in each of these areas are discussed separately, but solutions are brought together in section 4.6.5.

4.6.2 Air pressure

This variable is important to those who work at high altitudes, as in mountainous regions and aircraft, and below sea level, in diving and under-water construction.

4.5 Motion effect on the human body

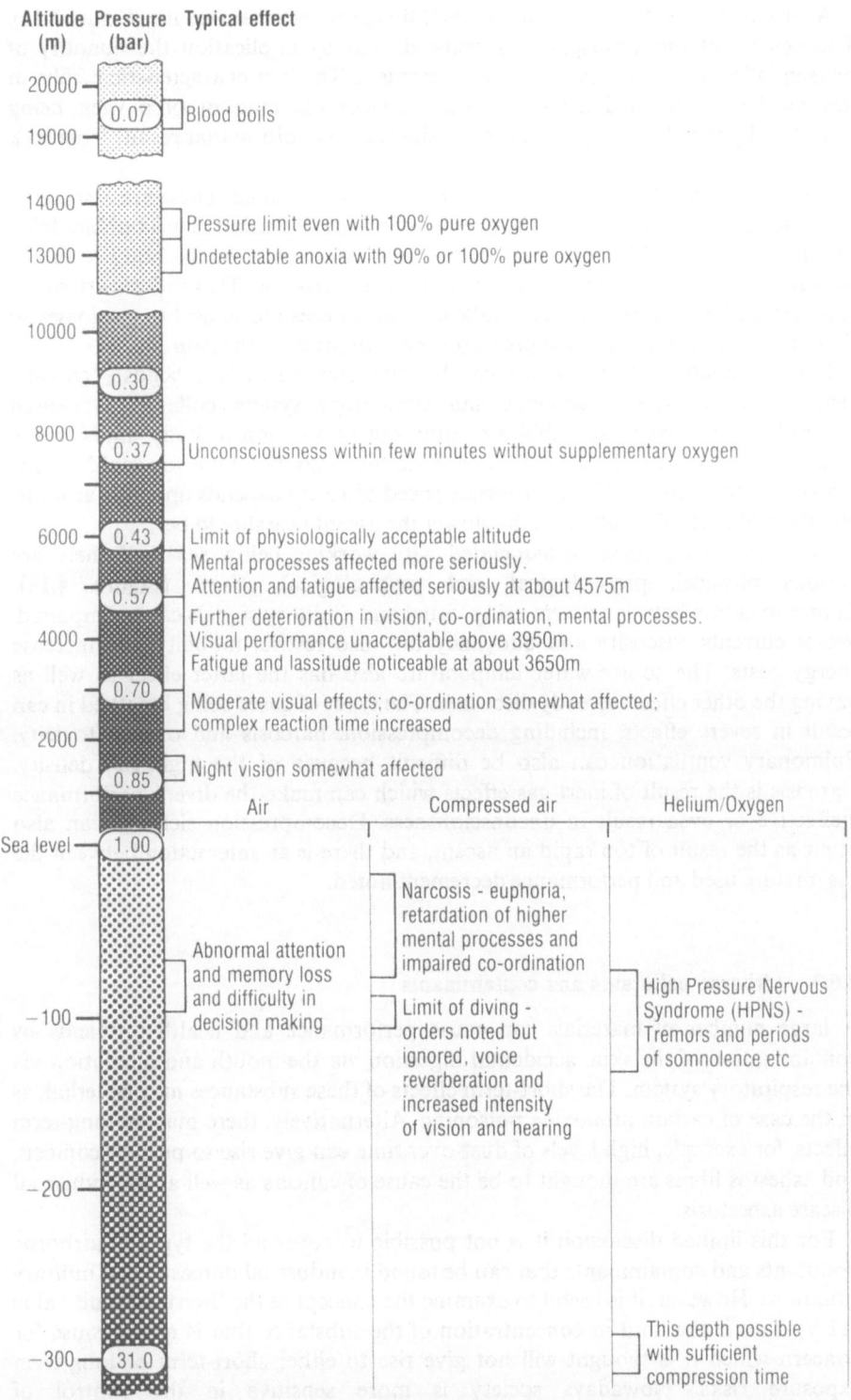

Figure 4.13 General effects of altitude and lack of oxygen, the effect of breathing pure oxygen and effects of increased pressure in diving with three breathing mixtures (after McCormick).

As the altitude increases above sea level there are two main results (Figure 4.13). The density of the atmosphere is reduced, and by implication the quantity of oxygen falls, and the air pressure also decreases. The first characteristic results in less air being breathed in; the second reduces the amount of oxygen being absorbed by the blood. Hypoxia (or anoxia) is the condition that results from lack of oxygen.

There are two other general effects which should be noted. These are associated with the expansion of gases in body cavities and the liberation of gas bubbles from the blood. As the altitude increases, so gases can expand in the body to create intense pain, the abdominal cavity being the most sensitive. These effects are rarely experienced below 8000 m. The middle ear and sinuses are subject to gas losses so that in descent it is the external pressures that can produce the pain.

Decompression sickness, which can be characterised by 'the bends', 'chokes' central nervous system collapse and circulatory system collapse, is caused generally by nitrogen gas bubbles coming out of solution in body fluids. If the ascent is slow enough, the gas may escape via the lungs, but if it is too rapid the gas is given off too quickly. The appropriate speed of ascent depends upon the altitude, and the higher the final altitude, the slower the ascent rate should be.

At increased air pressure associated with working below sea level there are various physical, physiological and psychological effects (Figure 4.13). Communication between divers using visual and auditory signals can be impaired. Water currents, viscosity and buoyancy can also restrict mobility and increase energy costs. The cooler water temperature also has the latter effect as well as having the other effects already discussed. The types of gases being breathed in can result in severe effects, including decompression, narcosis and oxygen toxicity. Pulmonary ventilation can also be difficult because of the high gas density. Narcosis is the result of inert gas effects which can make the diver's performance ineffective or even result in unconsciousness. Decompression sickness can also occur as the result of too rapid an ascent, and there is an interaction between the gas mixture used and performance decrement noted.

4.6.3 Airborne pollutants and contaminants

A large number of materials can cause performance and health problems by contamination of the skin, accidental ingestion via the mouth and inhalation via the respiratory system. The short-term effects of these substances may be lethal, as in the case of carbon monoxide poisoning. Alternatively, there may be long-term effects, for example, high levels of dust over time can give rise to pneumoconiosis, and asbestos fibres are thought to be the cause of cancers as well as the industrial disease asbestosis.

For this limited discussion it is not possible to cover all the types of airborne pollutants and contaminants that can be found in industrial domestic and military situations. However, it is useful to examine the concept of the threshold limit value (TLV). This is the limit in concentration of the substance that is giving cause for concern which it is thought will not give rise to either short-term or long-term exposure risks. Nowadays society is more sensitive in the control of carcinogens—those substances which research indicates are cancer-producing—so various TLVs are being recommended.

Research in America has indicated one area of concern—the control of these substances in coke ovens—where even with modern designs it has been difficult to control emissions. To rebuild existing coke ovens would be uneconomic, so a solution has been sought in two areas, namely through attempts to reduce the general atmospheric TLVs by coke oven redesign and to reduce the inhaled TLV by respiratory protection. The second approach, however, may affect performance in the coke oven environment where there are also climatic, auditory and visual effects (especially at night) to be considered.

4.6.4 Ionising radiation

Normally individual exposure to radiation is limited to that experienced during X-ray photography. However, those who work these machines, as well as workers in an increasing number of industrial situations and those in the nuclear energy field, may be exposed to various radiations. Ionising radiation penetrates matter and can therefore be a useful component of control systems where normal sensors are not feasible.

4.6.5 Design solutions to counter atmospheric effects

As with earlier attempts to minimise the effects of the environment on performance and health, there are four main approaches here. These are: to eliminate the effect at source; to provide localised protection to reduce the intensity; to provide refuges and design the work to reduce the exposure time and, finally, to isolate the individual completely from the conditions within a sealed structure or by the provision of an independent microclimate in the form of a personal protective system.

Elimination at source. Obviously this solution is difficult with atmospheric pressure, but it may be feasible to stop airborne emissions or radiations at source. However, there are many traditional industrial processes that have relied upon being able to release waste to the atmosphere, and in the short term redesign may not be economic. Redesign might also affect the tasks of the operatives and may also result in job loss, so there may be staff and union resistance to this type of solution.

Localised protection. Again, this is only feasible in the case of airborne contamination and radiation. A ventilated curtain may prevent the airborne problem from reaching the operative, or alternatively adequate rates of ventilation may reduce the likelihood of exposure to concentrations above the specific TLVs. Radiation-absorbent barriers may be a solution for the fixed operative but would have to be carefully designed. An example is a shielded area in X-ray rooms.

Refuge provision and work design. This approach is only really feasible with airborne hazards. Again the principle involved is to design the work so that, by spending some of the working time in a structure that contains normal unpolluted atmosphere, the overall cumulative effects of the hazard stays below the TLV. This may produce difficulties when there are problems with the production cycle and it can also be difficult to control worker exposure.

An example of a refuge design is the operating theatre (Chapter 18.13) which protects patients from airborne bacteria during an operation by providing a sterile working environment.

Work station isolation. The provision of a structure which protects against the effects of altitude, such as the pressurised cabin of an airliner, or against the pressure of the deep, such as in a submarine, is an ideal solution. Total protection of this form against airborne hazards may not be acceptable in industry because of the costs involved. However, in principle, if the operator can carry out his process control from such a structure then this would be a design solution.

Personal protective systems. This approach is favoured because of the relative cost savings and flexibility it affords. Personal protective systems range from the total protection afforded by pressurised suits for high altitudes, space flight and sea diving, to partial protective measures such as the simple face dust mask designed to filter out particulate matter.

The detailed design implications of the provision of such protection cannot be dealt with here, but a number of observations can be made. The first of these is that the degree of protection afforded greatly affects the man-machine interface. Secondly, the relevant aspects of the design of the microclimate produced as the result of the specific protection will have to take all relevant environmental factors into consideration, and should produce a total design solution. Thirdly, the wearing of the protective system may produce a weight and performance penalty. Fourthly, where respiratory protection is required, breathing air through a static barrier, such as a simple face mask, imposes a work load out of all proportion to its weight. Powered positive-pressure systems are normally preferable for this reason. Fifthly, the population who will be exposed to the various effects will vary in their tolerance, and they will also vary in the way that the protective system affects their performance. Care must therefore be taken in the use of data based upon populations not typical of the target population. Finally, expert help should be utilised when problems in the above areas are identified by the design engineer.

4.7 Conclusions

This chapter has indicated the complexity of human response to the physical environment and how human limitations interact with design considerations. When in doubt, consult the specialist. However, there are a number of general texts shown below which will fill in the technical gaps.

Discussion of environmental ergonomics as a separate entity is a convenient way of emphasising the environmental aspects, but in reality ergonomics is an integrated science, and this must be borne in mind when examining situations where man and work interact.

4.8 References

4.1 Leithead, C. S. and Lind, A. R. *Heat.Stress and Heat Disorders.* Cassell Ltd, London, 1964.
4.2 Graves, R. J., Leamon, T. B., Morris, L. A., Nicholl, A. G. McK., Simpson, G. C. and Talbot, C. F. *Thermal Conditions in Mining Operations.* Institute of Occupational Medicine, Edinburgh. NCB Technical Memorandum TM/80/9, 1981.
4.3 American Conference of Governmental Industrial Hygienists. *TLVs. Threshold Limit Values for Physical Agents Adopted by ACGIH for 1980.* ACGIH, Cincinnati, 1980.

4.4 Illuminating Engineering Society. *IES Code for Interior Lighting*. IES, London, 1971.
4.5 Morgan, C. T., Cook, S. C., Chapanis, A. and Lund, M. W. (eds) *Human Engineering Guide to Equipment Design*. McGraw-Hill Book Co, New York, 1963.
4.6 McCormick, E. J. *Human Factors Engineering*. McGraw-Hill Book Co, New York, 1976 (fourth edition).
4.7 Murrell, K. F. H. Ergonomics. *Man and His Working Environment*. Chapman and Hall, London, 1965.
4.8 Department of the Environment. *Code of Practice for Reducing the Exposure of Employed Persons to Noise*. HMSO, London, 1971.
4.9 International Organisation for Standardisation. *Guide for the Evaluation of Human Exposure to Whole-Body Vibration. ISO 2631*. ISO, Geneva, 1974.

4.9 Further reading

Burns, W. *Noise and Man*. John Murray, London, 1973 (third edition).

Edholm, E. G. *The Biology of Work*. World University Library: George Weidenfeld and Nicholson, London, 1967.

Fox, W. F. 'Human performance in the cold.' *Human Factors*, vol 9, no 3, 1967.

Fanger, P. O. *Thermal Comfort Analysis and Applications in Environmental Engineering*. McGraw-Hill Book Co, New York, 1972.

Gregory, R. L. *Eye and Brain. The Psychology of Seeing*. World University Press, 1966.

Hopkinson, R. G. and Collins, J. B. *The Ergonomics of Lighting*. Macdonald Technical and Scientific, London, 1970.

Kerslake, D. McC. *The Stress of Hot Environments*. Cambridge University Press, 1972.

5 Form

Peter Murdoch and Charles H. Flurscheim

Part 1—The theory of form

5.1 Introduction

Form is perhaps the most instantly perceived of all the elements of design, and in this lies an immense difficulty for the present-day designer. Whereas in the past form or shape evolved principally through straightforward functional considerations, this is no longer the case.

In engineering, improvements in both technology and techniques, instead of easing the problem, have made the designer's task more complicated. The reason for this is that many machines are now so complex that there are no longer simple functional clues for design techniques to follow, and there are now a host of considerations to be taken into account before the final form is achieved.

This chapter, therefore, is intended to provide not only a background to what form is, but also to give some practical guidance on how to interrelate design elements to provide a sound basis for design. Without first analysing the principles of form and the various factors dictating the design approach, the designer is working in a void, and runs the risk of arbitrarily seizing on one or other aspect as a starting point.

Prior analysis will show that form has evolved through a number of routes. There is an acceptable relationship between man and form which is important, and the psychological and ergonomic factors which are outlined elsewhere in this book give clues to this developed relationship, and should be examined before approaching the design problem.

5.1.1 What is form?

Put in its simplest terms, form is the image presented by the outer surface of an object or structure. But form is also the way in which the object is perceived or the manner in which it is able to be perceived.

Form cannot therefore be treated purely as a mathematical or functional problem. It must be viewed from the perspective of the potential user, and in relation to its environment. In this respect industrial design differs from conventional sculpture—the object is not viewed in isolation as a work of art, but in its context.

Like sculpture, however, form sets up a relationship or dialogue with the user or perceiver. It is based on certain factors such as a knowledge of what people like or dislike, what they fear or find threatening, what their aspirations might be and how they react to complexity or simplicity.

Form is also able to convey certain messages about the object in question—for example that it is fast, complex, dangerous, relaxing, modern or efficient. Some of these messages will be conveyed as a result of styling or fashion, but most will rely on the actual shape to convey this important immediate information.

Levels of perception vary according to the individual, but the initial perception is always of form, followed closely by colour and style. For example, the form of a racing yacht will look impressive to the layman, but it will mean much more to the experienced yachtsman, who is able to perceive the interaction of the elements within the particular shape chosen.

In engineering design a road bridge can also convey the same double message—to the layman it can be seen as a 'work of art', an example of symmetry and elegance. To the structural engineer it is a question of balance, stress, tension and materials contriving to do a functional job. If the design can also be elegant and appealing, this fulfils both aspects and creates the optimum response.

5.1.2 Basic psychology

There is, therefore, a fundamental psychological rationale that can be applied to form development, and the engineer can usefully employ this as a 'bench mark' when first considering the form of a product.

During the evolution of man, form has developed as a definite part of various functioning roles. For example, aggression is often associated with bright colours and/or large size; sexuality through curves and smell and certain colours; and protection by form and colour which blends with the background. It seems likely that certain forms have always provoked similar responses throughout the ages, and continue to do so today.

Thus, if the engineer first considers the effect that he wishes to achieve with the product, in relation both to users and to the environment, he has a clue to the development of its form. Does he want it to be anonymous, positive, aggressive, quiet, for men, for women, dangerous, relaxing? Does he want a combination of attributes? These are questions to be answered before design gets under way.

What then needs to be added to these basic concepts are other subtleties, such as use of materials, function, the competitive market-place and stylistic variations, and these form the subject matter of this chapter.

5.1.3 Development of form

As was mentioned in Chapter 1, three-dimensional shape has been developed and refined by man in the same way as other cultural and technological aspects. Originally, natural forms were utilised or taken as models for products. Nature has evolved its own forms, even the most aesthetically beautiful of which are basically functional. Thin, flexible forms sway with the wind and avoid damage; large bells or cups trap moisture or sunlight; broad leaves offer large surfaces for photosynthesis and protection—added to which are shapes important for both camouflage and attraction.

Man copies nature, but he did not have to use the same exact techniques once he had mastered the use of new materials. However, even today there are certain natural forms and constructions which man has only just been able to emulate: the wheat stalk and the spider's web illustrate some of nature's perfect solutions to problems of weight, tension, stress and material selection.

Man's first use of materials was for protection—mud, straw, leaves and boughs for structures, fibres for clothes and blankets. He utilised wood and then stone for weapons and tools, soon moving on to devise better ways of living, decorating his home and his clothes with paints and dyes, building better structures, and using fire and tools effectively.

The great Chinese, Greek and Roman civilisations all added to the new use of materials, and sculpture and architecture went hand in hand with intellectual and philosophical development. Plato, Euclid, Socrates, Galileo, Leonardo da Vinci, and many others brought society through to the modern age and the start of the Industrial Revolution. More recently another revolution in technology has brought robotics and miniaturisation of components, thus bringing industry more automation and less need for man-operated machinery.

During this time engineers have continued to study nature as a model for construction and development. An example is Louis Sullivan with his tension and compression structures based on the Morning Glory flower. Now, the new technique of continuous glass fibre structures, developed by Dr Math (Mathweb) of British Petroleum, go a long way towards helping man to emulate the spider.

Developments in rotational moulding, ceramics, glass, controlled crystallisation of metals and many other areas have all introduced new shape possibilities, so now the engineer is more often than not required to be the arbiter of shape and form, rather than being overtly constrained by necessity.

It has, however, become possible to distinguish three distinct elements in the design of form which can act as guidelines for the designer, and it is worth studying these in detail.

5.1.4 Form and function

Today one sees most evidence of 'form as an expression of function' in the industrial, as opposed to the consumer, sector of engineering. Apart from the reasons already outlined, this is also because peripheral styling or cladding material is very often seen as detrimental to the product concerned. To do anything to let form supplant function when designing such items as a plough, an aircraft, military hardware or power-generating equipment would be a folly, and the very nature of the functional requirement generally suppresses such action.

But function no longer stands as the sole criterion for the design of form or shape, nor need it be the primary aspect to be considered if other factors have a prior claim.

Where there is no direct relationship between man and machine, for example, it may well be possible for functional aspects to claim priority. But it may also be that secondary functional needs take over from the *primary* ones. An obvious example of this is in the world of electronics, where fitting electronic equipment into a compatible frame may produce a unit size which bears no relationship to the internal components. Similarly, where remote control techniques are available, the product may be of almost any size, with a console or module which is functional only in terms of the operator.

Design of form must therefore pay attention to the interrelated functions of machine and user. Where there is a direct relationship between man and machine, ergonomic considerations are at least as important as purely mechanical ones. The machine must not only operate efficiently, but be operated efficiently.

5.1.5 Form and fashion

As specifications, manufacturing techniques and materials become more standardised internationally, so also does the design of products. Given that this trend is no longer confined to items such as clothes, but is now apparent in engineering-based products such as cameras and cars, it is clear that fashion and style will become increasingly important. It is likely that the future use of form in engineering design will need the sort of understanding of the subject that fashion designers apply today.

Cars are perhaps the best example of this at present. It is not enough just to be a good design engineer working on cars, or in any other area where similar developments are occurring. It is necessary to be in tune with the current trends in styling and colour. As competition increases for the sale of technologically similar products, the engineer may need to be not only up to date in the various techniques required, but also to link his own personality with the design. In the world of cars, names such as Ghia, Farina and Issigonis have become associated with specific styles of design in the same way that Cardin, Gucci and Zandra Rhodes mean certain types of clothes.

5.1.6 Form and size

As the micro-chip and other forms of electronic miniaturisation are used more and more in engineering, less and less can the form of a product gain its *raison d'être* from either its mechanics or its basic operation. The result of this lack of inherent guidance can be a totally arbitrary form of development.

It is important that engineers understand that the rationale for form in these situations must be looked for in the relationship between the man and the machine. Because neither the man nor the device in isolation is able to give positive feedback that might be used to develop an appropriate form, the relationship becomes of prime importance (see Chapter 11).

A good example of this is in the design of watches. It is possible to design a watch so small that it ceases to work as a watch, both for functional and aesthetic reasons. The wearer could not adjust, let alone read, such a watch. Today's successful electronic watches are designed to take into account both use and mechanics, but above all the psychology of the purchaser.

Microprocessors have also taken over many of the functional roles within modern hi-fi or stereo systems, many of which could be designed to fit into much smaller units. But the needs of the consumer dictate the size, both in functional and aesthetic terms, since the equipment must suit the domestic environment. So the designer pays attention to form and produces the unit on a larger scale.

5.2 Balance and proportion

5.2.1 Historical ratios

One objective of good design of form is to make the product or assembled unit appear balanced. This can be achieved in a number of ways, most of which depend on the user's perception and understanding. Proportion and disproportion are aesthetic factors, the appreciation of which may be enhanced by knowledge, but which have their roots in ancient history.

Several ratios of dimensions have been developed historically in the world of

5.2 Balance and proportion

architecture that experience has shown to be aesthetically satisfactory. In engineering these ratios may be followed where they can be introduced without detriment to other aspects of design, but they are guides, not rules. These ratios include: the Golden Mean suggested by Euclid and Leonardo da Vinci, which calls for a ratio in rectangular dimensions of 5 to 8; the ratios developed by Palladio from the acoustical mathematics of music, such as 3 to 4, 3 to 5, 4 to 5, and including once more 5 to 8, which he used successfully in his villas; and the double cube, used by Adam, which as its name implies, defines a rectangular volume in which width and height are equal, the length being twice this dimension, a ratio which has been used in many large rooms.

None of these ratios was conceived in terms of engineering, but when they are appropriate the project will usually benefit.

The visual effect of some of them is shown in Figure 5.1, alongside a cube, which by comparison tends to look clumsy. They can be seen also in examples of

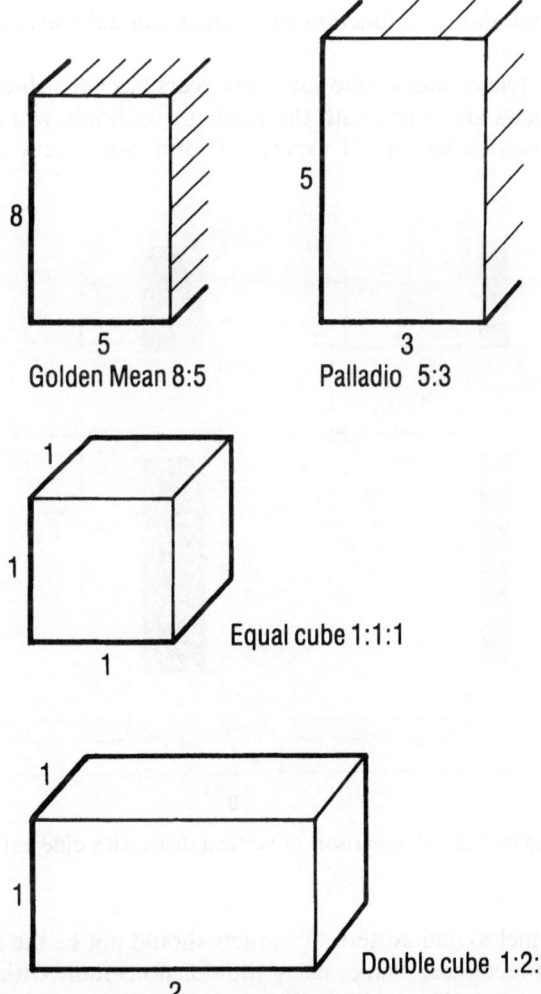

Figure 5.1 Comparison of three-dimensional shapes with different ratios of dimensions.

products which happen to conform approximately to these ratios. These include the Chubb Cash dispenser (Chapter 18.4), which has an outer panel in the ratio of 4 to 5 and an inner recess of 5 to 8; the True Data Computer System (Chapter 18.10), which has an outer panel of 4 to 5, the inset panel being 3 to 4; and the Penlon Ventilator (Chapter 18.6), which conforms closely to a double cube.

In most cases, however, when adjustment of proportions is possible, selection must rely on trained perception and what looks right to the designer. An example would be a small powered drill (Figure 5.2) in which the diameter of the cylindrical support column is dependent not only on the stiffness required to maintain accuracy, but on appearance. The design using a solid bar (Figure 5.2a) probably looks inadequate; with a large thin walled tube (Figure 5.2b) it appears gross; and the optimum visual balance is provided by a thick wall tube (Figure 5.2c).

More recently, the Bauhaus movement originating in Germany developed a complete philosophy for modern design. Designers and engineers collaborated to evolve methods for creating beautiful shapes, including various ratios. The principal object of the movement was to work towards a purity of form, expressing materials and function in an unemotional way, enhanced by a noticeable lack of colour.

Dieter Rams, who for many years has been chief designer for Braun, and whose views are in line with the Bauhaus tradition, works as a 'gestalt' engineer, which involves him in all aspects of form and function. He intends his designs to be

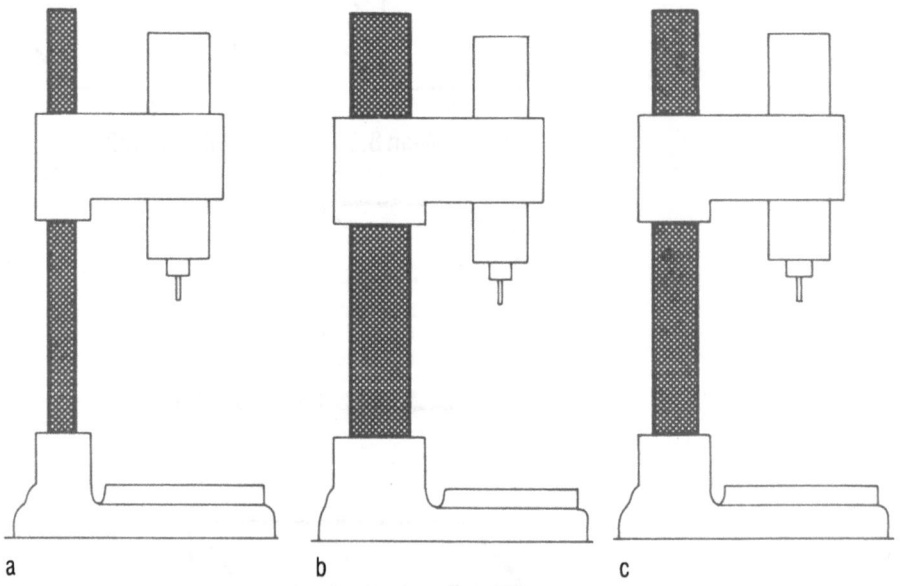

Figure 5.2 Comparison of vertical drills with different dimensions of support pillar.

timeless and austere. 'Products should not be the stars' in his view, hence his use of rectilinear shapes using little or no colour. Ultimately, though, despite the fact that he utilises various formulae for achieving his final design shapes, there are no such formulae for judging aesthetic quality. This is only learned from a great deal of experience.

Static forms are more suitable for the conventional ratios of dimensions outlined above, whereas movable form requires an enhanced appreciation of proportion and balance, and creates greater problems for the designer.

A movable object in some ways creates its own proportions and may be viewed from many angles, each giving a different perspective. A static object, on the other hand, must achieve the satisfaction of the viewer from its existing proportions. The impact of light, shade and colour in relation to its form are therefore important.

5.2.2 Stability

Another aspect of balance is stability, that is the use of shapes that generate an impression of equilibrium, individually or by their mutual relationship. Unstable forms upset the visual impression and affect the confidence of the user, especially when heavy components appear to be inadequately supported.

An example of how this can be overcome is shown in the Linear Accelerator (Chapter 18.12) in which a heavy outboard electron beam focusing unit is supported by a horizontal cantilevered arm. The GRP covers, which enclose the steel structure and electronic system, provide a satisfactory impression of stability because of their tapered form. Instability can also be suggested in quite simple rectangular structures of normal proportions, if they have horizontal divisions at or below half their height (Figure 5.3a, b) and this can be eliminated by raising the line to a position well above the centre line (figure 5.3c).

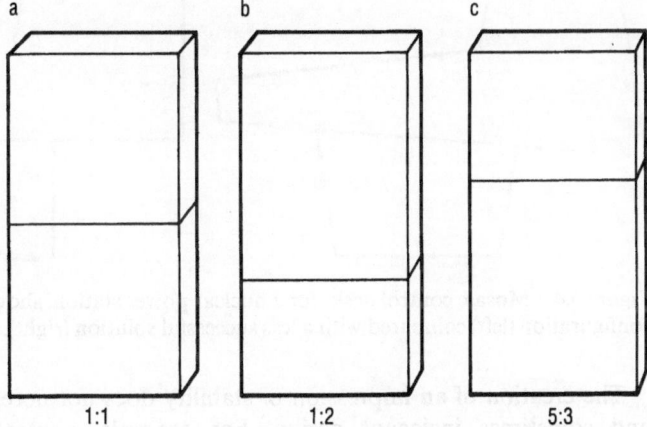

Figure 5.3 Comparison of rectangular panels with divisions at varying heights.

Another aspect of balance is shown in tall but slim free-standing structures. If these are constructed with completely parallel sides, they tend to look top heavy. The effect can be eliminated by the introduction of a *small* taper—too large a taper being worse than none.

Engineering examples of this include the gear measuring machine (Chapter 1 Figure 1.20) in which a subtle taper of the cast-iron vertical component of the structure contributes to an impression of stability and integration with the rest of the machine. The very slight taper on the porcelain insulators of the high-voltage circuit breaker (Chapter 18.20) contributes not only to functional requirements, but to the visual balance of the whole.

Figure 5.4 Mosaic control desks for a nuclear power station, showing imaginative configuration (left) compared with a less successful solution (right). (*Prolect Ltd*)

The creation of an impression of stability does not necessarily rely on massive and sometimes inelegant design. For example, control desks are normally constructed as rectangular, box-like structures, but the 'viaduct' format shown in Figure 5.4 can be used when the internal apparatus consists of items such as wiring harnesses. By careful selection of shapes, volumes and surface angles the arrangement has become ergonomically satisfactory and aesthetically interesting and balanced, despite the light supporting structure. Such design does, however, tend to be critical in visual response to the design parameters, and quite small changes could destroy the impression generated (Figure 5.4b).

This is one area where style may play a role in achieving a functional impression. To the engineer, the physical stability of the unit may not be in question, but what we are discussing here is not the physical qualities of the unit, but the visual impression of the form presented to the customer or user. The relationship between function and this aspect of styling is discussed later.

5.2.3 Symmetry

Symmetry is also a factor in the 'balance' equation. There is often a tendency to design equipment symmetrically and symmetry certainly suggests a state of order, but asymmetry can often create a greater sense of interest. Both have been used successfully in architecture, symmetry for example in the Taj Mahal, asymmetry in the works of Gaudi in Barcelona, or in the Château de Chenonceaux.

An example of symmetry in cubicle construction is shown schematically in Figure 5.5. The choice between the order of the Figure 5.5a and the interest of Figure 5.5b should be based on ergonomic efficiency and the relative cost of distributing power and circuitry internally, rather than on any aesthetic preferences. The symmetrical arrangement of identical switches on some car dashboards contributes to an impression of order and tidiness, but an asymmetrical arrangement, such as that used in the Leyland bus (Chapter 18.18) may well be ergonomically preferable. The computer system (Chapter 18.11) is another example where functional requirements have led to the use of an asymmetrical arrangement, which also increases the visual attraction.

Symmetrical arrangements, however, come into their own when multiple controls are associated with 'touch' techniques, as for example in typewriters, pianos, calculators and accounting machines, when symmetry generally contributes to the ergonomic requirements.

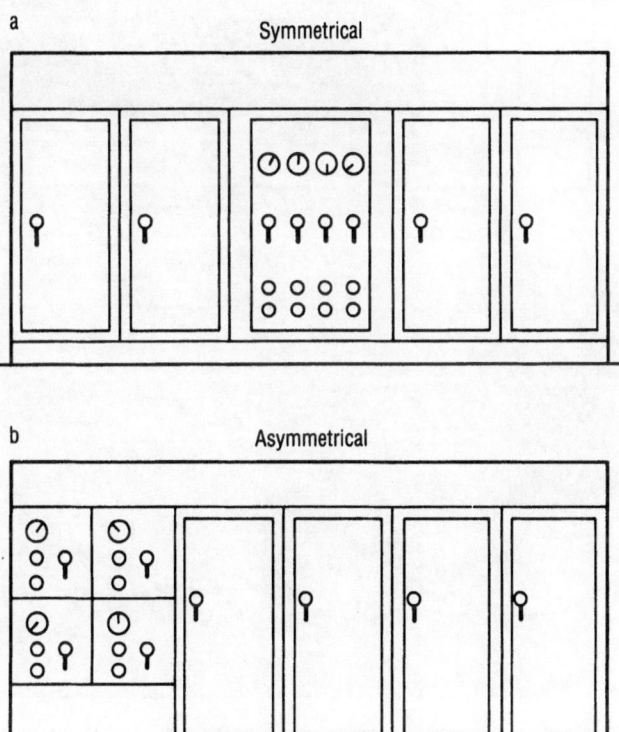

Figure 5.5 Comparison of symmetrical and asymmetrical cubicle construction.

5.2.4 Repetition

Repetition is yet another aspect of balance. Just as in music the repetition of a basic theme contributes to a sense of unity, so in engineering design the repetition of individually simple shapes or modules can help to create an impression of unified order and efficiency.

Repetition may be simply the copying of one form, as in the driving cab of the truck-mounted crane (Chapter 18.17), repeated in similar form and style at a higher level for the crane control; or a multiple of the form, as in the cylinder and valve gear module of the diesel engine (Figure 5.6).

It is important to realise that repetition creates its own form or pattern, distinct from the original components. In designing for multiples, therefore, it is important not to become too obsessed with the individual shape, but to consider how the necessary repetition of units can be used to best advantage. Modern examples include the solar screen, which consists of a number of repeated solar cells necessary for the creation of sufficient energy. The design strategy does not concern the individual cells, but the best form achievable through repetition.

The same holds true of individual components which are not so simple in design. An old roof tile appears irrelevant in isolation, but when fitted together with others

Figure 5.6 Repetition of form in an eight-cylinder diesel engine. (*Ruston Diesels Ltd*)

of a similar pattern, the logic of the design shape becomes apparent. So, too, does a new logic—ridges, gulleys and angles are formed by the constant repetition of the same rather ungainly design to form a new, aesthetically acceptable, pattern.

5.2 Balance and proportion

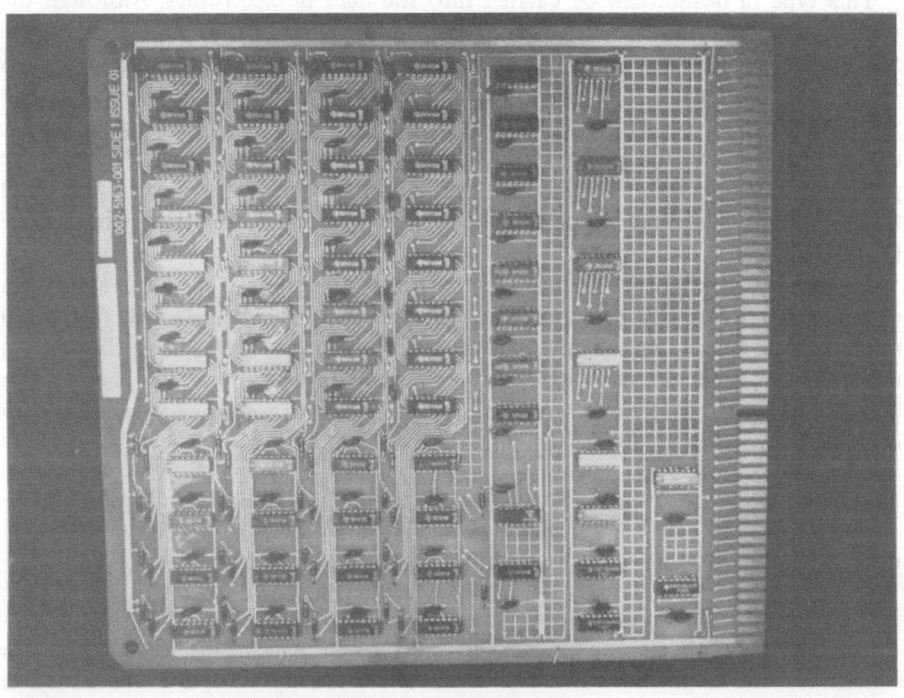

Figure 5.7 (above) Repetition of form in a digital switch PCB. (*The General Electric Co Ltd*)

Figure 5.8 (left) Repetition of form in a telephone exchange. (*The General Electric Co Ltd*)

This type of design can be carried into the field of electronics. Although the visual impact of printed circuit boards may not be a factor in their design, Figure 5.7 again shows how repetition, in this case of simple circuitry in a telephone switch PCB, adds to interest. The relevant construction here is the complete telephone substation, which presents an impression of order because of repetition of the mechanical detail of the PCBs, similar perhaps to books in a library. Visually the pattern formed is more interesting than a single book and has a quality of three-dimensional design which can support the objectives of efficiency and order (Figure 5.8).

When products involve repetition, therefore, detail design to accentuate the form of each module in a grouped and orderly display will probably give a better visual solution than softening the divisions between them, or introducing disguise to make the whole appear as a single entity. Nevertheless the impression of order can become overbearing with excessive repetition of identical components, and this can be offset by quite minor variation in detail.

5.3 Form and commercial viability

There can be two basic aims in design of form—to achieve the highest possible standards, and to provide what the market will readily accept. These aims are usually, but not necessarily, compatible. For some markets, notably consumer goods, styled ornamentation, or even the addition of aesthetically unnecessary and non-functional embellishment, can sometimes be of commercial value. For the more technical, engineering-based products, co-ordinated functional form is preferred, and ornament is likely to detract both from the aesthetic presentation and from commercial viability. An assessment of market reaction is therefore necessary before defining design policy in a development specification (see Chapter 16), covering issues such as the following.

Established practice in the market, which will influence the degree to which the form of a new product should conform or diverge.

The *character of user groups*, which tend to have their own specific ergonomic, psychological and educational characteristics.

The extent to which *innovation in style can be ephemeral or long lasting*. This will depend not only on the merit of the design, but on the extent to which change in style is acceptable to the user. Change will usually be more easily managed commercially if achieved by evolution rather than by revolution, because people relate best to form to which they are accustomed. Radical innovation in style is likely to be more readily accepted in products which have a short expectation of life.

In-house requirements for form, such as the use of standardised shapes, styles and details; and the need to emphasise the identity of the manufacturer, or to enhance 'pride of ownership' in users.

Part 2—Design guide for form

Form is an important aspect of the design of engineering products, but whereas it predominates in, say, sculpture it is only one of many design characteristics that affect quality.

In engineering, form has to comply with constraints imposed by performance and commercial requirements, to both of which it can contribute if it is co-ordinated with function. The development of product form depends largely on the art of three-dimensional design, and the following sections discuss some of the practical ways in which this can be approached.

5.4 Functional design

5.4.1 Expressing purpose

It is natural for engineering design to concentrate on the functional aspects of products. Many components, such as crankshafts, have their form dictated by the nature of their use and other considerations, such as stress. In this section, however, we are concerned with exposed components and the expression of purpose conveyed by the machine as a whole.

This is where considerations of styling may well overlap the purely functional considerations of the engineer. While the form of many components will be dedicated solely to their function, the designer will nevertheless have some latitude in how these are finally assembled.

Although it might at first seem logical that functionally designed elements will lead to functional overall design, this is not always true. Heath Robinson showed that effectiveness and presentation do not always go hand in hand. Functional design philosophy has to be directed towards emphasising and integrating the overall visual impression of function.

It may be helpful to introduce preferred forms that assist in the co-ordination process. These may have no functional value themselves, but they may assist the overall design and its appearance, and therefore represent a form of style. In practice, any functional design more often than not contains an element of style which assists rather than hinders the development of a good visual impression of the function of the machine.

Styled design has status of its own when it seeks to add detail shapes, ornamentation, embellishment and special finishes, ostensibly for purely cosmetic reasons. The more basic aspects of style likely to be used in support of function tend to have a long life, whereas these other stylistic additions, like fashion, tend to be transitory. If the latter are to be used at all in engineering products, it is important for the designer to consider the dating effect of this fashion-based styling.

5.4.2 Basic design

Functional design in its purest form will be achieved when all the visual components support a tidy and co-ordinated impression of function without the need for any imposed style. The design must then be natural, without disguise,

Figure 5.9 Liquid crystal display digital watch with (left) component details. Designed by David Edgerley. (*CRE Design Consultants*)

ornamentation, and non-essential covers used to disguise poorly designed details (which with good design should be proudly exposed), and the details must be inherently compatible. A number of examples can be seen in Chapter 1 of this book, beginning with the simple designs of the nineteenth-century microscope (Figure 1.9), handbrace (Figure 1.10) based on curves, and the lathe (Figure 1.8) using a compatible mixture of curves and straight lines.

An interesting concept in which function has dictated form is the liquid crystal digital watch (Figure 5.9) which won first prize at the Concours International de Dessin, Industrial Horloger, Chaud de Fonds, Switzerland in 1977. Such a watch has four formal elements: the battery, the electronic time chip, the display and the case linking them. Here the case was designed as a logical reflection of the system, carrying the three active elements in an adhesive-bonded, stainless steel, ladder-like frame, bent to improve visibility of the display, which is interrogated by pressing the protruding battery, thus operating a pressure switch mounted underneath the battery and reacting against the wearer's wrist.

5.4.3 Co-ordination

Because of their complexity, many products could present an untidy and unco-ordinated impression if a basic approach to functional design were taken. It therefore becomes necessary to introduce a styled co-ordinating medium. One widely used method is to base the design on straight lines with flat and rectangular forms; another approach, not as fashionable but equally valid, is to employ curved lines with two or three-dimensional curved forms.

The Yarn Top Detector (Chapter 18.2) is an example of a small component in which the untidy impression given by an early model has been corrected by the use of co-ordinated straight line construction. The lathe (Chapter 18.14) and the horizontal borer (Chapter 18.15) are examples of large machines co-ordinated by rectangular design.

Rectangular co-ordination is not appropriate for all products, especially when cylindrical or circular components are inherent in the function, when co-ordination with curves may be preferable. For example, the micrometer (Figure 5.10) is a well balanced design with cylindrical and semi-circular details. The linear

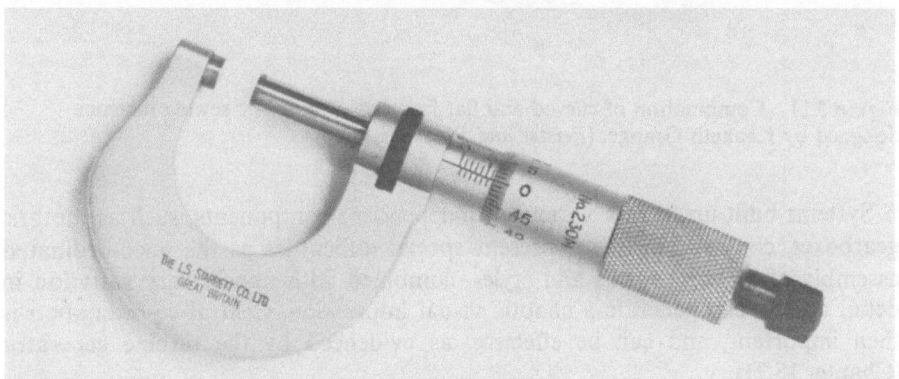

Figure 5.10 Cylindrical and curved co-ordination in a micrometer. (*The L. S. Starrett Co Ltd*)

accelerator (Chapter 18.12), which has a rotational movement on a large scale, has been co-ordinated on a curved surface basis. The associated control units display the variety of curved and rectangular styles that have been used over the years for similar equipment, reflecting the influence of fashion.

Rectangular and curved forms can also be co-ordinated successfully when they are justified in functional terms, and when detail is accommodated in a balanced manner. An example is the compressed air system (Chapter 18.1) in which cylindrical plug-in modules are accommodated by a separate rectangular air bus-bar. The sewing machine (Figure 5.11) also combines rectangular and curved forms in a single component. Ergonomic factors require a flat working area, and this has been co-ordinated with the body of the machine by large radii at the interfaces, which contribute to ease of handling of materials and cleaning. The sleigh type base, the re-entrant curvature of the vertical column balanced by the subtle taper of its further face, and the simple and matched shapes of the controls, all combine to generate an impression of efficiency and elegance.

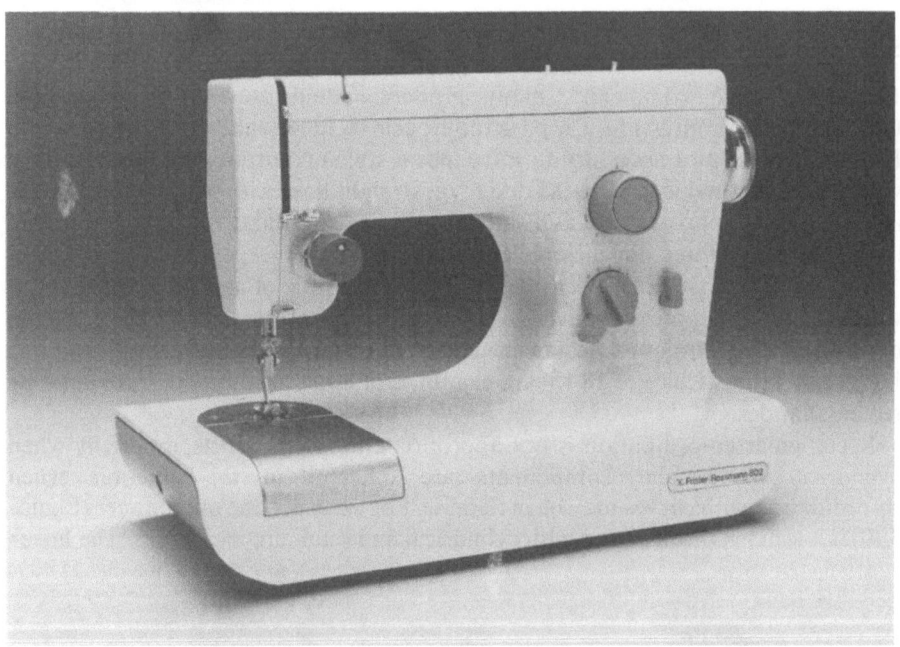

Figure 5.11 Combination of curved and flat forms in a domestic sewing machine designed by Kenneth Grange. (*Frister and Rossman*)

Systems built up from a variety of independent components, such as motors, gearboxes, control panels etc, present special difficulties as the unco-ordinated assembly of different forms and styles, combined with unnecessary variation in detail design, can generate a chaotic visual impression. Central co-ordination is then important, and can be effective, as evidenced by the turbine generator (Chapter 18.21).

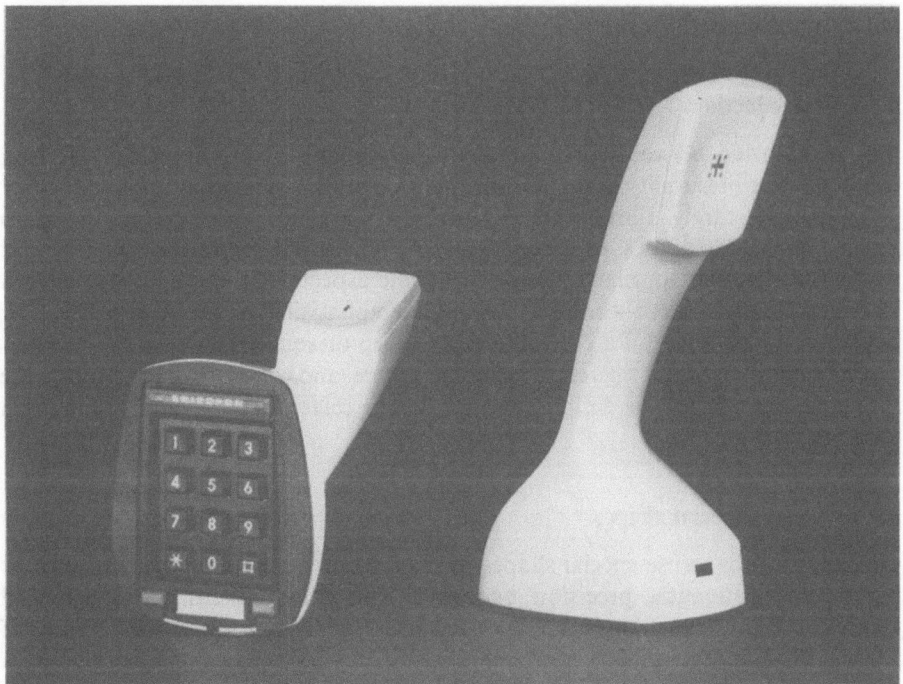

Figure 5.12 Solid-state electronics enable the form of telephones to be radically revised. (*Thorn Ericsson Telecommunications*)

5.4.4 Size

The miniaturisation that has resulted from advanced electronic techniques affects the way in which functional elements can be positioned in a design. Designers can now use previously unacceptable housings for integrated items, so freeing them from many of the design constraints which operated previously.

The design of telephones provides a good example of the way in which miniaturisation permits the use of entirely new shapes. In this case, it is now possible to integrate the entire telephone circuitry in a single component, permitting the designer to concentrate on balance, proportion, and ergonomic styling (Figure 5.12).

Clearly, in looking at any design in which microelectronics predominate, the engineer will be able to arbitrate on the final form of the object, unless miniaturisation has a complete and unchallenged advantage.

Where there are no overriding functional criteria to follow, the design can only be resolved by balancing the functional, material and psychological factors that do exist. This means that the engineer needs a greater understanding of marketing requirements, function, weight, compactness, need for visibility, relationship with user and so on.

This freedom of design—now manifest in the choice of fascia and unit size for hi-fi systems for example, where the electronics give no clues as to size or shape—compels the engineering designer to pay more attention to aesthetic criteria.

5.5 Style and fashion

5.5.1 Introduction

The relationship between form and style has already been discussed in Section 5.1.5. In terms of viability, both fashion and style play an important part.

Function is rarely allowed to exist without an element of styling, if only to distinguish a company's products. This may merely take the form of colour standardisation, but in many instances specific aspects of the design will have a strong fashion element—perhaps through emphasising a key component, or deliberately rounding off a conventionally sharp or square edge.

The designer should therefore examine where and when he can incorporate stylistic elements into his design and how he can relate these to purely functional elements.

5.5.2 Use of special shapes

One approach is to use special shapes to suggest some aspect of function, such as performance, strength, precision or speed. This may sometimes be achieved through the functional engineering design itself, or by introducing a measure of styling. The use of special shapes should really only be considered when the main shape or form cannot in itself create the correct expression.

Tilting the engine of a motor cycle forward to the extent that it can be employed as the front member of the frame was used in the 1920s and has now been developed by Honda (Figure 5.13), in an arrangement that gives an impression of

Figure 5.13 This motor cycle uses the engine as a load-bearing frame member, sloped forward to suggest speed. (*Honda*)

forward movement and speed using a purely functional shape. The tail fins springing out of the rear wings of cars of the 1950s were intended to suggest aircraft performance, but this was an exaggerated, unnatural and short-lived style. The microscope (Figure 12.3) in rectangular format is proportioned to suggest rigidity

Figure 5.14 Functional design is apparent in the appearance of the light alloy wheels of this motor cycle. (*BMW*)

and the sharply defined edges reinforce an impression of precision.

In reverse, form can be used to conceal function, for aesthetic or psychological reasons. In the deep therapy linear accelerator (Chapter 18.12) referred to above, the use of curves, and concealment of much of the elaborate electronic and high-voltage equipment, serves the additional purpose of reducing the severity of the visual impression, and contributes to alleviating a patient's fear of machine and treatment. Similarly, in the case of orthopaedic callipers the visual objective should be to reduce to a minimum the impact of the walking aid.

5.5.3 Styled ornamentation

By definition, functional design is free from detail ornamentation, and employs finishes selected to meet service conditions. Style, in terms of ornamentation, means the introduction of pattern, special finishes and sometimes graphics, usually associated with current fashions, with the objective of increasing product attraction.

Ornamental pattern may overlap with form if it employs a type of surface sculpture or if it requires components to be fabricated in ornamental shapes. It may on the other hand be superimposed by the addition of separate details such as chromium strips, or by using finishes such as plating to produce special visual characteristics. Styled ornamentation can also contribute to appearance if it is employed to break up and improve badly proportioned areas.

A type of product that exhibits the characteristics of both functional and styled ornamental design is the light alloy wheel for cars and motor cycles. This was developed originally by Bugatti in the 1920s, using aluminium alloy sand castings with eight outer spokes and eight inner spokes integral with the brake drum housing. Bugatti designed his wheel for purely functional reasons to reduce unsprung weight and improve heat dissipation from the brake system, but the result also contributed to the appearance and identity of his cars (Figure 1.12). With the advent of higher strength materials and diecasting, simpler forms of light alloy wheels with a single row of spokes are now widely used. Examples of interesting modern design include the **BMW** motorcycle wheel (Figure 5.14) in which cross-braced radial spokes provide a light, easy-to-clean and functional structure. The **SAAB** car wheel (Figure 5.15) employs four sets of styled flutes forming a ventilated disc. The Porsche approach (Figure 5.16) has been to co-ordinate the spoke format with the ellipsoidal style of the complete car.

Figure 5.15 A combination of functional design with a strong styling element in a light alloy car wheel. (*SAAB*)

Figure 5.16 The ellipsoidal styling links wheels and body elements in this car design. (*Porsche*)

5.6 Form and the visual environment

5.6.1 Designing in isolation

There is a great tendency to design individual items of equipment without any consideration for their interrelationship with other items or with their intended location.

If we return to our comparison with sculpture, it is accepted that a work of art may stand alone. Art galleries go to great lengths to ensure that the viewer is unobstructed and undistracted. In fact, the environment is deliberately made to be as unobtrusive as possible.

In engineering, however, designing in isolation is rarely, if ever, appropriate. Most products gain their *raison d'être* by being put into context, so the designer must start, in effect, by visualising a jigsaw puzzle with one piece missing. It is that piece which he must design, so that it completes the picture.

It can be argued that one can design in isolation and then adapt the end result to fit a specified environment. But many design errors could be avoided if consideration were first given to the way in which the item must co-ordinate with its surroundings, and appear or not appear to the user.

5.6.2 Visible or unobtrusive?

The form of products should generally be compatible with the visual environment of which they will be a part. For example, floor-mounted service equipment in kitchens should be modular, rectangular and clean in form. Telephone kiosks should be unobtrusive. Larger systems, such as factory assembly lines and process plants, involve the overall architectural presentation. In some products it is possible to develop functional forms that reduce the environmental disturbance they might otherwise create.

For example, if high-voltage transmission towers are co-ordinated in format with the catenaries of the cables they support, they can be graceful structures (Figure 5.17) which will minimise the adverse effect of transmission lines on the countryside in comparison with unco-ordinated designs.

When the product form has visual architectural consequences which affect the environment, the design alternatives should be one of the interface subjects agreed between engineering and architectural teams. The design for a submarine refit complex (Figure 5.18) is an example of such collaboration.

Figure 5.17 Transmission pylons designed to be more (left) and less compatible with a transmission line catenary.

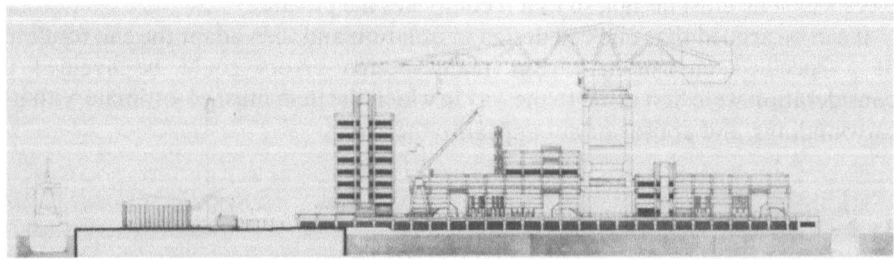

Figure 5.18 Design for a submarine refit complex, involving collaboration between engineering and architectural teams (Howell Killick Partridge and Amis, Architects in association with Sir Alexander Gibb and Partners, Engineers, for PSA, Ministry of Defence).

5.6.3 Environmental philosophy

The industrial designer Dieter Rams, mentioned earlier, takes the view that his company's destiny is best served if its products are 'timeless'. Most products in the Braun range are so designed.

Most of them use a rectilinear form with little or no colour, except for operating controls, in order to make them both unobtrusive and anonymous. Rams' view is that it is bothersome to live with such objects and that they should be in the background, rather than standing out as the 'stars of the show'. This somewhat elitist attitude is perhaps only relevant to intelligent people, but it sets a good example for others to relate to and is entirely appropriate for engineering products which may require this anonymity.

On many occasions, however, this is not appropriate. Strong colour, size and proportion may be used to achieve the required solution of visibility.

The current success of the Italian consumer goods industry, in contrast, has much to do with the creation of shapes that have a more pleasing, immediate appeal. The Italians certainly deploy more colour, use rounded shapes and pay more attention to fashion in their consumer items.

In the United States, mass production means the design of shorter-lived, fashionable, curved and coloured objects—an approach which has the commercial advantage of ensuring 'new lines' each year or so and is perhaps more appropriate for consumer-orientated products.

For capital goods and constructions, environmental considerations may play a greater part. The engineer who designs an electricity substation or process plant which fits into the surrounding countryside is more likely to gain the contract, especially where environmentalists have a say in the decision.

5.7 Form and safety

Form has a contribution to make to safety, and it should be studied as part of the design procedure and in the safety audit. In many cases, form will help to make the user aware of the functional purpose of the machine, rather than disguising it, and this can help in avoiding accidents (see also Chapter 13).

5.7.1 Prevention

Trapping can occur as the result of movement of an operator or user, or as the result of movement by machine components. Form can sometimes help to reduce or eliminate such risks. For example, if the operator of a large planing machine can accidentally get in front of the moving work table, the fixed bed-plate track should have a smooth shape that cannot trap his legs, so that at the worst he will be swept along without severe injury. In moving stairways the fronts of the steps are curved to prevent trapping of feet during the concertina closing action. (see also Chapter 18.6.)

The risk of injury due to impact can be reduced by appropriate shaping of components. For example, a flat area on car steering-wheels can reduce risks associated with collision or overturning.

The shape of machines such as chain saws, in which comprehensive guards are not possible, can help to ensure that the operator's hands are necessarily in a safe position when he is operating the saw. The risk of clothes catching in control handles and causing incorrect operation can be avoided if their open ends are shaped in a re-entrant form, to reach back towards the panel on which they are mounted.

Form should also contribute to reducing the risk of damage in service. For example, highly convex car wheels are bad because they will inevitably be dented or scratched when parking next to high kerbs.

By paying attention to shape and balance, and avoiding sharp contours and edges, the handling of large covers can be made both easier and safer (see Chapter 18.16).

5.7.2 Protection

Protection can be afforded by employing form to reduce risks arising from both internal and external sources. For example, machine tools can be shaped to retain and dispose of flying metal swarf (Chapter 18.14); control panels on aircraft can be shaped so as to assist the pilot to use them with precision in rough air conditions, thus reducing the risk of operational error (Chapter 18.8); windows on high-speed vehicles can be shaped to reduce risk from missiles (Chapter 18.19).

5.8 Detail design and form

There are a number of detail design issues, often rather mundane in character, which can have a disproportionate effect on the overall impression created by form. Some typical cases are discussed below.

5.8.1 Folded steel components

When several folded steel panels or components are assembled together to form a structure, there will be joints between them, and the design should ensure either that such assemblies will in fact line up accurately, or that a deliberate lack of alignment is introduced, large enough to prevent this appearing as a manufacturing error. Such joints also offer opportunities to modify the overall impression of balance, by creating vertical or horizontal lines that break up the surface. This effect can be enhanced by inserting a deliberate inset gap as used in the control panel (Chapter 18.9).

Mounting folded doors and similar components externally on the front face of panels can create a destructive domino effect, but this can be avoided by careful selection of proportions or by re-entrant folding of the housing and flush-mounted doors. The exposed edges of flat or folded components made from thin sheet always look untidy, and are often the points where corrosion gets a hold; re-entrant folding again offers a tidy solution.

Appearance will, in general, look taut and tidy if the internal folding radius approaches the minimum recommended in relation to plate thickness, but there are limits to this, as excessively sharp corners on very thin sheets can invite damage to paintwork.

Heavy sheet steel plate weldings should be used flat, folded or rolled to produce simple structures which emphasise their methods of construction. If the edges of thick plates can be left exposed they may help to define the form, as in the turbine casings (Chapter 18.21), rather better than if the edges are disguised by expensive corner welding of the adjacent plates.

5.8.2 Controls

The first priority for the three-dimensional design of controls must be to meet ergonomic requirements, but to the extent that this permits, their style should be co-ordinated with that of the machines they serve. For example, a squared wheel rim cross-section may reinforce the impression of efficiency of equipment designed in a rectangular format. When the machine form is simple the controls can be too, as they will still stand out. Colour is often usefully applied in these situations.

5.8.3 Services

Pipe systems used for lubrication, compressed air, steam, fire-fighting and hydraulic control, and electric wiring harnesses and power supplies, can all destroy an otherwise satisfactory form if they create an impression of disorder.

One aspect of this problem centres on flange couplings which so often look unco-ordinated and out of proportion, both on pipe systems and between services such as motors coupled to gearboxes. When flexibility of design exists (often unavoidably inhibited by the use of standards), it may be possible to reduce the size of flanges by the use of modern sealing methods, and flanges between services can sometimes be improved if their shape is matched with that of the machines they couple, round for round, square for square. The mounting details of securing pipes and harnesses are often crude, and if they are visible, care in their detail design can be rewarding.

Pipes and harnesses are often run without reference to the effect they can create. Within the limits imposed by their duty, service systems should be as inconspicuous as possible, and not be run in mid air or diagonally across rectangular forms. A possibility is to run them associated with existing divisional boundaries between machine components. When design permits, such services should be run internally with access doors for inspection purposes (see Chapter 18.21). A step further, on a smaller scale, is sometimes possible by substituting internal drilling for external pipework, an arrangement that can improve both reliability and appearance. An example of this approach 'before' and 'after' is the component (Figure 5.19) which was redesigned with all the connections drilled integrally within the castings. See also the ventilator (Chapter 18.6).

5.8.4 General shape detailing

The question of overall shape has been discussed earlier in the section on psychological factors. Rectangular shapes and curved shapes, for example, produce entirely different reactions in the user. A completely sharp-cornered square block will be as alien to the user as is possible. Softening the corners with a radius will make the object more 'approachable'.

In the opposite way, by giving an essentially rounded shape some 'flats', one can reduce the degree of obtrusion into the user's environment or activity.

The materials chosen for the manufacture of a machine can often commit the engineering designer to a final form which is incorrect. An example is the use of sheet metal for an object that is required to be omnipresent (represented in design terms by 'roundness'). Three-dimensional curves cannot easily be created in this material, and moulding or casting would be more appropriate.

Figure 5.19 Component with associated hydraulic pipework, before and after redesign to incorporate hydraulics within main casting. (*Hunting Engineering Ltd*)

5.9 Checklist

There are many aspects of the design spectrum that can contribute to efficient and aesthetic form for engineering products, and a checklist may assist in the selection and optimisation processes required in their implementation.

1. What are any special requirements of the user group?
2. What are in-house requirements affecting form?
3. Is form of product and control detail relevant to functional operation in terms of ergonomic and psychological criteria—such as operator concentration?
4. Does form contribute to safety in terms of trapping, effect of impact, handling during maintenance etc.
5. Would conformity with the past, divergence or innovation be appropriate?
6. Is 'style' relevant?
7. Is long-term or transient 'fashion' more appropriate in relation to the life expectancy of the product?
8. Should design be visible, anonymous, quiet, aggressive?
9. Should design express function?
10. Can design be made in isolation?
11. Should design be co-ordinated with associated apparatus?

12 Should design be co-ordinated with its local environment?
13 Should design be co-ordinated with its architectural environment?
14 Should the separate components of the product be co-ordinated for form?
15 Would component co-ordination using rectangular format, curved format or a combination of these be appropriate?
16 Would sculptural ornamentation be appropriate?
17 Would additional and non-essential ornamental details be justified and appropriate?
18 Are materials and form selected mutually compatible?
19 Do miniaturisation techniques require artificial adjustment of size for efficient use?
20 Are the product's proportions aesthetically satisfactory?
21 Is design balanced and stable in appearance?
22 Does the form take into account dynamic or static operation?
23 When both are possible, would symmetry or asymmetry be more efficient?
24 Could repetition of detail be used to enhance interest?
25 Would use of special shapes contribute to functional appearance?
26 Would use of taper contribute to balance?
27 Would greater exposure of detail contribute to efficient use or appearance, as by omission of non-essential covers?
28 Would co-ordination of form of control detail with structural form be ergonomically efficient and contribute aesthetically?
29 Is sheet metal detail satisfactory in proportion, exposure of edges, matching of joints?
30 Is structure optimised to avoid surface that will collect dirt, and for ease of cleaning?
31 Is form optimised to reduce damage in service?
32 Are services, as hydraulic, electric, compressed air, lubrication, fire suppression, optimised in layout as between their functions and their contribution to tidy and aesthetic appearance?
33 Does form assist maintenance?
34 Is form optimised in its contribution to function, appearance and cost through simplicity?

6 Colour

Allan Whitfield and Tom Wiltshire

6.1 Introduction

The principles underlying the selection of colours for engineering products are the same as those for other types of manufactured goods. In addition, they are appropriate for the treatment of any aspect of a product's visual appearance; be it shape, texture or proportion. There are two simple rules underlying these principles: first, identify the major functions of colour in the design of the product; and second, evolve a strategy that covers these functions. This chapter will attempt to describe these functions and illustrate design strategies that seem industrially viable. To begin with, however, it will be useful to digress slightly and say something about both colour perception and current methods of colour description and specification.

6.2 Colour perception

In general, perceived colour depends upon three things: the spectral characteristics of the light source illuminating a surface, the spectral transmission or reflectance of light by that surface, and the ability of an observer to detect this light and convert it into what is experienced as colour This triadic relationship underlies our perception of colour. For example, we are all familiar with the principle that the colour of an object is dependent upon the light source that illuminates it. In shops we often take an object outside to look at it in daylight in order to see its 'true' colour. Similarly, we have probably encountered individuals whose ability to discriminate between colours is impaired. Such impediments are fairly common, especially among the male population.

Major strides have been made over the past 20 years in our understanding of the mechanisms that underlie colour perception Much of this has resulted from sophisticated neurophysiological work carried out on animals that have similar colour vision to humans. The picture that emerges, however, is far from simple.

For convenience, it will be useful to conceive of colour vision as occurring in a series of processing stages within the visual system (eye-brain). In this an 'input' to the visual system triggers off a series of neural events that eventually result in the experience of colour being registered. The input to the system is energy in the form of light, and the system is designed to detect variations in both the quantity and distribution of energy.

At the first stage of processing, receptor cells respond to the presence of appropriate wavelengths. It has been demonstrated that there are two basic types of receptor in the human eye—rods and cones—which are located in the retina at the back of the eye. The rods are inactive in bright light, functioning only at low light levels, and are not involved in the processes of colour vision. The cones are operative at higher light levels and constitute the receptors that underlie the perception of the whole range of colours, including black, white and greys. By directly measuring the colour sensitivity of cones in the human eye, three different types of cone have been identified, each containing a specific pigment which is responsible for its selectivity. While each type of cone is sensitive to a range of wavelengths, the types differ from one another in reaching maximum sensitivity at different wavelengths. These maxima are conventionally, though inaccurately, referred to as corresponding to red, green and blue. The impulses generated by these three types of cone travel via a network of nerves to a layer of retinal ganglion cells. At this stage of processing, the impulses from the three types of cone are transformed into four chromatic signals, corresponding to red/green, yellow/blue, and two types of achromatic signal, corresponding to black/white. These signals are then relayed as impulses, via the optic nerve, into the brain proper. The third level of processing takes place in the lateral geniculate bodies which are located in the mid-brain region. Here the impulses appear to be sorted prior to their transmission to the visual cortex at the rear of the brain. Our understanding of this latter stage of processing in the visual cortex is, to say the least, incomplete. What is known, however, is that receptor cells exist here that are sensitive to the colour content of incoming signals.

It may be noted that all visual information about the outside world is transmitted by similar sets of impulses. Thus other receptor cells in the visual cortex are sensitive to features such as lines, edges and the movement of these. From the arrangement of cells encountered in this part of the brain it is thought that the visual cortex performs the function of extracting featural information, such as colour, from the initial input to the visual system and then assembling this information into the objects and events that we 'see'.

6.3 The description and specification of colour

In everyday conversation it is usual to describe colours using such terms as 'blue', 'light blue', 'greenish blue' and even 'navy blue'. While such simple descriptions may be adequate for everyday communication, they have obvious limitations when precision is required. For example, according to recent estimates a trained observer can detect approximately 7.5 million differences in colour under optimum viewing conditions. The term 'blue' could apply to literally thousands of these and the term 'light blue', while a little more precise, would nonetheless cover a vast number of discernibly different colours.

For scientific and industrial purposes it has been necessary to overcome this problem by devising systems of colour specification that describe the appearance of colour with a high degree of accuracy. The types of colour system that have been developed for this purpose can be broadly classified according to whether or not they are based on a collection of physical colour samples.

Sample-based systems are organised according to various principles; for example, equal perceptual spacing of small colour differences or the reflection of people's natural classification of colour. While these principles may appear fairly straightforward, a number of practical problems are encountered. There are difficulties, for example, in ensuring equal differences over the whole range of colours. There are also divergent views on how our perception of colour is structured, and consequently how our experience of colour can be appropriately classified. This disagreement is hardly surprising given the complexity of colour vision and the incompleteness of current knowledge. There is also the problem that colour perception is highly sensitive to even slight variations in viewing conditions (illumination, background etc) so that the colour appearance of, say, a paint sample can vary dramatically with changes in viewing conditions. A consequence of this is that a colour specification can only accurately describe a colour's appearance under specific viewing conditions.

Although a number of colour specification systems have been developed, few have achieved any kind of international acceptance. The one system that has achieved this acceptance—at least by certain influential industries and institutions in certain industrial countries—is the Munsell system. As this is probably the most widely used colour specification system in the industrial world and therefore of relevance to product engineers, it will be described in some detail. Also the terminology of the Munsell system will be used for the remainder of this chapter.

Munsell, a nineteenth-century American artist, developed a colour system which conceived of surface colour as possessing three major dimensions, and in which a complete description of a colour must specify its position on each of these dimensions. He termed the three dimensions 'hue', 'value' and 'chroma'.

Hue is a chromatic dimension and distinguishes between red, blue, green etc. There are 10 hue categories in the Munsell system, and these are denoted by letters (R for red, YR for yellow-red etc). The letter is prefixed by a number, usually 2.5, 5, 7.5 or 10, with 5 representing the 'centre' of the hue category. For example, 5R is meant to be a 'pure' red with no yellow or purple present. As the number becomes greater than 5 the hue inclines towards yellow-red (YR) and as the number becomes less than 5 the hue inclines towards red-purple (RP) and so on round the hue circle (Figure 6.1).

Value is an achromatic dimension and refers to the lightness or darkness of a colour. There are 11 categories of value from ideal black (0) to ideal white (10). In practice the range encountered is from 1 to 9, with divisions of each category being common. For example, in the BS: 4800 1972 *Paint Colours for Building Purposes*, the black has a value of 1.5 and the white 9.5. A rough percentage estimate of the light reflectance of a colour is given by the formula $V(V-1)$, where V is 'value'. Thus, colours of similar value have similar reflectances.

Chroma is a further chromatic dimension which describes the strength or saturation of a colour. It can be usefully conceived of as the 'amount' of hue present. The chroma scale goes from neutral grey (0) towards full strength at any given level of value. Numerical steps are usually in twos, thus 2, 4, 6 etc.

A complete reference for a colour in Munsell notation follows the order (a) hue, (b) value, (c) chroma. Thus the notation 2.5 YR 8/12 indicates:

(a) A hue (2.5 YR) in the yellow-red region inclining towards red.
(b) A high value (8) and therefore a light colour.
(c) A high chroma (12) and therefore a strong, saturated colour.

In fact, the notation 2.5 YR 8/12 would describe a strong, bright orange.

The one exception to this standard form of notation applies to achromatic colours which possess no hue and therefore no chroma. These neutrals—black, white and greys—are denoted by the value number prefixed by N, eg N2 or N9.

Figure 6.1 Colour circle.

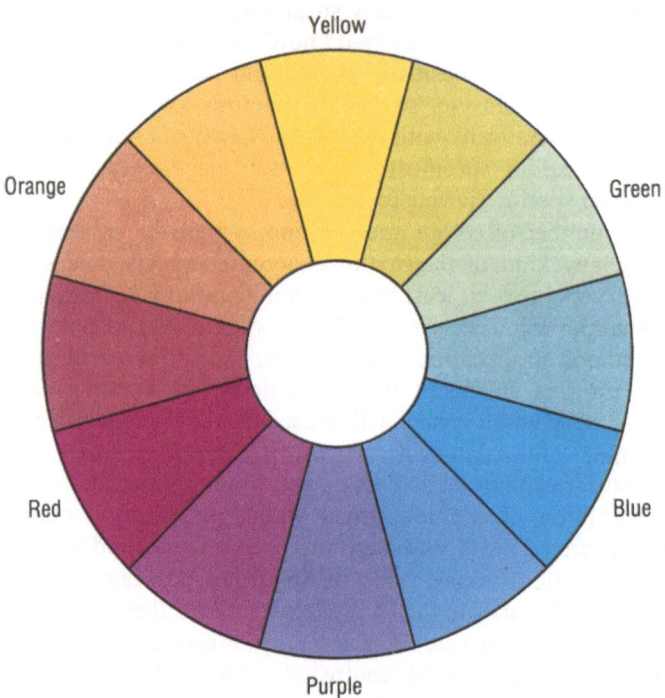

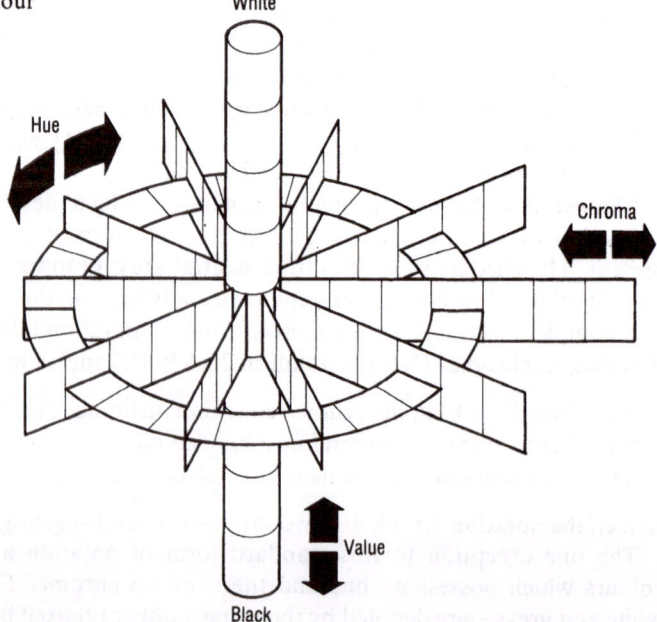

Figure 6.2 Munsell colour solid.

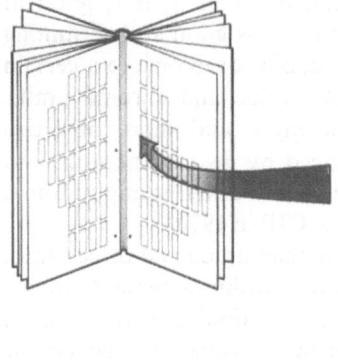
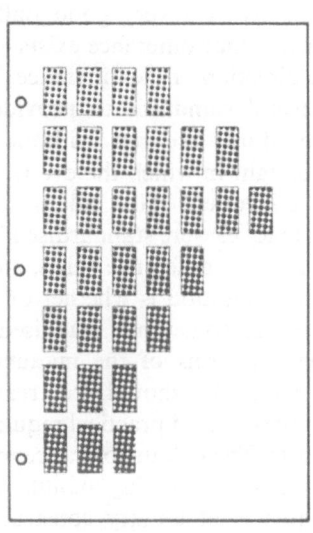

Figure 6.3 Munsell colour solid in 'book' form.

Figure 6.2 shows the way in which the three Munsell dimensions are co-ordinated to produce a 'colour solid'. This can also be usefully visualised as a widely opened book, each page of which contains colour samples laid out as shown in Figure 6.3.

The great advantage of using the Munsell system is that the imprecision of everyday colour communication can be avoided. Accurate descriptions of colour can be achieved which are meaningful to other users of the system, and which can be visualised by following the recommended viewing procedures attached to the system.

While the Munsell system is probably adequate to meet the needs of the designer, it is important to know that there are systems that are not associated with physical samples. The most important of these, and the one which is usually employed in connection with instruments for colour measurement, is the CIE system developed by the Commission Internationale de l'Eclairage. Although the CIE system is rather difficult to understand, the basic concept underlying it is simply that of additive colour mixing, in which a given colour can be matched by a mixture of coloured lights. From a knowledge of the quantity of the coloured lights used to match the colour it is then possible uniquely to define that colour. Since the system is instrument based, it does not rely upon people viewing coloured patches. The observer is, in fact, effectively replaced by a numerical description of a 'normal observer's' response to colour, and in this sense the system can be considered as objective rather than subjective.

A particular colour for a product can be specified using either the Munsell or the CIE system, or any of their derivatives. However, as this colour can only be manufactured to a certain level of accuracy, there will inevitably be some variations between the product and the specified colour. Because of the importance attached in manufacturing to the question of tolerances, considerable effort has been devoted to the measurement and evaluation of colour differences. In all cases a two-stage procedure is necessary, involving, first, detection of the difference in colour between the product and the specification and, second, determination of whether this difference is acceptable for the product in question.

Colour differences can be established using both direct visual and instrumental methods. The human eye is a very good null detector—that is, it is good at determining whether a colour difference exists—but it is less good at determining either the size or the direction of the difference. This can be overcome to a certain extent by having several standards to provide hue, value and chroma limits. However, these procedures are not without difficulties, and more objective methods have been sought in which the eye is replaced by instrumentation, and measurements are converted into colour differences using computational techniques. These techniques were standardised by the CIE in 1976.

Having established that a difference exists, and the magnitude and direction of this difference, the next question is whether or not it is within acceptable limits. Acceptable tolerances are sometimes inadvisedly set by stipulating the tightest possible limits, either in terms of the manufacturing processes or the colour difference measurements. It should be realised that such tolerances are inappropriate in that they exceed practical requirements and will almost inevitably result in increased costs. The colour tolerance must therefore reflect the nature of the product and its function, including manufacturing aspects and marketing. For example, if the product is of unitary construction and is an isolated piece of equipment then the colour tolerance can be quite large. However, if the product is assembled from several components which are manufactured at different industrial locations then close control over the colour of the components must be exercised if undesirable colour differences are to be avoided. In such circumstances it may be desirable to use distinct colours for each of the components so that the colour tolerances can be relaxed.

Finally, it does not make sense to specify tight colour tolerances on objects where fading may take place or where the accumulation of dirt and grime will obscure the surface. Equally, it is of little use specifying a colour tolerance if other aspects of surface appearance are neglected. For example, a mismatch will result if two surfaces are of the 'same' colour but one is matt and the other gloss. Although the determination of colour difference is internationally standardised, the permissible colour differences for a specific product are not. The exceptions tend to be where the interests of the general public are concerned, as in the case of traffic lights, or where many manufacturers are involved in supplying the same standardised product.

6.4 The functions of colour in product design

While a vast literature is available covering many aspects of colour, its application in product design has received relatively little attention. The nearest body of relevant information comes from the building field, where much work of a theoretical, practical and experimental nature has been carried out. Here, the requirements of colour can be reduced to three basic functions, which can be summarised as 'structural', 'ergonomic' and 'aesthetic', and a similar approach can be followed in the case of products.

6.4.1 Structural functions

The structural function of colour concerns its capacity to affect the way in which an object is perceived. Essentially, colour is conceived here as a means of influencing the perceived structure of an object in a controlled and predictable manner. Gloag and Keyte (1957) refer to this as a capacity to 'express form or breakdown of surface', while Hardy (1966) refers to it as the ability to accentuate a satisfactory three-dimensional structure or 'to correct unsatisfactory proportions by altering their apparent form and creating centres of attention'. While information on the effects that can be generated remains on the qualitative level, there is sufficient consistency to give some credence to the ideas expressed. This is given some support by Albers (1963) who demonstrates a series of striking colour effects that can be achieved in two or three-dimensional design, and Birren (1969) who displays a similar acumen with building exteriors. In addition, basic design courses in art, design and architecture in Britain generally include instruction in this and related aspects of what may be loosely termed 'visual language'. In general, however, this field remains underdeveloped, with little empirically based information available on the perceptual effects that can be achieved by using colour in a design context.

Despite the above limitations, certain general principles can be exemplified.

Centres of attention. Attention can be directed to a product, or some aspect of it, by colouring it in such a way that it stands out from its background. In general this is achieved by using strong, vivid colours that contrast well with those around them. If, however, the surrounding colours are also strong and vivid it will be necessary to use some other device. This could involve the use of stripes, fluorescent colours and even black or white. The important point is that a contrast occurs.

Separation/Association. The separate parts of a product can be visually associated or separated by appropriate use of colour. This can have the effect of making the product appear visually more unified or fragmented, more simple or more complex. One of the most useful devices for achieving separation or association is through manipulation of the strength of the boundaries between areas of colour. As the boundaries weaken, the areas of colour will converge, or associate, and vice versa. It is the value difference between the areas of colour that is most crucial in determining the strength of boundaries.

Proportion and orientation. The visual proportions of a product can be enhanced or altered by breaking down its surface into areas of colour. The same is true of its orientation; that is, whether it appears essentially horizontal or vertical. Colour schemes can also be used to disguise the form, using techniques of camouflage.

A diagrammatic representation of the use of colour or tone to alter the visual form of an object is given in Figure 6.4.

6.4.2 Ergonomic functions

The ergonomic function of colour designates its role in the achievement of safety, good visibility conditions, durability of appearance and product identification. It may be noted that the structural function tends to underlie the ergonomic function—that is, structural effects are used to achieve ergonomic goals.

The ergonomic function has been investigated in a number of studies. For example, Hardy (1966) has considered the problem of colour selection for floor coverings in terms of light reflectivity and maintenance, while Hopkinson (1952) has considered the colouration of chalkboards with reference to visual discrimination. Colour has also been investigated as a means of minimising strong

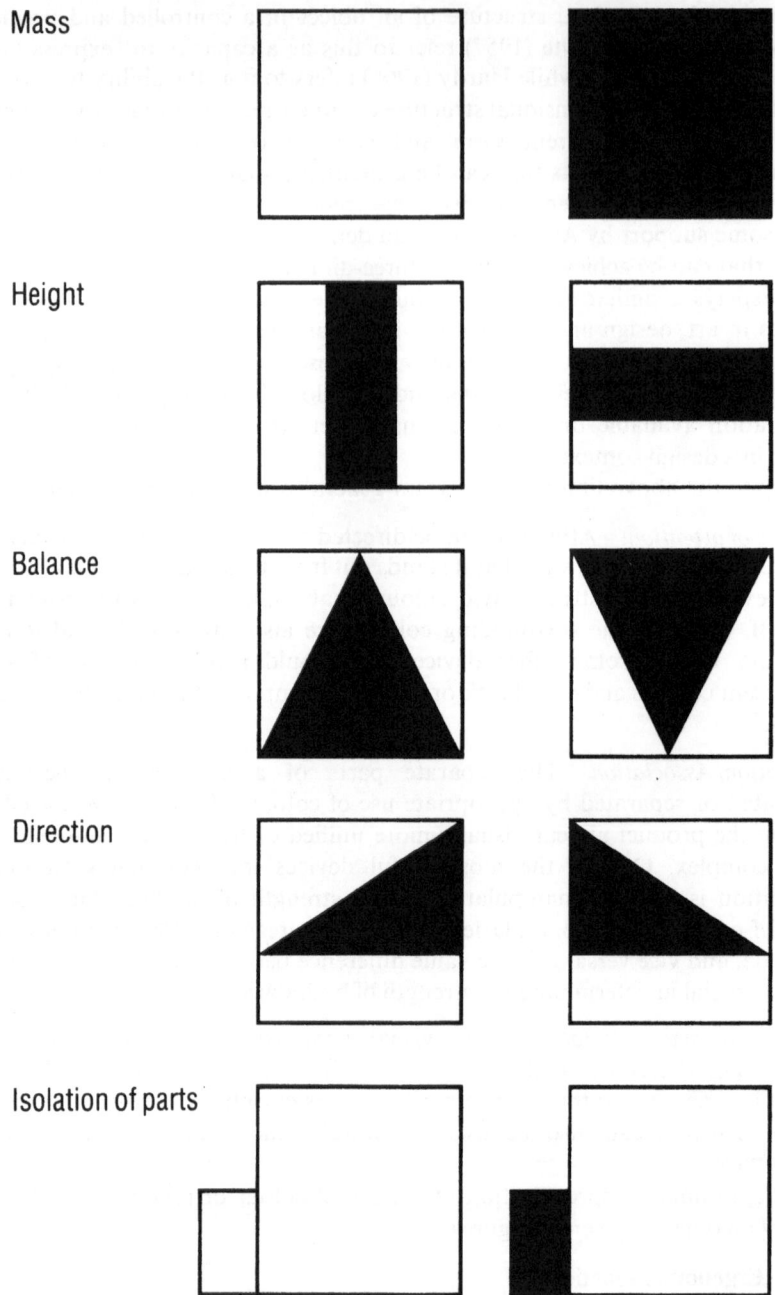

Figure 6.4 The use of colour and tone to alter the apparent physical properties of an object, in diagrammatic form.

brightness contrasts between sources of illumination and their surroundings, and as an aid to controlling the level and distribution of illumination in a room. This has led to value, the light-reflecting component of colour, being regarded as an important parameter in calculations for both daylight and artificial illumination design.

To return to product design, the ergonomic function of colour is easily exemplified.

Safety. This can be aided by careful selection of colours, either to improve the visibility of potentially dangerous equipment or to draw attention to dangerous moving parts. The usual method of achieving this is to create centres of attention—in other words to apply strong, vivid colours to such equipment or components, so that they contrast with their surroundings.

Visibility. In tasks requiring fine discrimination it is important to maximise brightness contrasts at the level of the task, but minimise those in the immediate surroundings. For example, reading at a desk ideally requires maximum contrast and therefore legibility at the level of the text, but reduced contrast between the page and the desk top. For this reason desk tops should be fairly light in colour. Where equipment is to be manually operated, colour coding of controls can improve their visibility. Colour coding can also be used to help distinguish between the various functions of a piece of equipment, and even to facilitate its dismantling and reassembly.

Durability of appearance. Many products are subject to contamination of various forms, quite apart from normal weathering, and the appearance of a product will deteriorate if it is permanently exposed to oil, grease, corrosive materials and dirt of one sort or another. Where such contamination is unavoidable and the results of it are unacceptable, three strategies can be followed. The first is to choose a colour for the product that matches the colour of the anticipated contaminant, so that the product's general appearance does not actually alter much in use. The second is to choose a colour that may still appear acceptable when discoloured by contaminants, such as the traditional grey or dull green of many types of metal-working machinery. The third approach is to choose a colour that will actually emphasise the effect of contamination in the hope of encouraging users to keep the product clean. It is important to remember that service conditions are not the only ones to be considered: product colours should also be chosen with regard to possible damage in transit or while awaiting sale, either to minimise the effect of damage or to enable easy repair or replacement if this is likely to be a problem.

Identification. Colours are associated with various standardised products. At times this is for commercial reasons to help provide a 'brand' image, and at times for public reasons to help with identification. Examples of the latter would be telephone boxes and electrical cable connections to appliances.

Compatibility. A product's colours should be compatible with those of its service environment. In some cases this will mean that colours should be chosen so that the product will have the least possible impact and merge into its surroundings, as, for example, with some electricity sub-stations. In others it will be necessary to make the product stand out from its surroundings. It should be noted that the trend in some hitherto rather bland and traditional fields, including capital goods, is towards brighter colours, and the modern production plant is a much more colourful place than it used to be, both architecturally and in its equipment. The environmental conditions under which the product will operate are, of course, important when selecting colour, and temperature, lighting and weathering must all be considered. Light colours (that is, those with a high Munsell value) will reflect light and heat, while dark ones will absorb them. Care must, however, be taken to avoid reflected glare, which will affect operator comfort and efficiency, as mentioned above.

Maintenance. Many aspects of the use of colour to aid maintenance are very similar to those concerned with day-to-day operation of equipment. Colour can again be used to identify particular areas and systems that are associated functionally, or to indicate particular diagnostic sequences. A particular maintenance aspect is that of identifying service systems, and of coding connections in such services to help ensure correct reassembly. It should be remembered that quite a large proportion of the population, males especially, exhibit some degree of colour blindness. It may therefore be necessary to consider some form of pattern in addition to colour (as in the case of the earth wire in domestic electricity supplies) and, of course, critical components should ideally be designed in such a way as to make misassembly impossible anyway.

Not surprisingly, British Standards have been compiled to cover some aspects of the ergonomics of colour. These tend to focus on identification, and include standards for pipelines (BS 1710: 1960), safety colours in industry (BS 1919: 1967) and electrical cable connections to appliances (BS 4410: 1969). More recent standards have been directed towards co-ordinating colour ranges for various materials used in buildings, such as paints and plastics. Given the importance of these developments and their general industrial and architectural acceptance, the designer should be familiar with them. An outline of the rationale and workings of these standards is given in Gloag and Gold (1978) and in the 'master standard' itself (BS 5252: 1976). The 'satellite' standards currently available are for paints (BS 4800: 1972), vitreous enamel (BS 4900: 1976), plastics (BS 4901: 1976) and sheet and tile flooring (BS 4902: 1976).

6.4.3 Aesthetic functions

The third function of colour is to provide an agreeable stimulus, appropriate to the product, its style and desired image. The problem of this, the aesthetic function of colour, has proved notoriously difficult to resolve, possibly owing to the scarcity of reliable information on the subject compared with the quantity and variety of speculation.

A feature of this area is the variety of advice available, particularly the contrast between what may be termed the 'popular' and 'professional' literature. The essential difference would appear to lie in the assumptions each makes regarding what people actually like. The popular literature lays emphasis on fashion and change, while the professional literature inclines in emphasis towards balance, order and consistency. An example of the latter is the preoccupation with laws of 'classical' colour harmony in the tradition of Goethe, Chevreul and Munsell. Evidence of this preoccupation is found, not only in texts offering guidance to architects and designers, but also in the rationale underlying recent industrial colour standards. Given the architectural and design professions' historical interest in principles underlying 'good' design and their heavy reliance on the philosophic tradition, a concern with objective and hence timeless values is perhaps understandable. Unfortunately, grave problems arise when such principles are tested. This has been demonstrated for 'classical' colour harmony on both a theoretical and empirical level, though it applies equally, for example, to the 'golden section'.

A further important influence on design theory, including colour, has been that of 'functionalism'. Following the famous dictum (attributed to the architect Sullivan) that 'form follows function', this doctrine has come to embody the notion that the aesthetic merit of an object is determined by considerations of practical and structural convenience. The inference is that products that are 'easily' perceived, in terms of being relatively uncluttered, well defined and ordered, will also be nice to look at, and the functionalist doctrine is clearly an appealing one in the case of engineering products, where performance aspects tend to predominate. It should be realised, however, that this distinction between 'functional' performance and 'aesthetic' performance fails to acknowledge that aesthetic performance is necessarily part of the overall function of a product. The idea that aesthetic requirements can be satisfied by paying attention only to utilitarian requirements is as impractical as the notion that utilitarian requirements will be satisfied simply by taking care of aesthetics.

In the context of this argument it may be noted that a design strategy that solely encompasses the ergonomic and structural functions of colour must necessarily exhibit a particular aesthetic (for example, 'jerry-built' and 'utility' styles). In fact, it can be argued that the creation of an aesthetic is unavoidable, and that by considering only non-aesthetic design criteria an 'aesthetic by default' is achieved.

6.5 Design strategies for colour

In considering approaches to the use of colour in product design it will be useful to deal briefly with certain practical problems which arise, prior to focusing upon perhaps the more awkward problem, namely, an approach to the aesthetic function.

The first decisions to be made should probably concern the number of colours to be used on the product and whether the product will be manufactured in a range of colours. There are clear economic advantages in keeping the number of colours as small as possible, both within and across the product range. However, if the product is to be manufactured as a range containing major or minor alterations, then colour can be usefully employed to distinguish between members of the range. Colour can help in this way to maximise what may in reality be minor differences in the product range.

The second set of decisions concern the existence, or absence, of ergonomic and structural requirements of the colour used. For example, the product may require frequent cleaning, or it may be desirable to distinguish between its components for reasons of maintenance, visual acuity or simply to 'break up' its visual mass.

A further set of decisions revolve around marketing and production considerations. For example, in the case of export markets it will be necessary to take account of cultural or ethnic differences, and these may also influence the packaging and advertising material associated with the product. Similarly, for all markets, the particular social and economic group or groups likely to buy the product must be considered. It is, of course, also important to consider how the colour of a particular product will fit in with the manufacturer's general policies in this area—or indeed whether it can be used to introduce a 'new look' for the company as a whole. In many cases it may benefit the company considerably to create a recognisable 'house style', in which colour can play a major part, but the need for future changes must also be kept in mind.

There are also the practical problems of production, on which a number of points have already been made. The main aims here are to ensure that appropriate tolerances for colour matching have been established, related to the nature of the product and manufacturing techniques; that colours have been selected that are either easy to match in the various finishes and materials involved, bearing batch production in mind, or are sufficiently distinct to avoid the possibility of an apparent mismatch; that allowance has been made for future changes of colour and for the effect of different colours on stock levels and costs; and that overall costs of manufacture, assembly, fault rectification, servicing and so on are not being pushed up unnecessarily by the choice of inappropriate colours for the product or product range. Monitoring and quality control are also important aspects of production and, if colour matching is critical, it is important that all those involved in production should understand the limits that apply and should be provided with control colour samples for checking purposes.

Finally, a set of decisions is required to ascertain the relative weightings to be given to the ergonomic, structural and aesthetic parameters. In the context of this problem a caution may be offered, namely, that it is important to recognise that the weight a designer may attach to the practical performance requirements of the product's appearance may vary markedly from those of the customer. This is probably more true of products for the domestic market. For example, the popularity of metallic finishes on cars is increasing in Europe, despite the problems of repair and maintenance. This would seem a case of aesthetic considerations outweighing practical considerations. In fact, the use of gleaming paint finishes and chrome on cars only makes sense aesthetically. It is after all difficult to conceive of any other requirements being met. A further ergonomic absurdity from the car industry is the use of vinyl on roofs. As the roof is the area of external paintwork least subject to corrosion and damage, to cover it with a coat of durable vinyl affords it extra protection it does not require. How much more practical to cover the lower and more vulnerable parts of the car! Again, one can only assume the involvement of aesthetic considerations.

It is important, therefore, carefully to consider the priority status of the functional requirements to be met in the use of colour and not to assume automatically that a practical function must take precedence over an aesthetic function. It is also important to bear in mind that products are purchased new. In this condition the aesthetic can have considerable appeal. It is notable that even agricultural machinery is increasingly being manufactured in gleaming colours, despite the fact that such machinery will quickly soil and may never be cleaned.

At this point some engineers—and product designers—may object and claim that surely a machine should be designed to perform the practical functions required of it and not be subject to irrelevancies of colouring and styling. Unfortunately (or fortunately, depending upon one's moral persuasion) a requirement for most artefacts is that they should look good. Accordingly, a good-looking machine will have a sales advantage over a not-so-good-looking machine. In addition, an attractive appearance should help offset minor practical disadvantages, though this will depend upon the nature of the product and the extent of its practical disadvantages. In effect there would seem to be no advantage in neglecting the aesthetic appearance of a product, only disadvantage.

Having considered problems of production, the ergonomic and structural

6.5 Design strategies for colour

requirements of the colour and their weightings relative to the aesthetic, and ascertained the constraints imposed by these, the problem remains of meeting aesthetic requirements. In effect, the designer must decide what constitutes an attractive, marketable colour for the product. It is suggested that a solution to this problem can be pursued at two distinct levels: first, by a consideration of colour itself (a typical question here would be 'are some colours inherently more attractive than others?'); and second, by a consideration of market forces, fashion trends and the industrial methods used to harness them.

A useful framework for conceiving of colour is to distinguish between its connotations (ie associations) and its desirability, and to regard its desirability as a function of its connotations. For example, a product that appears masculine and heavy may be considered more desirable than one that appears feminine and light.

Connotations are such qualities as:

Perceptual	*Evaluative*
'warm — cool'	'expensive — cheap'
'hard — soft'	'subtle — vulgar'
'heavy — light'	'masculine — feminine'
'strong — weak'	'fashionable — unfashionable'

These connotations can be divided into the perceptual, which are fairly tangible and on which most people will agree, and the evaluative, which are less tangible and on which there is generally less agreement. As examples of perceptual connotations, colours in the red, orange and yellow hue range appear warmer than colours in the green and blue hue range. This has been demonstrated experimentally and appears as true of the British as of the Americans, Swedes and Japanese. Furthermore, colours of high Munsell value (light) appear lighter in weight, larger in size and spatially nearer (assuming a neutral background) than colours of low value (dark).

There are also a number of generally accepted connotations associated with different colours which may be mentioned. These include the link between red and aggression or masculinity, as opposed to pink for softness or femininity; blue for tranquillity; orange or yellow for cheerfulness and warmth; green for growth; white for purity, and so on. However, too much reliance should not be placed upon these associations when selecting colours. The reasons are that the links are somewhat tenuous, and variation occurs from culture to culture.

It is interesting that people can extract such detailed connotational information from colour and also that these connotations, particularly those of a perceptual nature, remain reasonably stable from one coloured object to another. In this sense the connotations of the colour appear to be perceived independently of the connotations of, say, the shape, texture and even the identity of the object. This useful property of colour, therefore, can be exploited to influence the overall perception of a product. In effect, a product can be made to appear warmer, lighter in weight and even smaller in size by the appropriate use of colour. It can also be made to appear expensive, subtle and sophisticated, though these latter goals will be more difficult to achieve than the former. Unfortunately little work has yet been carried out to describe these latter, rather nebulous connotations.

In the absence of detailed information, the designer is obliged to discern the connotational properties of colour himself, unless he is prepared to carry out or commission a proper study in relation to his particular product. If he simply assesses the connotations himself it is important that he checks his judgements against those of others—ideally those of potential customers.

Having decided upon the appropriate colour connotations for the product and selected colours with these connotations, it is advisable to check that these are accurate and influence the desirability ratings of the product. A simple procedure here would be to photograph the product in a range of colours chosen to vary in the appropriateness of their connotations. These can then be rated in terms of both their desirability and connotations. While this approach may seem rather elaborate, and to some designers perhaps superficial, it is worth noting that major manufacturers of domestic products, including foodstuffs, go to great lengths to ensure the visual acceptability of their products. It is just too risky to trust a designer's or food technologist's judgement alone.

With regard to the problem of colour selection, a common error is to attach undue importance to the dimension of hue (ie whether a colour is a red, green or blue etc). Many people seem to regard hue as *the* important dimension of colour, and even as being synonymous with colour. This can perhaps be traced to early school education where colours are (or used to be) classified into 'colours' (hues) and 'tones'.

A consequence of this misunderstanding is the assumption that colour connotations and preferences are largely a function of hue. This is now known to be untrue. Much recent work has convincingly demonstrated that connotations and preferences are a function of the interaction between the three dimensions of colour. The value (lightness) of the colour, however, emerges as a highly influential determinant. This is as true of 'single colour' judgements as of colour relations. For example, available evidence points to the value difference between two colours as being the most important influence on people's preferences for colour pairs.

A possible design strategy to follow where more than one colour is to be used on a product is to 'match' one or two dimensions of the colours—for example, to vary the hue while keeping the value (lightness) and chroma (strength) constant. Probably the most obvious and frequently used example of this is the 'one hue scheme'. In this the hue is kept constant, and possibly the chroma, while the value is varied. If the value steps are kept fairly equal, assuming that more than two colours are used, then the result is a very 'controlled' colour relationship. Now, why should such a relationship appear attractive, assuming that it does? Well, there may be social and historical reasons why matched colours appear so. For example, it costs money to match colours; with poverty anything goes. In 1899 Veblen coined the phrase 'conspicuous consumption' and argued that in order to gain respect and recognition it is not sufficient simply to have wealth, it must be demonstrated; and a very obvious form of display, or demonstration, can be achieved by a costly attention to detail in personal visual appearance (such as clothing and home furnishings). This need not necessarily be ostentatious and therefore too apparent, but can be achieved by subtle means. The matching of colours could be interpreted in these terms, as may the whole sphere of 'colour co-ordination'.

It may be noted that objections may be raised here and the argument offered that surely there are some colour relationships that are inherently pleasing and that those described above, if pleasing, are found so because of their natural lawfulness. While this may be true, unfortunately such arguments tend to appeal to the very existence of natural lawfulness as a justification in itself, while failing to specify the general principles of natural lawfulness, to define them operationally, or account for why such lawfulness should be appealing anyway.

Further examples of apparent lawfulness may be exemplified, using the 'colour wheel' of the Munsell system. This shows variation in the two dimensions of hue and chroma, and provides the starting point for selecting colours which some think will 'harmonise', according to the following schemes.

Monochromatic. Monochromatic colour schemes (6.5a) have been discussed above. These use colours of the same hue but which vary in their value and chroma.

Apposite. Such a scheme (6.5b) employs closely related or adjoining hues, again with variation in value and chroma.

Complimentary. Two hues that are diametrically opposite each other on the colour circle are said to be complementary as (in theory) they make a neutral grey when added together (6.5c).

Triadic. This scheme uses three hues at roughly equal intervals around the colour circle (6.5d).

Split complementaries. This uses a three-way division of the colour circle, but with two of the hues close together (6.5e).

Double split complementaries. This is a more ambitious variation of the complementary method, in which two adjacent hues are used with their complementaries (6.5f).

In concluding this section the question may be posed: 'Are some colours inherently more desirable than others?'. The simple answer to this, from the standpoint of product design, is 'No'. While some colours may be regarded as desirable at a given time this does not mean that they will continue to be so. It is also becoming increasingly apparent from experimental work that the judged desirability of a particular colour will change according to the object to which it is applied. For example, while light yellow may be considered appropriate for a shirt it may be considered inappropriate for a suit. Similarly, the desirability of a colour will probably vary with the style of the object. For example, the wall colours considered appropriate to a room furnished in the 'Modern' style are unlikely to be considered appropriate to that same room furnished in the 'Georgian' style. This should act as a caution against treating colour in isolation. Colour is but one attribute of an object and must be considered in relation to the other attributes, be they perceptual (such as line and form) or cognitive (such as style and identity).

It was stated earlier that a solution to the problem of colour selection for products can be pursued at two distinct levels: first, by a consideration of the nature of colour itself and, second, by a consideration of market forces, fashion trends etc. It is the latter to which we now attend.

In industries in which the achievement of a desirable product colour appearance is important, steps are taken, or should be taken, to maximise the likelihood of success. These can be considered as three separate processes.

(a) The collection of own-product sales data to ascertain the popularity of the different colours offered—assuming that customers are given the choice.
(b) The collection of data on competitors' colour ranges, and sales, if possible.
(c) The collection of speculative data on future colour trends.

Both (a) and (b) are self explanatory. Basically they provide information on past and present colour trends. While this information should prove useful when colours are to be dropped from a range, it has limited use when new colours are to be added. The problem is that colours for products have to be selected long before they are actually manufactured. The time delay will vary with the industry and the

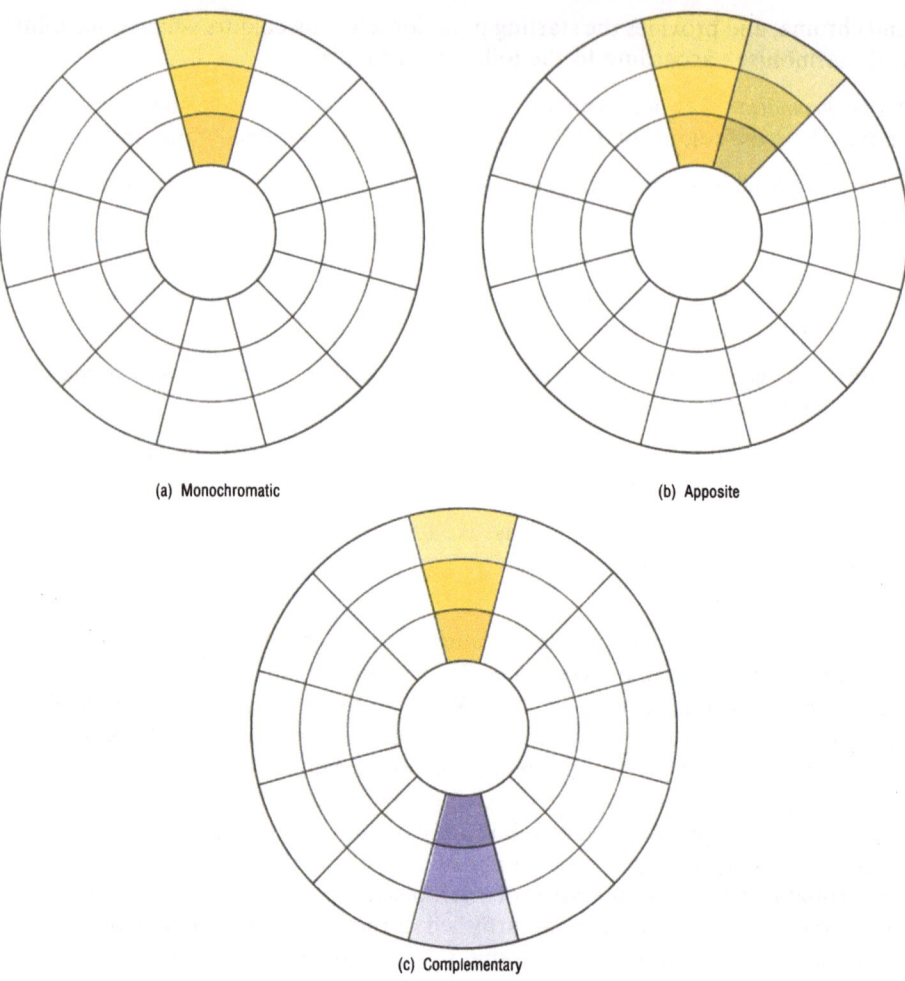

Figure 6.5a to f (above and opposite page)   Strategies for selection of colour schemes, based on the Munsell system.

nature of the product. A delay of two or even three years may not be unrealistic. A consequence of this delay is that current trends will be past trends at the time of manufacture. However, the collection of data on colour sales over a number of years should give some general indication of possible future trends.

It is at this point that the benefits of (c) are apparent. Various specialist bodies predict future colour trends, generally for the fashion/clothing industry. The workings of these bodies are rather mysterious, but it would appear that groups of individuals, well endowed with predictive faculties, decide amongst themselves what is going to happen in the colour scene. How successful are their predictions? It is difficult to tell. If, however, we assume that there *is* some validity in their predictions, which is by no means clear, then the questions may be asked: 'How do they arrive at them and why are they valid?'. A highly speculative answer to these questions runs as follows. If we assume that prominent designers influence less prominent designers, who in turn influence industry, then assuming the industrial dread of being caught out commercially with the wrong product appearance, it would seem sensible in this frighteningly uncertain area to attend to

6.5 Design strategies for colour

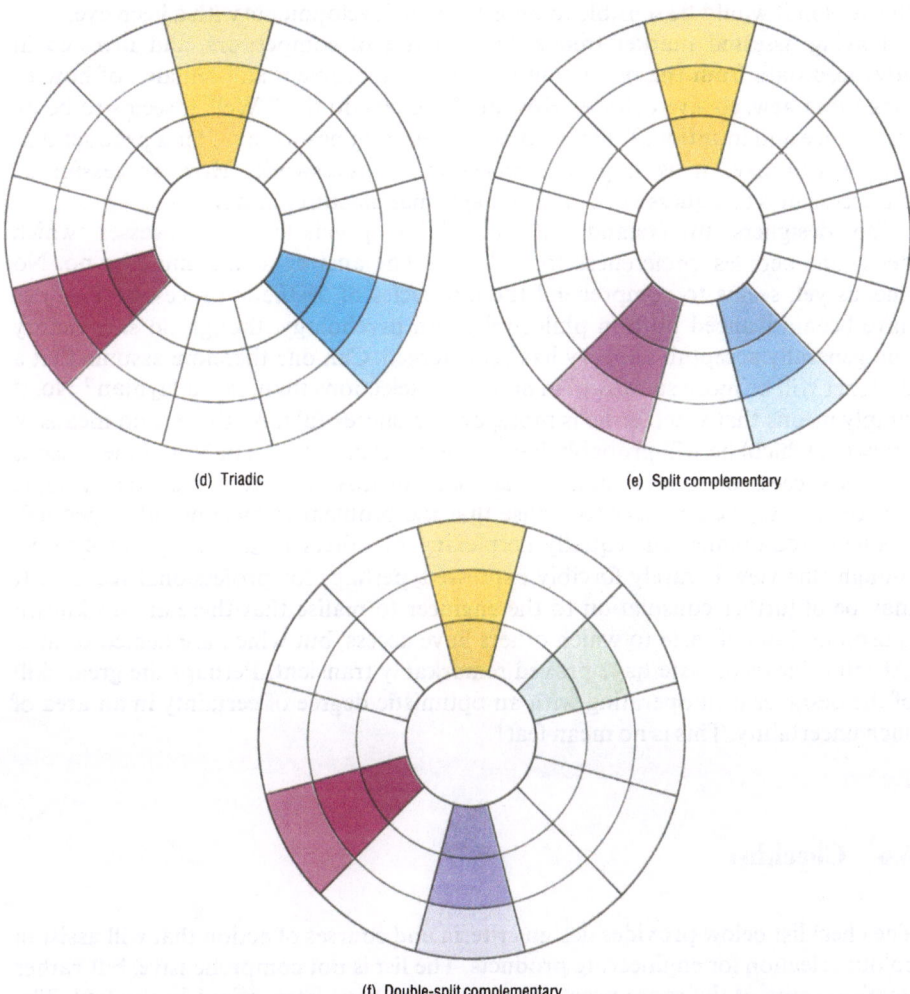

(d) Triadic (e) Split complementary

(f) Double-split complementary

what the prominent, trend-setting designers are up to. The 'prediction' industry probably acts in a liaison capacity by detecting trend-setting developments for industrial consumption. In the absence of any more reliable information to act upon, industry manufactures in accordance with these trends, which it accompanies with promotional advertising. As these trends now correspond fairly closely with those of 'top' designers, they are sanctioned and thus given credibility. The customers then, perhaps, learn to regard these new types of appearance as desirable. What this speculation suggests is that we are not so much dealing here with prediction as with self-fulfilling prophecy.

At this stage in the discussion the reader may well feel that developments in the fashion/clothing industry are rather irrelevant to engineering products. This is not necessarily so, however. The reason is that the timescale for manufactured goods varies considerably, so that fashion clothing will be marketed ahead of domestic products, which in turn will precede engineering products. If we accept that fashion is an important market force, then developments in shorter timescale industries are potentially important due to the influence they can exert on public taste. For this reason it would be sensible to observe such developments with a keen eye.

Having assessed market trends, the practice of competitors, and invested in advanced data from the prediction industry, the problem still remains of how to pick those new, elusive colours. How do designers do this? Well, it seems to be by experience and intuition. When it comes to selecting new colours for a product that people will like, it is largely a question of guesswork. How successful or unsuccessful is this guesswork? Ask a major manufacturer and see!

Do designers understand the underlying psychological processes which determine peoples' preferences for colours? The answer to this must be no. No one, as yet, seems to comprehend the intricacies of aesthetic processes. Theories have been advanced both in philosophy and psychology, though no satisfactory and generally acceptable theory has yet emerged. Can one therefore assume that a designer will be more successful in his colour selections than, say, a layman? No, it simply means that whether he is more, or less, successful (and there is no means of knowing which) he will probably have no more clear understanding of the reasons for his success or failure than the layman. In this context it may be of some consolation to the engineer to realise that the problem of meeting other people's aesthetic requirements is equally perplexing to others in the design profession, though this view is rarely forcibly expressed, perhaps for professional reasons. It may be of further consolation to the engineer to realise that there are no known 'absolute' laws of taste to which others have access, but which are denied to him. 'Absolute' laws of taste have proved remarkably transient. Perhaps the great skill of the designer is in operating with an optimistic degree of certainty in an area of such uncertainty. This is no mean feat!

6.6 Checklist

The checklist below provides design criteria and courses of action that will assist in colour selection for engineering products. The list is not comprehensive, but rather itemises some of the more practical points that have been raised in the text. The points are interrelated and must be considered in parallel, although different weight will be given to the various aspects, depending on the nature of the product and its market.

The photographic illustrations that follow (Figures 6.6 to 6.13) have been chosen to demonstrate some of the ways in which the principles identified in this chapter can be applied in the design of actual products.

Figure 6.6 (opposite, top)   The isolation of components in this Slimtherm boiler is carried out by using a bright, contrasting colour. (*Butler, Dennis, Garland and Partners Ltd*).

Figure 6.7 (opposite, bottom)   The form of this Hydrovane compressor is broken down and made more interesting visually by the use of a contrasting colour scheme (designer John Payne).

6.6 Checklist

Figure 6.8 (above) The proportions of the windscreen of the Leyland T45 are altered by including a black panel beneath it, giving the impression of a deeper area of glass and therefore a lighter cab structure. (*Leyland Vehicles*)

Figure 6.9 (left) The moving guard portions of this Dualform press are highlighted in bright yellow, for safety purposes. (*Dualform Press and Shear Manufacturing Co*)

6.6 Checklist

Figure 6.10 (above) An ergonomic test model of S. Davall and Sons submarine metal detector, showing the very unusual colour coding devised for use in extremely poor working conditions (*designer Alistair Ewen/Bill Moggridge Associates*).

Figure 6.11 (right) Colour coding of services in the BHRCA Fluid Engineering Laboratories.

Figure 6.12 The colour scheme employed in the ICL 2903 computer system (see Chapter 18.11) was selected to reduce its apparent bulk and be compatible with office surroundings. (*ICL*)

Figure 6.13 (above) A design study model based on the Triumph TR7 showing the use of colour and tone to enhance the styling and provide an impression of speed. (*Royal College of Art*)

6.6 Checklist

A Colour and product marketing

Consideration	Aim	Action
User/specifier needs	Satisfy needs and compete with other products on the market	Consult specialists to establish appropriate practice and standards and select colours accordingly
Geographic/ethnic group requirements	Take account of different cultural requirements for all forseeable markets	Consider cultural influences for all markets for all aspects of colour in product design and promotion
Socio-economic group requirements	Take account of buying habits and changing conditions	As above, but include special consideration of market change using surveys such as MINTEL and examine competing and related products
Company policy	Enhance existing and future policies	Choose colours to support existing company practice if any, or consider establishing an appropriate 'house style' while allowing for future changes

B Colour and ergonomics

Consideration	Aim	Action
Operator efficiency and safety	Reduce fatigue and errors and present information clearly	Reduce glare, high reflectivity and excess contrast Isolate and identify functions such as control Draw attention to, and clarify, graphics Identify system functions and safety aspects in particular Use appropriate colour coding for functions
Maintenance	Enhance and encourage proper maintenance and reduce risk of incorrect reassembly	Isolate and identify areas and systems involved, including diagnostic sequences Identify services for checking and reassembly by colour coding

Based on marketing and ergonomic requirements, decide on total number of colours required for main components of product

C Colour and environment

Consideration	Aim	Action
Structural and aesthetic appearance	Emphasise good form and enhance other characteristics while maintaining an interesting overall scheme	Use colour and tone to alter or emphasise apparent size, weight, proportion, interest and orientation of product while ensuring that colours are compatible with service environment and with one another
Psychological and emotional effects	Minimise ill effects on all those involved with the product and encourage its proper use	Consider effects of product, especially in critical situations with 'captive' audiences, and choose colours to produce harmonious overall effect Consider camouflage of vandal-prone products

Consideration	Aim	Action
Contamination	Enable product to resist weathering and contamination without detracting from appearance	Select colours to reduce apparent contamination and minimise effect on colour matching, or alternatively consider use of light colours to encourage regular cleaning and maintenance
Handling and storage	Enable product to reach end customer in good condition	Choose colours to reduce apparent damage in transit or deterioration in storage, or to simplify rectification Consider use of materials with integral colouring to reduce effects of damage or wear

D   Colour and production

Consideration	Aim	Action
Colour matching	Avoid unintentional mismatching and facilitate economic production	Choose colours that can be matched easily in appropriate materials and finishes, or alternatively ensure adequate deliberate contrast between colours
Manufacturing quantities and techniques	Allow efficient production in appropriate quantities and with economic stock levels	Consider chances of colour change or changes in type of finish and the effects of wide colour range on stocks for sale and repair
Monitoring and quality assurance	Effective and economic control of colour quality	Establish appropriate tolerance levels and provide proper colour samples for all staff concerned

6.7   Further reading

No book is currently available that gives a serious and comprehensive coverage of this area. Information on most of the subjects covered in this chapter is available, but finding it means searching through a number of books and articles. A very readable and well illustrated account of colour perception is given in Lindsay, H. P. and Norman, D. A. *Human Information Processing* (Academic Press, New York, 1977). As the title suggests, this book adopts an 'information processing' approach to perception, and it covers not only colour but various facets of perception and general psychology. A useful treatment of colour specification systems (including Munsell), colour measurements and tolerances, and industrial aspects of colour technology is given in Billmeyer, F. W. and Saltzman, M. *Principles of Colour Technology* (Wiley, New York, 1966). Again, this is a readable and well-illustrated book which provides a fairly comprehensive introduction. Those wishing to advance beyond this level should consult Judd, D. B. and Wyszecki, G. *Colour in Business, Science and Industry* (Wiley, New York, 1963). Information on the ergonomic and structural functions of colour in design is extremely limited. For the ergonomic see Hardy, A. C. (ed) *Colour in Architecture* (Leonard Hill Books, London, 1966 and Gloag, H. L. *Colouring in Factories* (HMSO, London, 1961). Probably the most thorough treatment of the principles underlying the structural use of colour is given in Albers, J. *Interaction of Colour* (Yale University Press, Newhaven, Connecticut, 1963). A paperback version of this book is also available, but without the range of superb illustrations found in the original. See also Birren, F. *Light, Colour and Environment* (Reinhold, New York, 1969). There are numerous texts of varying quality dealing with the aesthetics of colour, though few of these are explicitly directed towards product design. For overviews of the psychological literature on colour and aesthetics see Kreitler, H. and Kreitler, S. *Psychology of the Arts* (Duke University

Press, Durham, North Carolina, 1972) and Pickford, R. W. *Psychology and Visual Aesthetics* (Hutchinson, London, 1972). Those wishing to sample the philosophical literature may consult Sparshott, F. E. *The Structure of Aesthetics* (Routledge and Kegan Paul, London, 1963) for a discussion of issues relating to aesthetics in general, and Whitfield, T. W. A. and Slater, P. E. 'Colour harmony: an evaluation' (*British Journal of Aesthetics*, vol 18, 1978, p 199–208) for a discussion of certain issues relating to colour. Coverage of the rationale underlying building colour standards is given in Gloag, H. L. and Gold, M. J. *Colour Co-ordination Handbook* (HMSO, London, 1978) and in the master standard itself *BS 5252: framework for colour co-ordination for building purposes* (BSI, London, 1976). For information on other standards relating to safety etc, it will be necessary to consult the British Standards Index. Finally, examples of the strategies that industries employ in ensuring the acceptability of product colours can be found in Gantz, C. M. 'Mass-market colour selection' (*Colour Research and Application*, vol 3, 1978, p 137–140) and Hall, J. 'Colour popularity in automotive finishes' (*Colour Research and Application*, vol 2, 1977, p 105–107). These short articles provide an interesting glimpse into the world of product marketing methods.

6.8 References

6.1   Birren, F. *Light, Colour and Environment*. Reinhold, New York, 1969.
6.2   Gloag, H. L. and Keyte, M. J. 'Rational aspects of colouring in building interiors.' *Architects Journal*, vol 125, 1957, p 399–402, 443–448.
6.3   Hopkinson, R. R. 'The selection of suitable chalkboard colours'.*RIBA Journal*, vol 59, 1952, p 377.

7 Machine graphics
Ronald Easterby

7.1 Introduction: The man/machine interface as an information device

Designing devices for communicating between man and machine is a nice blend of technology and aesthetics. The technical requirements of the machine, the ergonomics of the display, and the relationship of displays to other visual aspects of the machine all interact with one another. Machine displays provide information on the current state of the machine so that it can be properly controlled; control elements enable the operator to influence machine behaviour. Machine graphics are required to label and identify these control and display elements reliably and to provide other forms of supplementary information. In addition, other machine graphics such as borders, outlines, and coloured and contrasting areas can be used to enhance the man/machine interface, and facilitate the operator's task by making it easier to identify specific control and display functions.

Two issues confront the designer: deciding on the form of the display elements and specifying their detailed characteristics. Ergonomics dictate the choice of the *form* of the display or control element; performance and cost determine the *detailed* technical aspects of their design. The effectiveness of displays in communicating with the operator is of prime importance; hence the significance of ergonomics in the choice of machine graphics for display design.

Machine displays divide naturally into either static or active displays and each type has a distinctive role to play. Static displays label or provide instructions, or present fixed numerical information; active displays provide discrete status indications, indications of steady-state values, indications of dynamic-state values (or rate information), indications of spatial relations or simulations of the real world.

The choice between these different forms of display is an important ergonomic decision, governed by human factors research which provides objective data on the influence of display formats on speed, accuracy, user stereotypes, user preferences and overall task performance. But the final detail design of displays also depends on the practicalities of specifying and producing a display appropriate to a given machine. So this chapter will consider not only the ergonomics of machine displays, but how the related disciplines of graphics, typography and industrial design affect the final display design.

7.2 The function of the man/machine interface

Machine graphics can only be considered within the context of the total interface—including both controls and displays (see also Chapter 3). A variety of psychological processes are used by the operator in controlling his machine. Designers must be aware of these, since they determine to a large degree the limits of control display formats.

Detection: *Determining the presence of an interface element*
Control and display elements must be made the right size for adequate detection.

Discrimination: *Defining the differences between one interface element and another.*
Labels, symbols, shape and colour contribute to the discrimination process.

Identification: *Attributing a name or function to an interface element.*
Fitness for purpose and function of an interface element enable the operator properly to identify controls and displays. Control and display elements must have explicit characteristics that enable the operator to identify them reliably.

Classification: *The grouping by the operator of elements with similar purpose or function.*
Spatial grouping, juxtaposition, shape coding and colour coding all facilitate classification.

Recognition: *Knowing what a display or control element purports to do.*
The form of the display must be chosen so that there is no conflict between its apparent function and its actual function.

Scaling: *Assigning values to variables.*
Scales, indicators and tabular displays all provide scaling information.

Ordering/sequencing: *Determining the relative order and priority of using control and display elements.*
Spatial arrangement and relative positioning of elements and groups of elements facilitate the ordering and sequencing process.

Not only is the detailing of the elements important, but their pattern and structure become important too. The spatial configuration of the interface forms an important part of the machine display, and the total pattern provided by the spacing of elements creates a visual order to which the operator will respond. A formal uniform layout may in fact militate against effective detection, discrimination, identification, classification and recognition. The result will be ambiguity, longer operating times, operation errors, and the possibility of serious errors leading to damage to the product, the machine, or the operator.

7.2.1 The interface as a display: basic principles

We therefore need to know how the operator perceives his machine and the interface when operating it, because this ultimately determines the most effective way to organise the display aspects of the interface.

Operators organise their tasks in hierarchies; each section of the task has sub-tasks, which in turn have more detailed components, and we can continue to break a task down until we have analysed and defined it in terms of individual visual and control activities. We can also analyse the *order* in which the task components occur, which in turn gives us some clues as to the best spatial configuration of the interface elements. Lastly, we can examine information required to perform the task components; this determines the form of interface which will be required.

7.2.2 Mapping the operator's task

A valuable approach is to think of displays as a form of map. We often use a map in the literal, everyday sense of that word, but if we extend the term and use 'map' in a more general sense, we can consider the control panel of a machine as a map of the operator's task. Any task can be considered as psychological territory with both familiar and unfamiliar features, and displays should give a map of the task in both the literal and figurative senses. The display must therefore define relationships that exist between the machine elements, and give some clues as to what to do next and when. The human factors specialist has traditionally used some basic heuristic principles to develop the design of the display interface. This has involved some form of task analysis, followed by the grouping of display and control elements according to one or other of the following four principles: function, sequence, frequency of use, and relative importance. The difficulty of applying these time-honoured principles is that they are not independent of one another. Appealing as this basic approach is at first sight, it simply does not work for anything other than the simplest of machines.

However, we can instead use a new set of principles for the design of the interface, based on the notion that we are devising a map of the task, and that the structure of any interface is a display in itself, incorporating information displayed in the constituent interface elements.

First, *group the elements by function.*

Then *organise the sequence of use within these functions* based on a hierarchy of the sub-elements of each functional task.

Finally, *organise the major functional groups* (now much fewer in number than the number of individual elements) *according to the task structure*, arranging them according to the traditional criteria of sequence, frequency of use and relative importance.

These design principles are much more tractable in this new form, and the number of items to be dealt with is greatly reduced. Equally important, the visual structure of the interface now assumes the characteristics of a map of the operator's task, an appropriate objective for any man/machine interface.

With these general principles in mind, we can now consider some aspects of typography, the use of symbols, how to select text, the exploitation of colour coding and the design of tabular displays, all of which relate to the larger problem of designing an optimum man/machine interface.

7.3 Fundamentals of typography

Ergonomics and technical design issues interact when considering typography for

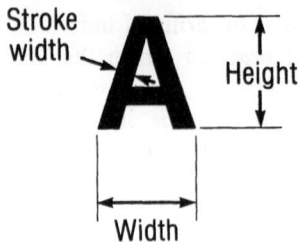

Width/height ratio 0.7:1

Stroke width/height ratio

Black on white 1:6
white on black 1:8

Figure 7.1 The geometrical specification of letters and numerals in a typeface.

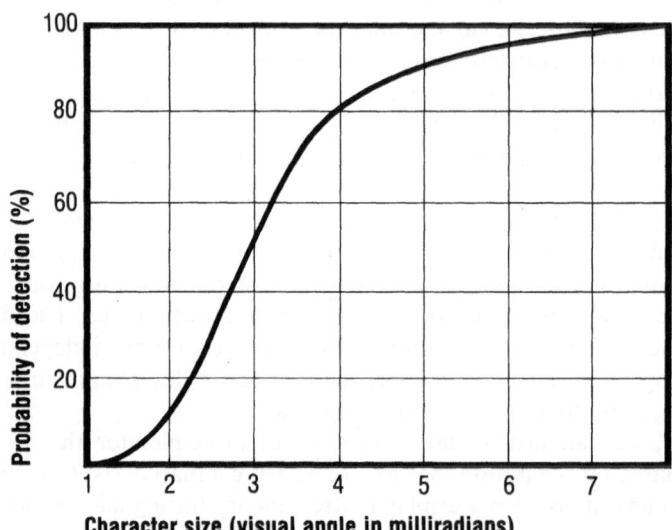

Figure 7.2 The probability of correct identification of a typeface with variation in visual size.

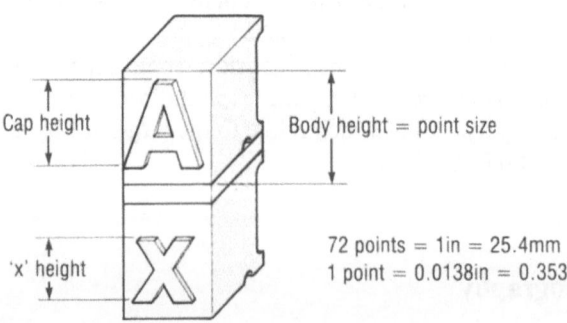

72 points = 1in = 25.4mm
1 point = 0.0138in = 0.353mm

Figure 7.3 A typeface and its formal specification.

machine displays. Objective ergonomic characteristics include legibility and discriminability, and technical characteristics are defined by the exact way in which the typesetting process operates. Both sets of characteristics must be taken into consideration.

A major difficulty is that the measurement systems for letterforms as geometrical shapes differ from those applied to pieces of type. Ergonomists have not usually used the typographers' specification methods in designing their research material, but happily it is possible to relate the two.

7.3.1 Geometrical specification of letters and numerals

The optimum legibility of letters and numerals has been defined in terms of the relationship between character width, stroke width and character height (Figure 7.1). Systematic research, using the formal geometrical specification of characters, has yielded an optimum character width/character height ratio of around 0.7 : 1 and (for black characters on a white ground) an optimum stroke width/character height ratio of 1 : 6. For reverse contrast (white letters on a black background), the optimum ratio needs to be 1 : 8 to compensate for the irradiation of the white characters. In terms of operational character size, however, we must relate the size of the letter to the distance from which it is viewed. We therefore specify size in terms of the visual angle in milliradians subtended at the observer's eye (1 milliradian is the angle subtended by a 1 mm high character at 1 m viewing distance). It will be seen from Figure 7.2 that characters can be reliably identified on 90 per cent of the occasions when the visual size is 3.5 milliradians, and that by increasing the visual size to 7 milliradians we can ensure 100 per cent certainty of identification.

A decision can thus be made on how large the character size must be depending on the viewing distance, and how important it is in operational terms for the characters to be reliably detected. Alternatively, the *loss* of detectability with reduction in size can readily be estimated and used when establishing the size of the machine graphics.

7.3.2 Principles of typographic form

To translate the optimum legibility of geometrically defined forms to real displays it is necessary to understand the system of metrication used in the printing industry, since high-quality machine graphics can only be achieved by using proper typographic composition and setting methods. However, with the demise of traditional techniques of hot-metal typesetting, the old methods of measurement will become redundant. The introduction of new photocomposition systems has been accompanied by the introduction of metric systems of typographic measurement.

7.3.3 Traditional type measures

Traditionally, type has been produced by casting characters on an individual slug of metal (Figure 7.3). In specifying type manufactured in this way, reference is made to the following.

Body height: the height of the slug of metal. This height is measured in points, the 'point size'. For all practical purposes a point is equal to 1/72 inch (72 points = 1 inch = 25.4 mm.) There are slight differences between the European point and the American point, but this need not concern us here.

Cap height: the height of the capital letter 'X'.
x height: the height of the small (lower case) letter 'x'.
Leading: the space between the body of successive lines of type, measured in points.

Thus, for a 36 point type, for instance, the body size is 36 points high, (0.5 inches or 12.7 mm). But the height of the letters is less than the body size and, for each different typeface, the characters may be a different height, even though the point size may be the same. Thus two other factors, over and above the point size, are needed to completely specify the letter form:

Typeface: the style of letters to be used specifying its form (eg Times Roman, Helvetica, Univers) and its proportion (eg Condensed, Standard, Extended).
Type weight: the relative weight of letters within a given typeface, basically but not exclusively related to stroke width (eg Bold, Medium, Light).

Thus, a typical complete type specification could read:

14 point—Helvetica, Upper and Lower case, with 2 point leading.
        ie point size: 14 point
        Typeface: Helvetica
        Typeface specification: Upper and Lower case
        Leading: 2 points

7.3.4 Metric type measurement

Modern photocomposition techniques have eliminated the need for solid metal setting and its outdated measuring system. It has been replaced by rational methods of typesetting using metric measures which refer directly to the dimensions of reference letters in the alphabet for each typeface. Thus, body height is now irrelevant, and the interline spacing can be directly specified, rather than derived from considerations of body size and type form (Figure 7.4a, b).

Figure 7.4a Reference lines in metric type measurement.

Figure 7.4b   Reference dimensions in metric type measurement.

The key reference dimensions used are:

Base line: a line drawn through the base of the capital letters
Cap line: a line drawn through the top of the capital letter 'X'
x line: a line drawn through the top of the lower case 'x'
Ascender line: a line drawn through the top of the ascender on the lower case letter 'k'
Descender line: a line drawn through the bottom of the descenders on the lower case letter 'p'.

The dimensions between these reference lines then yield the following reference sizes:

E height: Base line to cap line
Ep height: Base line to descender line
x height: Base line to x line
kp height: Ascender line to descender line
Interline space: Base line to cap line
Pitch: Base line to base line

7.4   Symbols for labelling machine functions

The use of symbols to identify machine functions has grown in the past decade. Few machine systems are now without symbols of some form for the labelling of machine control and display functions, and a range of generally accepted standard symbols is currently under development.

7.4.1   Psychological principles in symbol design

In the design of individual symbols, one set of important principles derived from psychological theory can be seen as helpful. These suggest that any graphic symbol should tend towards the following ideal characteristics:

Figure/ground: Clear and stable figure to ground articulation is essential.
Figure boundary: The figures should be bounded by a contrast boundary in preference to a line boundary. Where more than one

graphic element is required for a symbol, the most important should have a solid contrast-bounded element. Rationalisation of these principles leads to the following consistent allocation between solid and outline elements of the symbol:

    Symbol dynamic: solid
    Moving or active part: outline
    Stationary or inactive part: solid

This allows a consistent approach to the design of complex symbols, since the stationary part (solid figure) will, by definition, never require the dynamic part (also a solid figure) to be superimposed. Problems of overlapping solid parts of a composite symbol should therefore not occur.

Geometrical forms: Where simple geometrical shapes are used, it is preferable to have a solid rather than an outline figure. For these solid contrast-bounded figures, optimum discriminability will be achieved if the following rules are kept in mind:

    Triangles and ellipses: maximise area
    Rectangles and diamonds: maximise one dimension
    Stars and crosses: maximise perimeter

Closed figures: Line contour figures should always form a closed figure unless it is essential to the meaning of the symbol for the outline to be discontinuous. In this case the discontinuity should be unequivocal to avoid misleading interpretations.

Continuity of figures: The smoothest continuous outline for the figure should be used. Again, as in closure, if a discontinuity is essential it should be unambiguous.

Simplicity: The symbols should be as simple as possible. Fine detail makes no contribution to unambiguous and rapid interpretation.

Symmetry: Symbols should be as symmetrical as possible, providing that asymmetry adds no further meaning to the figure.

Unity: Symbols should be as unified as possible. This can be achieved by consistent use of the same size and proportions of individual elements when they repeat. Secondly, when solid and line outline figures occur together, more unity is achieved when solid figures are integrated by enclosing them within the line outline figures.

Orientation: The prevailing outlines of the symbol should follow as far as possible the main horizontal and the vertical axes.

These basic principles of visual form are illustrated in Figure 7.5 with references to some typical symbols designed on this basis for use on machine tool control panels.

7.4.2 Symbol standards

Continued efforts of the International Standards Organisation Technical Committees, notably through its co-ordinating committee TC145-Graphic Symbols, have resulted in the development of some useful guidelines for symbol

7.4 Symbols for labelling machine functions

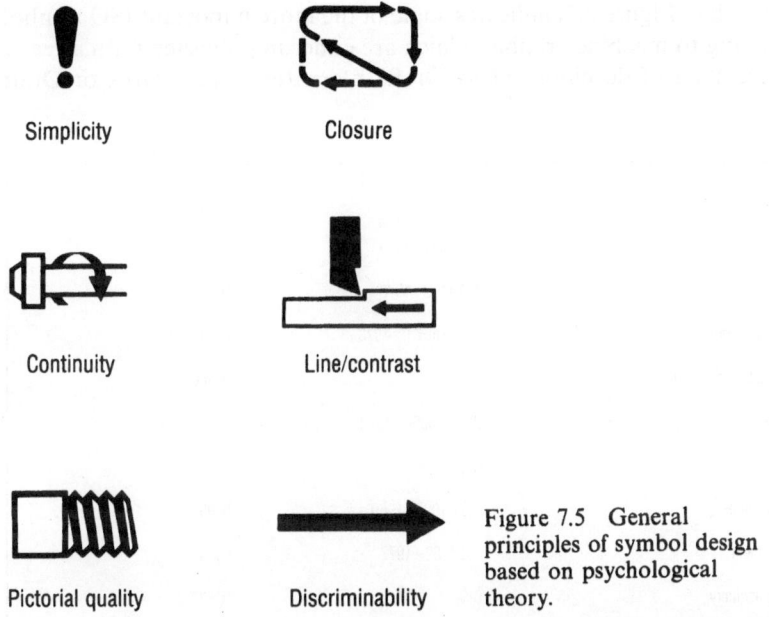

Figure 7.5 General principles of symbol design based on psychological theory.

design, especially for use on machine controls. *ISO 3461: Graphic Symbols—general principles of presentation* elaborates these in some detail. It has two main aims: first to facilitate the graphic design of symbols, and second to achieve some uniformity of presentation of standardised symbols.

While not completely effective (the rigid grid design suggested sometimes conflicts with the psychological principles outlined earlier), with careful implementation the two sets of ideas can be usefully harmonised (see Figure 7.6). The outcome of this ISO work is a series of standards in a number of technical areas which are in turn complemented by a more elaborate and detailed national

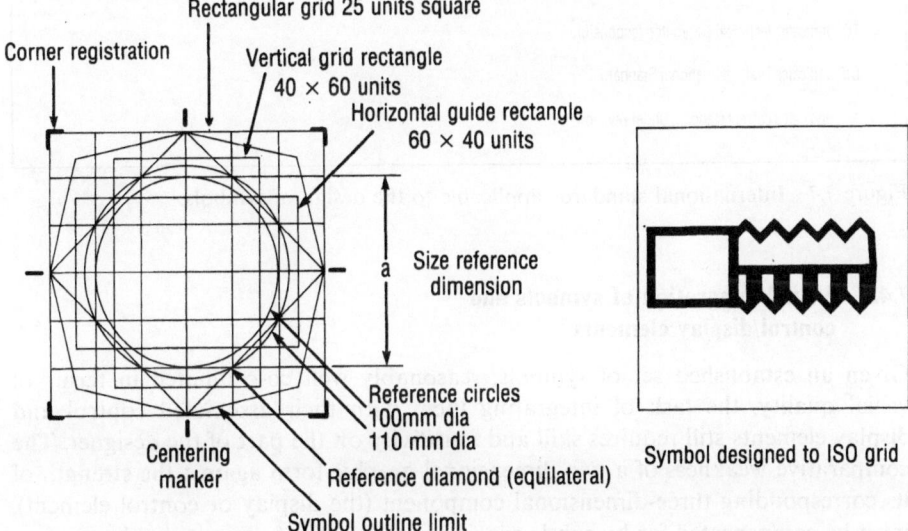

Figure 7.6 The ISO/IEC grid for the design and co-ordination of graphic symbols.

standard. The table (Figure 7.7) indicates some of the more important ISO symbol standards relating to machine graphics which are either in published form already or in an active state of development as Draft International Standards or Draft Proposals.

Graphic symbols	ISO DIS 7000	TC145
Graphic symbol design	ISO 3694 - 1973	TC145
Use of the arrow	ISO 4196 - 1978	TC145
Co-ordination procedure	ISO GUIDE II - 1978	TC145
Symbols for electrotechnology	IEC 41C	TC4(a)
Symbol review	DIN 30 6000 - 1978	
Agricultural machinery	DIS	TC23
Calculating machines	ISO R1093 - 1969	TC95
Dictation equipment	ISO 4062 - 1977	TC95
Earthmoving machinery	DIS	TC127
Industrial trucks	DIS	TC110
Ionising radiation	ISO R303 - 1963	TC80
Machine tools	*ISO R369 - 1964	TC39
Numerical control	ISO 2972 - 1975	TC93
Safety signs/colours	*ISO R557 - 1967	TC80
Textile machinery	DIS	TC72
Typewriters	ISO R1090 - 1969	TC95

Brief titles only shown

TC Indicates technical committee responsible

DIS Indicates Draft International Standard

*   Indicates current standard under revision

Figure 7.7 International standards applicable to the design of symbols.

7.4.3 Visual integration of symbols and control/display elements

Given an established set of symbols, reasonably well co-ordinated in terms of visual quality, the task of integrating them with their associated control and display elements still requires skill and sensitivity on the part of the designer. The comparative weakness of a two-dimensional graphic form against the strength of its corresponding three-dimensional component (the display or control element), must be compensated for by subtle spacing and the use of various graphic devices to relate symbolically coded labels unambiguously to their associated elements.

7.4 Symbols for labelling machine functions

Here a simplified form of visual syntax can be applied: combinations of symbols, by superimposition or juxtaposition, to make combined symbolically defined functions (Figure 7.8a); superordinate symbolic labels which then divide hierarchically into sub-functions (Figure 7.8b); visual integration of symbolically coded labels with the corresponding control and display, using outline graphics with a key pointer (Figure 7.8c). Note that the preferred position for symbolic labels is *above* the associated mechanical/electrical elements. Vertical spatial separation of the visual elements, and suitable asymmetry of the enclosing boxes, then ensures that no ambiguity exists as to which control the label refers to.

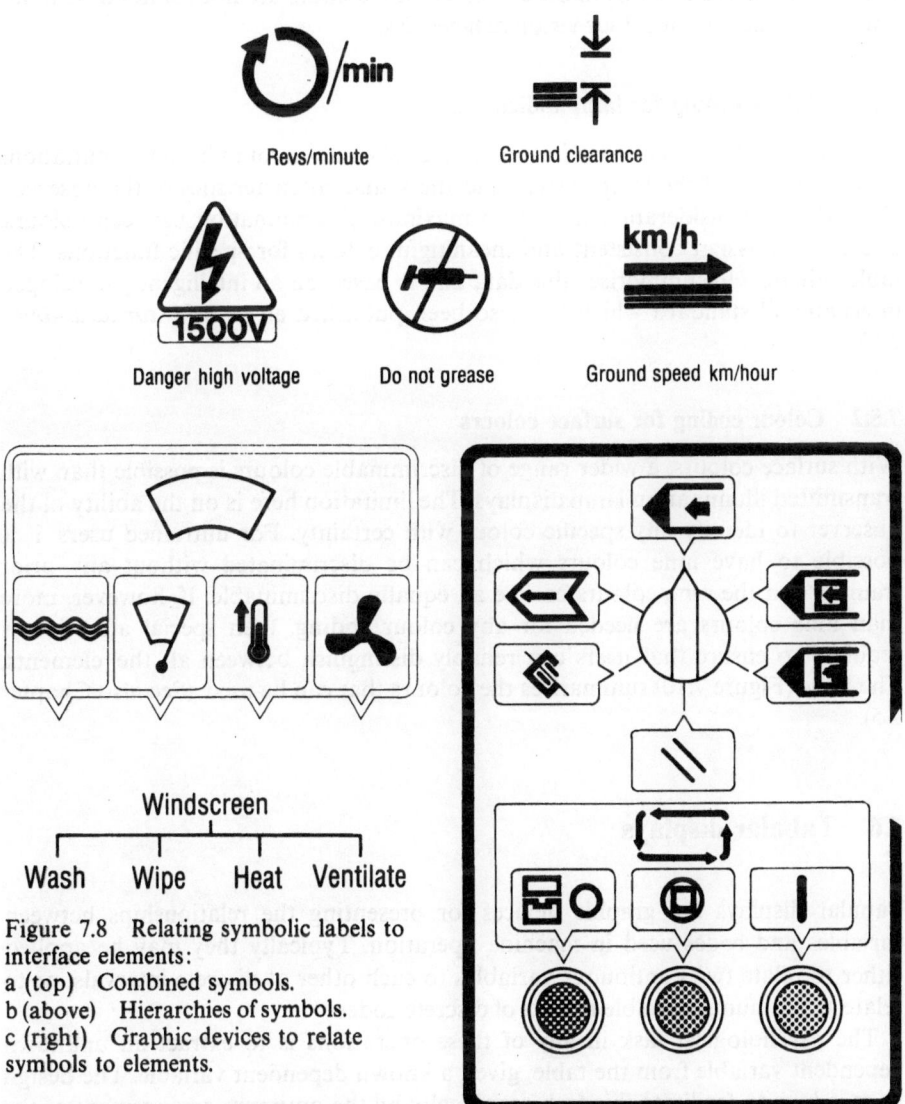

Figure 7.8 Relating symbolic labels to interface elements:
a (top) Combined symbols.
b (above) Hierarchies of symbols.
c (right) Graphic devices to relate symbols to elements.

7.5 Colour coding

Colour coding of machine graphics can enhance some important features of the display, mainly by drawing attention to specific elements. The colour can be supplied by the use of lamp displays, or alternatively as surface colour, as part of the applied graphics, in the form of text, symbols or other graphic devices.

Given that a proportion of the user population is subject to colour blindness, the range of colour codes that can reliably be used in this way is limited. Reasonable control should also be exercised over the specification of the colour, especially in relation to surface colours, in order that certain colours are not confused with one another by those with colour vision deficiencies.

7.5.1 Colour coding for lamp indicators

The colour coding of lamp indicators should take into account both the limitations of performance of the lamp system and the visual characteristics of the observer. The primary consideration here is to maximise discrimination between colours, and then to assign consistent and meaningful colours for specific functions. The table (Figure 7.9) summarises this data and is based on an intelligently developed international standard which has also been published as *British Standard 4096: 1976*.

7.5.2 Colour coding for surface colours

With surface colours, a wider range of discriminable colours is possible than with transmitted illumination lamp displays. The limitation here is on the ability of the observer to identify any specific colour with certainty. For untrained users, it is possible to have nine colours which can be discriminated without any prior training and the nine colours can be all equally discriminable. If, however, more than nine colours are needed for any colour coding, then special attention is required to ensure that users can reliably distinguish between all the elements. The table (Figure 7.10) summarises the colours that can be used. (See also Chapter 2.5).

7.6 Tabular displays

Tabular displays are graphic devices for presenting the relationships between variables and codes used in machine operation. Typically they may be applied either to relate two continuous variables to each other at discrete intervals, or to relate a continuous variable to a set of discrete codes.

The psychological task in any of these operations is to extract an unknown dependent variable from the table, given a known dependent variable. The design issue is how to facilitate this task graphically by the optimum arrangement of the table and the use of graphic devices such as lines, rules and spacing. The variables that predominantly affect the design are:

1. The *number of independent variables* (with more than one independent variable, special design strategies are needed to enable the table to be easily searched).

7.6 Tabular displays

Function	Lamp colour	For untrained users including colour blind	9 Equally discriminable colours	
Danger – alarm Action needed	Red			
		Red	3R	
Caution – impending change	(Yellow)	Orange	9R	For more than nine codes special training required
		Yellow	9YR	
		Blue	16Y	
Safety – proceed, equipment safe	Green	Purple	3G	
		Grey	7BG	
		Buff	9B	
		White	9PB	
Specific functions – eg reset recalibrate	Blue	Black	3RP Select those farthest apart if less than nine codes needed	
Non-specific function	(White)			
SOURCE BS 4096: 1976	( ) indicates confusion if used with colour-blind users	SOURCE Fed. Spec. 11.C.595	SOURCE Conover (1959)	SOURCE Chapanis, Halsey (1959)

Figure 7.9 Colour codes for lamp displays.

Figure 7.10 Colour codes for surface colours.

2 The *range of the independent variables* and their form (text, codes, order, number, continuous variables, both integer and real numbers).

The graphic variables available (as detailed in earlier sections) are:

Type style
Type size
Type weight
The use of space—horizontal and vertical spacing
Grouping
Rules and lines
Headings/labels—symbols and text.

The preferred ways in which these graphic devices can be used in tabular display design are as follows:

Type style: It is good graphic design practice not to mix typefaces within a given table, unless deliberate differentiation of some major element is needed, as in a heading.

Type size: The use of different type sizes *within* the main data sections of a table is not usually beneficial, but changing the type size can enhance the emphasis needed between headings, labels and the main data in any table.

Type weight: Type weight is very useful for differentiating between variables within a table, especially between independent and dependent variables.

Space: Spacing is probably the most powerful graphic tool available for the design of tabular displays. Most tabular displays can be greatly improved by the intelligent use of spacing. A reduction in type size to create more flexibility in spacing can give benefits which more than compensate for any loss of legibility (but refer to Section 7.3 to determine any loss of legibility for a given reduction in visual size).

Horizontal spacing: This allows functional grouping of related elements, thus facilitating visual scanning *down* the columns. When scanning data grouped in this way, only the first entry in each group needs to be examined during the search process and the remainder can be skipped until the right approximate area of the table is found. Grouping in either twos, threes, fours, or fives is a very useful way of organising horizontal spacing in tabular displays.

Vertical spacing: This form of spacing allows for separation between elements for easy and unambiguous transverse scanning.

Alignment/indentation:: The use of aligned and indented lines is very useful for creating visual pattern, which again eases the search and reading task. Indentation is more useful when used with a continuous text, but it can be used to advantage in some other applications. Thus the alignment of common elements enhances the identification of individual elements and develops coherent relationships between them, while indentation can contribute visual pattern and structure to the display.

Rules: Rules should be used sparingly and their implementation should be considered carefully in relation to the requirements of spacing as outlined above. Often a space together with appropriate alignment of variables can create a vertical division without the necessity to resort to a rule. The vertical rule is best used to prevent adjacent unrelated columns of data being read together when scanning horizontally across a line.

Horizontal rules: Horizontal rules should be used sparingly and restricted to the separation of one form of displayed material from another.

Figure 7.11 demonstrates a number of these principles.

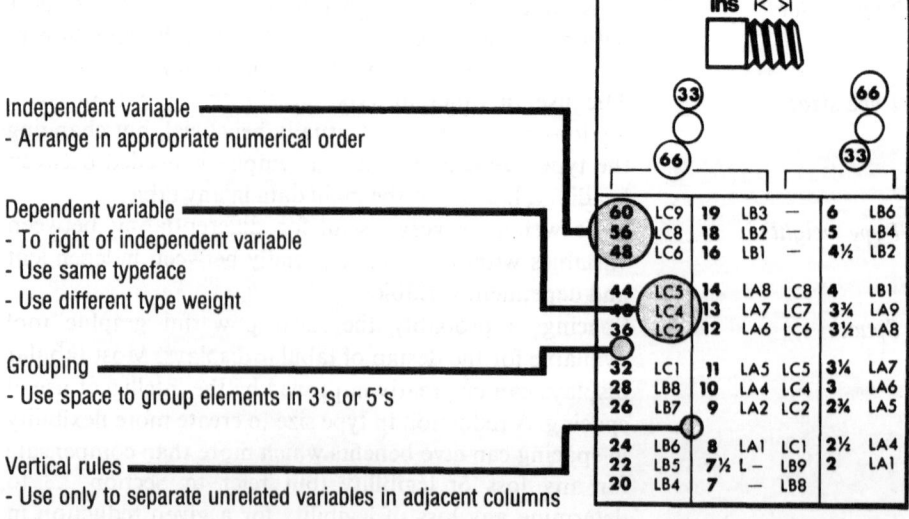

Figure 7.11 Tabular display principles.

7.7 Design and instrument scales

Instrument scale design involves the application of comparatively straightforward principles. What is required is a scale that achieves the optimum compromise between speed of reading and accuracy of reading. A detailed design guide is given in Chapter 3.4.2.

8 Selection of materials for the man/machine interface
Alec B. Kirkbride

8.1 An approach to defining material requirements

8.1.1 Philosophy

The act of designing is complex, involving designers in both intuitive and rational judgements. The choice of materials is basic, and can provide a starting point for the rest of the design. This aspect of design is, however, sometimes considered inadequately, with the result that the overall design concept is distorted and the most is not made of the materials available.

Here we are concerned with the many areas in which men and machines interact, and the criteria for defining the properties that are needed can be classified under the headings of 'basic design' and 'environmental factors'. The aim of these brief checklists is thus to enable the designer to arrive fairly quickly at a broad picture of what properties are needed in the man/machine interface area.

The specification of need must then be compared with a picture of what is available. To assist in this, there follow here some checklists which include most of the materials suitable for interface applications. Since descriptions of their 'engineering' properties are widely available, the notes here have been restricted to descriptions of the characteristics that are of special interest in the interface context.

When a shortlist has been drawn up, the physical properties, fabrication methods and costs of using the short-listed materials should be discussed in greater detail with specialists, such as suppliers, in order to make a final selection.

8.1.2 Basic design factors

In choosing materials the designer needs to analyse the product or system as a whole, and the following are some of the relevant areas for consideration.

Function. The product may be a compressor unit in a quarry, a lung ventilator in a hospital, or an audio unit in a home. A consideration of what it must do, and of what its expected life must be, will give general guidance on the class of materials which will be appropriate.

Economic background. Including selling price, work costs, maintenance costs, amortisation of tools, raw material costs, frequency of replacement, and an acceptable value of work in progress in relation to output which may influence both manufacturing processes and materials.

Production quantities and methods. The quantity of products to be produced in batches per year will affect the economic methods of production and hence materials. Modular design may be relevant and will affect production levels and hence materials.

Basic structural requirements. Balancing the design requirements for the engineering, ergonomic and aesthetic aspects for performance, for reliability, and for mechanical structural needs against production quantities, processes and their consequential cost is the central dilemma in material selection. Endurance may be more valuable than initial aesthetic beauty, or than cost savings based on low-quality materials that may lead to premature deterioration in use.

Marketing. The designer must assess customer expectations, and if possible produce a product that exceeds these, but not too far. So far as materials are concerned customers' attitudes are affected by all types of products they are in contact with, and not just the one the designer is working on. Expectations about materials are affected by history in some fields, and by conventional usage in others, so to some groups 'new' does not automatically mean 'better'—often rather the opposite.

Safety. Safety is fundamental, and material selection should always comply with the relevant standards and take into account foreseeable risks. Examples are the possibility of electric shock from hospital equipment unless its tyres are anti-static, the chance of plate glass being broken in exposed positions, the possibility of vital controls becoming jammed if they incorporate moisture-absorbing plastics bearings, and deterioration associated with hidden corrosion.

Appearance. Appearance is an important aspect of interface materials, since it is affected by the immediate visible surface of components and by the forms in which the components can be manufactured. In terms of surface, materials affect the natural colour, the light reflection and 'grain', additional finishes that can be applied and their appearance, and the degree and rate of deterioration in service. As regards form, the manufacturing process used—casting, moulding, pressing, folding or cutting—will depend very much on the material used, and affects the character of the form that is economically practical.

Control. There may be requirements in this field relevant to material selection, for example appropriate manual contact friction grip levels, impact resilience, or suitability for graphics display by forming the material or by superimposing a graphics process.

Corrosion resistance. Resistance to corrosion is a major criterion for selection of materials, affecting performance, appearance, life, safety and maintenance. Corrosion-resisting materials tend to be expensive, and the extent to which they can be used will be related to their value, as determined by market assessments, except where safety is directly involved. In such cases, a choice must be made between preventing dangerous corrosion by material selection, and clear definition of inspection, maintenance and testing procedures needed to maintain safety—for example with lifting devices or vital controls.

Weight can be an important factor, more especially with all forms of transport. It can also cause problems if large components have to be handled during maintenance, and in such cases light alloys or plastics may be preferred.

8.1.3 Environmental factors

Environmental factors in materials selection include the following:

Shock forces. Machines may be subject to a wide range of shock forces, for example 'roll-over' and falling rocks can affect earth-moving equipment, rapid acceleration can result from heavy shunting of railway stock. Products may also be dropped during handling or in transit.

Vibration. Vibration can lead to disintegration if, for example, heavily loaded studs are threaded into soft alloy, or friable plastics. It can also lead to fatigue failure of components when vibration produces stresses in excess of the fatigue limit for the material, which may be surprisingly low, as in the case of GRP.

Temperature and humidity. Known extremes of temperature always present a problem in material selection. For example, materials which are poor conductors of heat will be more suitable for manual control handles in extreme temperatures. However, it is often the unexpected changes in conditions that tend to be overlooked. For example, the control window on an air compressor housing may need to be 'double glazed', with glass on the inside and polycarbonate externally. Humidity also causes problems, for example with dimensional changes as a result of moisture absorption by some plastics.

Vandalism often receives inadequate attention. Equipment exposed to the public is vulnerable and it may be cheaper in the end to use expensive materials, such as stainless steel, rather than to keep replacing less expensive alternative materials or finishes.

Fire. Fire hazards must be considered when selecting materials from the points of view of safety and continuity of service. Criteria include flammability, rate of flame propagation, generation of poisonous fumes, generation of gas which can contribute to the creation of an explosive atmosphere, and the length of time a material will resist high temperatures without failing to perform its function. Even the heat from transformers can cause some materials to give off noxious fumes and accelerate the ageing and cracking of certain plastics.

Solvents. All materials, finishes and paints should be tested using all solvents to which they are likely to be exposed under the most extreme environmental conditions for which the product is designed.

Animals. Some materials, such as wood, rubber and some plastics, are eaten by insects or birds and other fauna, especially in warm climates.

Dirt. Surfaces may be abraded by some materials when they are cleaned, and this should be considered. If exposed to electrical stress, there may be an increased risk of tracking once a surface is eroded.

Weathering. Ultraviolet light, frost and rain will weather products and bleach graphics, such as instructions, if unsuitable materials are used.

Misuse. Material selection should take into account both normal and rough use, and the misuse to which the equipment is likely to be subjected during delivery and in service. Manual controls may in practice be operated by kicking—the man side of man/machine interfaces can be very unreliable.

Noise. When noise is a problem, and expecially when it is sudden or unexpected, operators may need the protection of sound-deadening materials. For example, composite sheet materials can reduce transmitted noise, whereas plain metal sheets may aggravate the problem.

8.2 Metals

8.2.1 Copper-based alloys

The copper-tin bronzes are exceedingly resistant to corrosion and combine this quality with rugged mechanical properties. They are therefore excellent for exposed, stressed interface components, such as marine fittings, when their high cost can be justified.

Specialised manufacturing processes permit bronze nameplates to be cast with very precise graphics with stroke widths as small as 0.3 mm, giving characters about 2.5 mm high, and these can be contrasted with their backgrounds by painting the nameplate and wiping the paint off the raised letters.

Brass, or other copper-rich alloys in cast, extruded, rolled or sheet form, can perform some interface duties that might otherwise indicate less expensive materials such as steel or aluminium. Their special properties—hardness, low friction levels, relatively good resistance to corrosion, colour and ease of fabrication—may be of value. However, unless protected, brass forms a patina which limits its use in cases when this will affect performance. It is frequently used in interlock mechanisms, levers, linkages, locks and hinges.

8.2.2 Zinc alloys

Die-cast zinc alloys can reproduce fine detail on complex components, at low cost if production quantities are high. They are not suitable for heavy loads on sliding surfaces, and friction levels are higher than with copper alloys, nor do they possess the same resistance to corrosion. They can make excellent handles for tools, and light-duty mechanical components, but many of the components traditionally using die-cast zinc can now be better from the operator's point of view, as well as cheaper and often lighter, if they are made from engineering plastics.

8.2.3 Steel

Rolled sheet steel is the most commonly used material for cubicles, control panels and desks. It is easily fabricated by press, welding and folding techniques. Zinc-coated sheet is frequently used to increase protection and life of painted components, and plastics-coated and coloured sheet steel is becoming widely used. It provides excellent corrosion protection combined with flexibility, which often permits coated sheet to be formed to the finished shape without damaging the finish. Another advantage is that the colour is embodied in the coating, and is not just on the surface, so that minor scratches in service are less obvious. Sheet steel is also available with factory-applied paint finishes, and both plastics and paint can be supplied with imitation wood grain or leather effects. Sheet steel is also available in perforated or expanded form for use as grilles or ventilators, and in the form of modular louvres.

8.2.4 Stainless steels

Almost indestructible, stainless steels are used in bar, cast, forged, sheet or strip form where their properties of high strength, high resistance to corrosion, and the ability to accept an excellent polished or matt finish offset the high cost. Examples are rigging for boats, fixing devices such as screws, nuts and bolts, toggle catches etc, or window frames and trim for cars, kitchen utensils, instruments and medical apparatus, and so on.

In sheet form, stainless steels can be fabricated to provide complex shapes, and can be marked by chemical etching or engraving, and colour filled. Their uses can therefore extend into areas of information display.

Non-magnetic stainless steels are available for applications where heating or compass deflection might result from the use of magnetic materials. They are also available in woven forms for use as screens or filters.

8.2.5 Aluminium and its alloys

Aluminium, which is available in cast, forged, rolled, sheet and extruded forms, is corrosion resistant, but less so than bronze or stainless steel. In alloy form it can provide mechanical strength comparable with that of mild steel at about one third of the weight. In pressure die-cast form it offers the possibility of precision reproduction of detail. The versatile extrusion process permits aluminium to be shaped accurately, and is used in a very wide range of applications, eliminating many machining, fabricating and assembly operations, especially for framework structures.

Being a good conductor of heat, aluminium is used to transfer heat, and sections incorporating fins are widely used to disperse heat to the atmosphere. As a good conductor of electricity it is used for bus-bars, but special processing, such as silver plating, is then needed at all joints to prevent the build-up of electrical resistance.

Except in the most porous forms of sand casting, aluminium can be anodised to provide a hard-wearing lustrous finish which will accept a wide variety of coloured dyes. Integral graphic markings can be included below the anodic finish. It is, however, unwise to expect exact colour matching of anodic films when seen in the same light and plane, as on control panels for example. Aluminium will accept a high surface polish, but it should be remembered that this may give rise to dazzle.

In terms of cost, aluminium is about twice as expensive as steel for a given mechanical function.

8.3 Glass

Optical transparency, more especially in the ground plate glass or float-produced forms of glass, and brittleness are the properties immediately associated with this material. It is also extremely inert and impervious to most chemicals, so it is of great value in chemical process plants.

Glass can, however, be etched by suitable acids, or engraved, to record information or to make large panels visible for safety purposes.

Like all ceramic materials, glass is relatively weak in tension, except in drawn fibre form, but will accept high compression loads and is a good electrical insulator.

In plate form, glass is widely used to protect humans and to transmit light. Safety considerations may dictate the use of heat-treated 'armour' plate glass because of its resistance to shock. Such toughened glass will shatter, but the small crystals resulting from fracture avoid the lethal shards produced by ordinary plate glass when it breaks. Laminated glass, comprising a glass-plastics-glass sandwich, will remain in place even when broken, and may be even safer, though more expensive. Glass in windows etc can be located by the use of flexible extruded seals, or may now be fixed using suitable adhesives.

Glass fibres have very high tensile strength, and are used in composite plastics (see Sections 8.6 and 8.10).

In the form of vitreous enamel, glass is also used as a finish to permit less amenable materials such as steel and cast iron to take on the surface characteristics of glass. This finish is much used in food-processing and laboratory interfaces.

8.4 Rubber and synthetic rubber

These materials constitute a whole range of special technology. Their properties depend on the particular compound mix and process used, and because these are not easily identifiable from a sample, advice from suppliers is usually essential to establish properties and tolerances in production.

Rubbers offer flexibility, extensibility, the ability to absorb shock, the ability to dissipate dynamic energy by coverting it into heat through distortion, energy storage by extension, and the retention of load when distorted, for example as gaskets. Rubber is waterproof, fairly resistant to most chemicals, and can be used in intimate contact with humans. It can also provide a very high coefficient of friction—of the order of 1.0 in contact with a rough surface.

Rubber is easily moulded without expensive tools, or can be rolled or extruded. It can be bonded to itself, to metals, to glass or to plastics, and can therefore be reinforced with a range of fabrics or threads. It can also be foamed, with or without an external skin.

8.5 Wood

Possibly one of the most versatile materials known, wood is now used more often for psychological and aesthetic reasons than for mechanical or economic ones at the interface level. This is because most people have grown up surrounded by wooden artefacts and feel 'at home' with it as a natural material.

While timber is still much used in buildings, the machine designer is more likely to find himself thinking of wood as a possible finish rather than as a constructional material—hence the veneers, printed effects and general simulation of wood on metals or plastics that are often encountered.

The properties of wood can be modified by lamination or compression, or impregnation for stressed duties.

8.6 Thermosetting plastics

These include phenol formaldehyde, melamine formaldehyde, urea formaldehyde, epoxies, and some polyesters used for DMC (dough-moulding compound) and

GRP (glass-reinforced plastic). These plastics do not soften or melt when heated but will eventually char; however, some will support combustion unless specially treated. The characteristic of not softening under heat is their great virtue over thermoplastic plastics, they also tend to be harder but more brittle.

Melamine laminated with paper is much used for surfacing other materials such as chipboard and is available in solid-grade forms for arduous tasks such as laboratory bench tops. These solid-grade melamine laminates can be machine engraved and the graphics colour filled with catalytic paints to produce extremely hard-wearing control panel surfaces for rugged applications such as rock drills, mining machines and ships deck equipment. A melamine sandwich called 'Traffolyte' in which one colour is engraved or cut away to reveal a different coloured layer underneath is also used for such applications. As dirt is darkish (especially coal dust) it is as well to cut through a light colour to a darker layer beneath so that dirt in the grooves thus formed will seem similar in tone.

DMC is a glass-filled thermosetting polyester 'dough moulding compound' which can be compression moulded to form strong compound panels with wall thickness of 2 to 4 mm which, because they do not soften with heat, can be used on engines and boilers.

Urea and DMC have good dielectric strength properties which, together with the fact that they do not melt, make them excellent for switchgear enclosures such as circuit breakers.

Unlike urea, DMC is difficult to colour evenly and is often finished by painting, which can be an advantage if an exact colour and texture match is required between plastic and metal parts. It can be stove enamelled at the same temperatures as metals and so can use identical paints.

8.7 Thermoplastics

8.7.1 Polyacetal

One of the engineering plastics, acetal is available in two forms, acetal copolymer and acetal homopolymer, which is somewhat stronger. Both are very resilient, stiff, resist high temperatures and abrasion and have low coefficients of friction. With these properties they are used for gears, pipe fittings, light springs, bearings etc. They can be injection or blow moulded or extruded.

8.7.2 Polyester

Thermoplastic polyester resins are often combined with glass fibre and in this form are commonly used for printed circuit boards as they resist soldering temperatures and flux solvents. They can be injection moulded with or without reinforcement and have excellent dimensional stability in changing temperature conditions such as the engine compartments of cars, boats etc.

8.7.3 Polyethylene

Available in two types, LDPE (low density) and HDPE (high density), polyethylene has very good chemical resistance. LDPE is flexible and tough while

HDPE is stronger and stiffer. Both can be extruded, rotationally moulded, blow moulded or injection moulded, they do not taint food including milk and are commonly used for containers; they also have excellent electrical insulation properties.

8.7.4 Polypropylene

Strong and stiff, polypropylene has outstanding fatigue resistance and good temperature resistance. It can be injection or blow moulded, extruded or thermoformed (vacuum formed). In man/machine interface terms, polypropylene is perhaps best known for chair shells, control knobs and safety helmets. Its fatigue resistance enables it to be used for 'living hinge' type applications—hinge action by bending a thin membrane rather than pivoting.

8.7.5 PTFE

Polytetrafluoroethylene is tough, has an extremely low coefficient of friction and is exceptionally resistant to chemical attack and harsh environments. It is not possible to mould by normal techniques and is processed by sintering and machining. Its unique properties mean that it can be used for unlubricated bearings running at high temperatures, coatings for saucepans and other items handling sticky substances and, of course, chemical pipework, bellows, O-rings etc. As it is not rejected by the human body, some grades of PTFE are also used for 'replacement parts' surgery.

8.7.6 Nylon

Although rather moisture absorbent, which can upset their dimensional stability, nylons are extremely tough and resistant to abrasion and are used in many engineering applications calling for injection mouldings capable of surviving repetitive impact shocks. Chains, cams, propellers and lock mechanisms are typical items. Nylon is also used for dip coating of metals.

8.7.7 Cellulose acetate

Rather poor on dimensional stability because of moisture absorption but hard and tough, it resists human skin acids and perspiration, and is often chosen for tool handles and as coverings for much-handled objects such as car steering-wheels.

8.7.8 Cellulose acetate butyrate

CAB is tougher than cellulose acetate but not as stiff as cellulose acetate propionate—the other cellulosic thermoplastic. Both have excellent weather resistance and are used for outdoor illuminated signs as well as tool handles, typewriter keys etc. Business machine parts subject to great wear, such as typewriter keys, can be injection moulded in two colours.

8.7.9 Polyurethane

Commonly used as a basis for structural foam in a great many interface and machine housings, Polyurethane can be stiff or flexible. A relatively little-known use for solid polyurethane is in cast form for extremely tough components required in small quantities with low tooling costs, such as complex guards on machines, muffler mouldings on pneumatic tools etc. In suitable cases, injection-moulded polyurethane structural components can be designed to replace mild steel at little higher cost.

8.7.10 PVC

Polyvinyl chloride comes in three common forms: unplasticised, plasticised and as a copolymer with vinyl acetate. PVCs can be extruded, injection, blow or rotationally moulded or thermoformed. The copolymers are easier to vacuum form and plasticised PVC is much used for dip and extrusion coating.

PVCs have very good chemical and weather resistance and can be transparent; the unplasticised form is tough, resistant to abrasion and self-extinguishing.

8.7.11 ABS

The name is abbreviated from acrylonitrile-butadiene-styrene. It is stiff, tough, will resist abrasion and staining and has the unusual property of being readily metal plated. It can be thermoformed, extruded and, of course, injection moulded. In interface areas it is commonly used for instrument and fascia panels and for many moulded components on modern machines.

8.7.12 Polystyrene

In its general purpose form polystyrene is somewhat brittle, but is very transparent and often used for packaging applications.

High impact polystyrene is a compound of styrene and rubber and is much tougher but more opaque. It can be injection moulded, extruded, thermoformed and does not taint foodstuffs. It is not suitable for outdoor applications.

8.7.13 SAN

Styrene acrylonitrile has many of the properties of high impact styrene but retains the transparency of general purpose styrene.

8.7.14 Acrylics

Optically transparent and very immune to weathering as they are not affected by ultraviolet radiation, acrylics can be thermoformed, cast, injection moulded or extruded. They are used for machine guards as they do not shatter easily, but solvents can cause stress cracking.

Because of their weather resistance and thermoforming properties, they are used for illuminated signs and bathroom fittings being resistant to hot water and, of course, warmer to the touch than metal or vitreous enamel.

Acrylics are frequently used for visual display areas using 'black-out' techniques, and their 'light piping' properties enable them to be used with remote sources of illumination. The physical basis for this piping effect is that any ray which encounters an acrylic/air interface at an angle greater than 43° from the normal will be totally reflected back into the acrylic material.

8.7.15 Polycarbonates

Transparent, stiff and tough (more so than acrylics), polycarbonates are used in situations requiring optical clarity with a high degree of impact resistance. They can be injection or blow moulded, extruded or vacuum formed. The fact that they are often specified for windows in schools gives a good indication of their usefulness on safety guards, clear moulded covers for control areas that need to be seen but are tamper proof etc.

8.7.16 Polysulphones

Because they have good flame resistance and do not present any smoke hazards, polysulphones and polyethersulphones are often chosen for use in aircraft. They can be injection moulded, thermoformed or extruded. Some can withstand temperatures used for stoving processes of about 160° to 170°C. Typical applications are ducting, passenger service units and integrated circuit boards.

8.7.17 Polyphenylene oxide

Known as PPO it has very good electrical properties over a wide range of temperatures. In foamed form it is used in office machine housings and TV sets. It is robust and does not burn or produce fumes at high temperatures.

8.8 Foamed plastics

Nearly all thermoplastics can be foamed, the results of which divide into soft foams, commonly found in upholstery, and hard foams, some of which have become known as 'structural foams'.

8.8.1 Soft foams

Softness is of vital interest in relation to interface situations in protecting operators from bruising or more severe damage in situations involving speed or violent movement. Soft plastics foams have revolutionised the design of furniture. For our present purposes it is perhaps enough to note that the following tend to form the basic plastic cushioning materials.

> Foamed rubber latexes—very resilient but relatively expensive.
> Foamed polyethers—widely used in furniture, they usually incur a fire risk.
> Foamed polyesters—these have been superseded to quite a large extent by polyethers as they have a 'dead' form of resilience.
> Crumb foam—a loose foam filling.
> Reconstituted foam made from crumbs bonded together is harder but suffers from similar fire hazards to ordinary polyether foams.

Polyester fibre—used for wrapping around foam core in luxury upholstery. Self-skinning injection-moulded polyurethane and polyethylene foams are still flexible.

Designers need to check the degree of risk involved from heat or fire when specifying foamed plastics or rubbers or combinations of these materials in interface conditions.

8.8.2 Structural foams

Enormous strides in the use of structural foams have been made in recent years because of their economy in manufacture, their light weight, their ability to replace large complex assemblies with fewer parts, good appearance and, of course, their structural properties.

A variety of techniques are used in manufacture, but generally lower moulding pressures and less costly tooling have meant that the number of interface items made from structural foam has grown at a considerable rate.

Structural foams can be moulded with integral solid skins ('self-skinning'), can have metal fixing inserts incorporated and can be reinforced with other materials such as metal or wood.

The following tend to be the most commonly used plastics for structural foam components.

Polyurethane
Polypropylene
Modified PPO (polyphenylene oxide)
Polycarbonate
High density polyethylene

Some other thermoplastics such as polystyrene are frequently used in a foamed condition for non-structural uses—notably for insulation and packaging—and others, such as polyurethane in various degrees of flexibility, for soft or resilient upholstery components and crash pads.

Structural foams usually show swirl patterns and are generally painted with compatible paints or given a textured surface finish.

8.9 Floor coverings, steps etc

Control rooms always require floor coverings, and machines themselves occasionally also have floor areas, steps and gangways.

So far as control rooms and other near-architectural applications are concerned, sources such as the Barbour Index, The Building Centre or Commodity File No. 8 of the Building Documentation (UK) cover materials such as cement, ceramics, thermoplastic tiles, rubber, wood, non-woven felt tiles and traditional carpeting—the range is very large.

More mechanical or 'engine room' floor coverings include rubber, which is often good looking as well as being practical.

Metal tread plate is made in both steel and aluminium expanded metal and is useful if dirt is to be shed through the floor. Both rubber and expanded steel are used for flooring in the JCB unit shown in Figure 8.3.

8.10 Composite and compound materials

The growing number of materials in these categories are, by their nature, difficult to define. Alloys or substances such as GRP or nitrile rubber are compounds, while (say) stainless cladding on aluminium, sandwich panels and laminated glass are composites. Some of these are listed below with their principal properties.

8.10.1 GRP glass-reinforced plastic

Often termed Fibreglass (in fact a trade name), GRP can be machine or hand moulded into very tough compound shapes, it can be integrally coloured and has excellent weathering properties so is useful for outdoor applications of a semi-rugged nature.

Some resins used in GRP will support combustion unless specially treated—the most frequently used resin is thermosetting polyester, which is light, strong, translucent and capable of being moulded and fabricated into extremely intricate and quite big structures. It is very useful for smooth, well rounded coverings where operators are subject to violent movement, such as in small boat cockpits.

GRP items are often low on tooling costs but high on labour costs. Orientation of the glass fibres in a particular direction can greatly enhance the tensile strength of components, but the fatigue tensile level is lower than might be expected.

8.10.2 GRC glass-reinforced cement

With a minimum wall thickness of approximately 8 mm, GRC can be moulded from low-cost tools to form strong but rather heavy components which have good fire and heat resistant properties. GRC is usually given a textured finish and can be made to simulate the appearance of stone and similar materials. GRC will probably be used mainly for non-load-bearing architectural applications.

8.10.3 Carbon fibre in substrates

Carbon fibres when oriented carefully in resin substrates, usually polyesters or acrylics, can imbue components with enormous flexural strength in specific directions. Carbon fibre may also be used to reinforce metals.

8.10.4 Superplastic alloys

The traditional alloys are compound materials, and the new super-plastic zinc/aluminium/plastic alloys are an extension of this area. They have the unique ability to be formed like thermoplastics at temperatures above 260°C yet behave like metal thereafter—including having electrical screening capabilities.

8.10.5 Epoxy resin with additives

Epoxy resin mixed with sand can be cast to form electrical enclosures. This material is also used for certain tools as is epoxy resin mixed with aluminium powder. Mixed with bronze powder, 'cold cast' bronze can imitate lost wax cast bronze forms.

8.10.6 Sintered metals

In powder form many metals may be alloyed under pressure without melting to produce sponge-like structures capable of absorbing fluids such as oil—so they are frequently used in self-lubricating bearings. More solid varieties can be formed into items such as gears.

8.10.7 Sintered ceramics

Compounds such as carbon-graphite can be moulded or machined to produce bearings where no lubrication or maintenance is available—they are not affected by most chemicals.

Sintering, which is partial fusion, is used to formulate many compounds in semi-conductors, ferromagnetics, piezoelectrics etc, and for high temperature uses.

Flame spraying is another technique which can be employed to deposit both ceramics and metals on other materials—thus imparting properties well outside their normal range.

8.10.8 Glass-reinforced compounds

Glass is a ceramic and is frequently used, as previously noted, as a reinforcement in GRP and GRC. In addition, glass is used to reinforce other thermosetting and thermoplastic plastics such as 'dough-moulding compound' DMC, which is a glass/polyester compound. Glass is also used to fill and reinforce nylon, polypropylene etc.

8.10.9 Resin-bonded paper

The well known melamine-impregnated papers such as Formica are widely used in interface situations as surfacing materials. 'Solid grade' materials of this type are available up to 35 mm thick and are much used in laboratories.

8.10.10 Resin-bonded cloth

'Paxolin', 'Tufnol' and similar materials are composed of phenol and other resins with cloth. They are still frequently used for machined plastic parts and electrical baseboards.

8.10.11 Resin-impregnated woods

Materials such as 'Permali', jigwood etc, have the pores of the wooden substrate filled with plastic resin under pressure, which gives them enhanced properties in terms of strength and dimensional stability, and electrical characteristics.

8.10.12 Sandwich panels

A very great number of composite sandwich panels are possible and they form a new family of materials for use by designers. A typical example would be, for instance, a caravan wall which is composed of a panel of hardboard or plywood 2 to 3 mm thick on the internal face with a 0.3 mm PVC vapour barrier film, then

25 mm of styrene or polyurethane foam, then a 22G aluminium external skin which will itself usually be prefinished with acrylic paint, using curtain coating techniques.

8.10.13 Melamine-faced chipboard

A composite material with almost universal applications in current furniture construction.

8.10.14 Honeycomb structures

Some frequently used composites of this kind are aluminium on aluminium honeycomb for stiff panels used in scientific instruments and hardboard on paper honeycomb used for domestic doors and similar panelling applications.

8.10.15 End-grain wood sandwich composites

Great strength allied to lightness is produced by a sandwich of end-grain balsa wood between two sheets of aluminium or GRP.

8.10.16 Bimetal laminates

Often used as thermal detection and switching devices due to their differential expansion and consequential deflection.

8.10.17 Clad metals

Stainless clad aluminium is a composite which has uses beyond mere finishing (such as PVC-coated steel or aluminium etc, see Chapter 9). The thermal conductivity of aluminium, or copper, combined with the hardness and corrosion resistance of stainless steel make ideal cooking utensils for frying etc.

8.10.18 Plastic cladding

Composites such as PVC-clad ABS extruded sheet have been used to protect the ABS from ultraviolet light degradation outdoors, but new ABS formulations are now making this less necessary. Many other such clad materials can be formulated and can often be useful as finishes, (see Chapter 9).

8.10.19 Metal cladding of non-metals

This technique is commonly used as a finish to give plastic components such as taps, car grilles etc a plated appearance, but it is also much used in printed circuit board production to form the conductive paths—usually copper on polyester-impregnated cloth substrates.

8.11 Examples of materials selection

In the first of the examples that follow, the suggested procedure for materials selection has been followed. Other examples are included in precis form to show solutions that are acceptable within their contexts, or to demonstrate advanced ways of improving the man/machine interface.

8.11.1 EMI complex switching panel

This was designed to carry out the complex video mixing needed in television studios. The three main requirements were for completely flexible specification on the part of the television company; manageable production techniques for EMI; and simplicity of use and good ergonomics.

For flexibility, simplicity and ease of repair in case of damage in transit, the panel was designed as a number of functional units, such as input selection and mode selection, and these were produced as standard modules, which produced benefits in terms of wiring and assembly. The need was then for a design that supported these modules in an ergonomically sound way and was attractive from a marketing point of view.

The design criteria checklist thus read as follows:

Function—complex switching with extreme flexibility.
Economic background—modular construction must save costs to compete with large-scale production in the USA and with non-modular panels.
Production quantities and methods—short run, therefore no expensive tooling costs. Plastics or metal extrusions would permit variations in module length and provide extra material where necessary. Final assembly must not involve complex fitting or wiring.
Basic structural requirements—need to make a large, flat panel from small pieces, hence interlocking planking and longitudinal tie-rods secured in grooves in extrusions with easy access to underside for servicing.
Marketing position—Must be flexible, look up-to-date and be competitively priced.
Control—graphic problems arise with linking of modules. The surfaces in a studio must show the location of controls but not produce glare, hence screen printing on a semi-matt finish.
Environmental factors—Must withstand shocks in transit, be stable in spite of changes in temperature and humidity, comply with all electrical safety standards, withstand cleaning using methylated spirit, resist termites in tropical conditions, resist writing on control surfaces. Vandalism, dirt, weathering and noise presented no special problems.

The final design consisted of an assembly of extruded aluminium planks of varying length, interlocking with one another and held together by steel tie-rods in tension, supported on an outer frame and 'tub' of sheet steel, as shown in Figure 8.1a, b.

8 Selection of materials for the man/machine interface

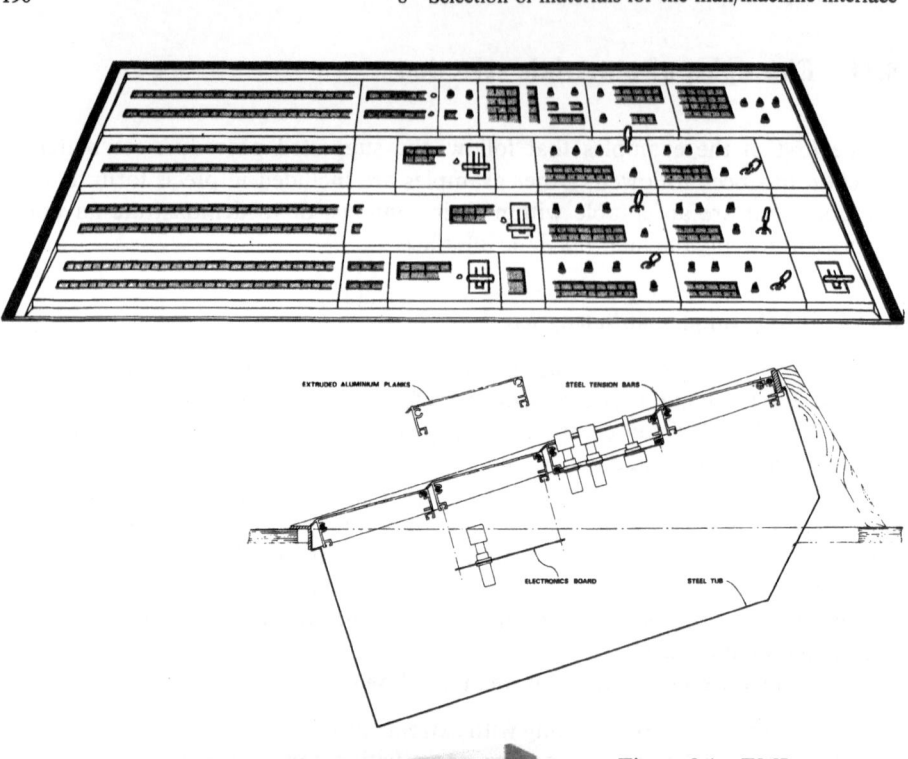

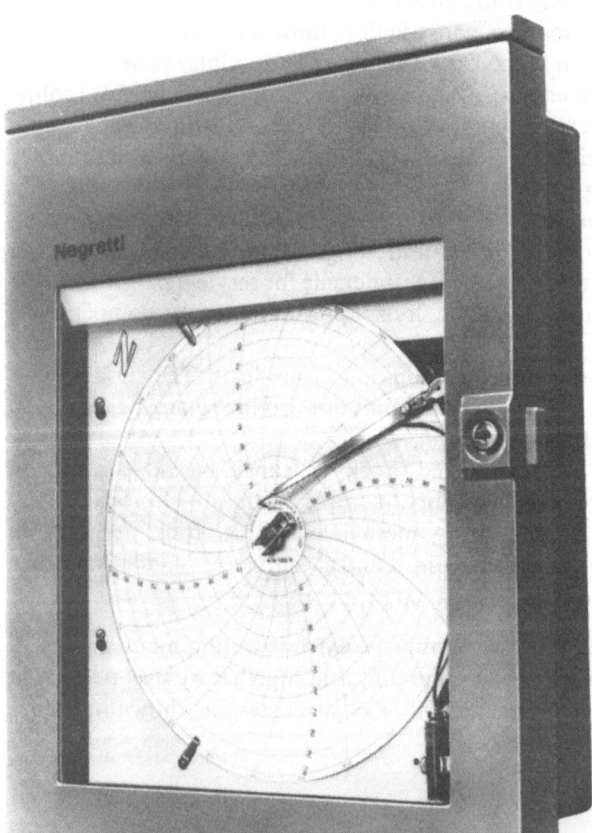

Figure 8.1 EMI video switching panel, general view (top) and (above) section showing construction details.

Figure 8.2 (left) Disc chart recorder. (*Negretti and Zambra Ltd*)

8.11.2 Negretti disc chart recorder

The case of this piece of equipment protects a visual recording mechanism. The environmental conditions are harsh, including hosing down with hot water or even steam in milk-processing plants. Production quantities are small, with older designs using cast aluminium, which is expensive to finish. Sealing arrangements meant that matched tools were needed to produce the intricate shapes necessary for gaskets, hinges and a lock. Thermosetting plastics allowed low tooling costs, and DMC was chosen as a material, being formed in aluminium tools with the resulting mouldings then being stove painted at 160°C, as shown in Figure 8.2. A feature of DMC is that large changes in section can occur without producing sink marks. The glass-filled polyester is very strong, and a matching range of instruments, also with DMC cases, can be mounted on, for example, diesel engines to record exhaust temperatures.

8.11.3 JCB articulated cab

This machine (Figure 8.3a) operates in very arduous conditions and must have a roll-over cage to protect the driver complying with the SAE J394 test. Welded steel tubes were chosen for the energy-absorbing structures as their strength to cost ratio comes out well ahead of any other materials. The configuration shown in Figure 8.3b, with two posts at the front and one larger one immediately behind the driver was chosen to give the best possible vision when reversing.

Figures 8.3a and b   Articulated loader cab with (insert) detail of cab framework construction. (*JCB Ltd*)

8.11.4 Bang and Olufsen tuners and amplifiers

(These notes are included with the kind permission of Mr Jens Bang. The author wishes to emphasise that he has not been personally involved with the design of these units, unlike the other examples in this chapter.) Touch switches, which operate by capacitive coupling with the human body or pressure and avoid moving parts are used in this equipment and, with membrane switches, are likely to find many more applications in the future. Figure 8.4a, b show the exterior and a cross-section of the arrangement used in the Beomaster 1900 unit, which uses an injection-moulded ABS front panel. The hollows indicate where to place the fingers and also eliminate sink markings. Figure 8.5a, b show the different arrangement used on the Beomaster 2400, using a PVC adhesive strip which is screen printed with the necessary graphics. Both these units use a contact spring to connect with a pin on the PCB, but it is possible for the sensors to be copper areas on the PCB itself (Figure 8.6). Modern touch keyboards will operate behind acrylic panels of up to 6 mm thick, or behind other non-conducting materials such as glass.

The above examples suggest the sort of approach to be used when selecting materials, but many other examples are to be found in other sections of this book, and especially in Chapter 18.

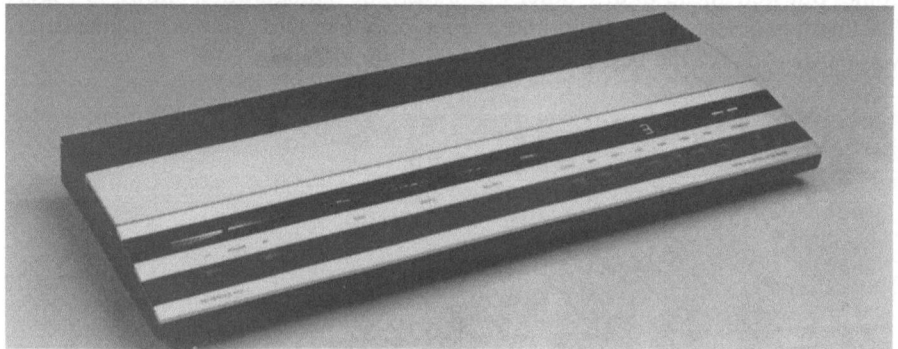

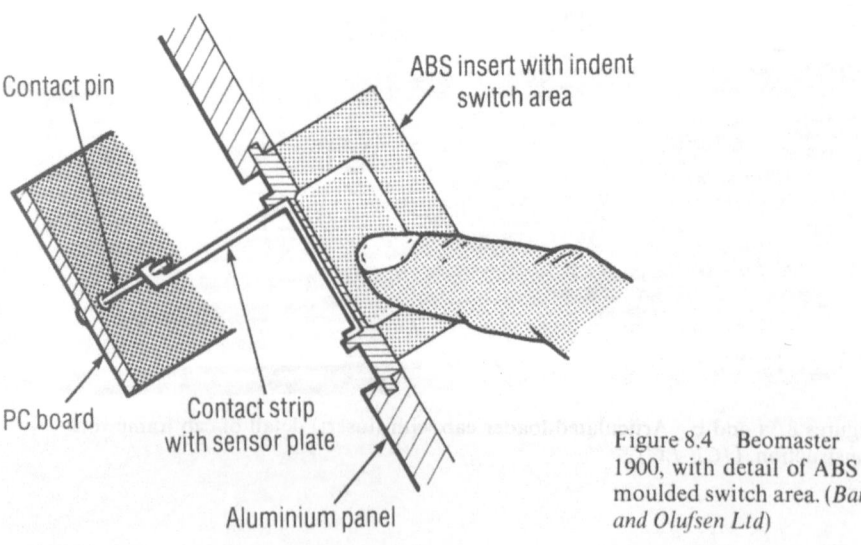

Figure 8.4 Beomaster 1900, with detail of ABS moulded switch area. (*Bang and Olufsen Ltd*)

8.11 Examples of materials selection

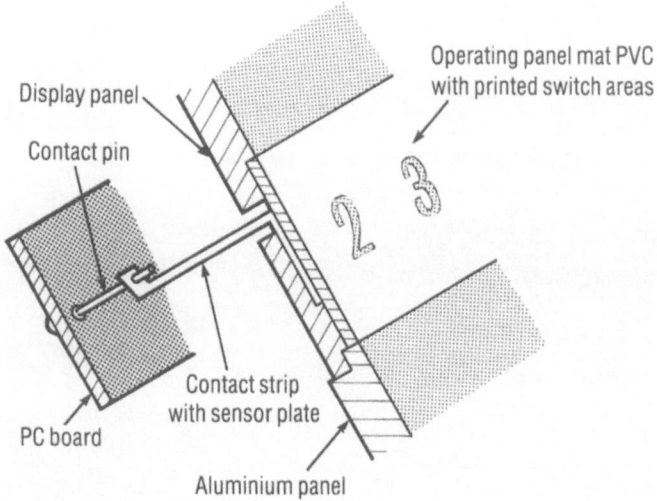

Figure 8.5 (top and above) Beomaster 2400, with detail of printed PVC switch area. (*Bang and Olufsen Ltd*)

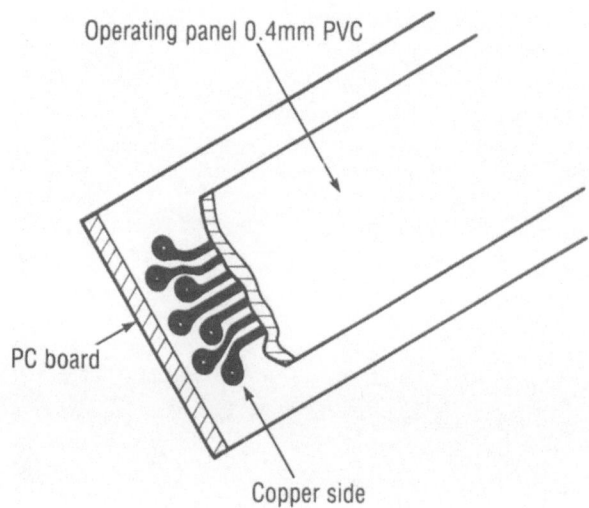

Figure 8.6 (right) PCB used as switch area.

9 Selection of finishes for the man/machine interface
Noël London and Howard Upjohn

9.1 Introduction

Broadly speaking, we have come to accept the meaning of the word 'finish' as the impression we receive from looking at or touching the surface of a manufactured item. Since most materials available to the designer have to undergo some form of processing before they become part of a product, it is essential to understand the part that finishing plays in order to use it to the best advantage. In practice, finishes may be chosen according to their contributions to market appeal, functional efficiency, life and reliability, safety and cost.

9.2 Market appeal

Many factors influence what we buy. While a designer's decision to use a particular finish or combination of finishes may never be the deciding factor in persuading a prospective customer to buy a particular item, a finish that is badly chosen or poorly executed may prejudice the chances of an otherwise excellent product.

Some companies have established a reputation for quality and reliability. We often hear these two words used together; delete either one and the firm has lost its reputation. Quality is, in fact, often measurable in terms of the way a company finishes its products, and the higher the standard, the more it suggests good performance. Sadly, however, this vital aspect of product design is too often denied its fair share of the budget when a product is being developed. Capital goods are bought by people who are also domestic consumers, and they would no doubt be most upset if they were accused of having one set of standards for their personal purchases and a poorer set when acting on behalf of their companies.

9.3 Functional efficiency

Finishes contribute to functional efficiency in many ways. They may be chosen according to the way in which they affect performance, by controlling friction between sliding surfaces, for example; or for their contribution to economic durability in terms of resistance to wear, corrosion and misuse.

Economic durability can have implications for safety. For example, if because of inappropriate finishes an emergency cut-out seizes up when it is needed, having been unused for years or months, safety will be at risk. More generally, however, economic durability implies that product finishes are evaluated in terms of the likely use that will be made of the product throughout its lifetime, keeping their costs in mind.

Ergonomic factors also overlap with safety to a considerable extent, particularly in the man/machine interface area, where visual and tactile responses are involved. Clearly, different finishes will be appropriate for laboratory measuring equipment and earthmoving machinery. The former may be best served by plating, polishing or matt black surfaces; the latter may require corrosion and impact-resistant paint and perhaps galvanising.

Ergonomic considerations will be uppermost in the design of control systems and associated visual displays. Manual and foot controls may need high or low-friction surfaces to promote grip or allow slip, and wear resistance may also be important. Visual displays must have finishes that prevent reflection and glare, and enhance contrast.

9.4 Safety

Finishes can affect safety in many ways, such as by providing non-slip and non-reflective surfaces and by differentiating between safe and unsafe areas through colour. The choice of appropriate finishes is a matter of common sense provided that safety requirements are well defined in the brief.

Fire risks often demand special consideration, and although this mainly concerns structural materials, finishes are also important. Some plastics in particular will burn and melt, assisting the spread of flame. Others may not support combustion, but give off highly toxic fumes when heated. Painted surfaces are generally less of a risk as the material is spread thinly, and some finishes are flame retardant. Some plated finishes can themselves create toxic risks in contact with food or children.

Safety goes hand in hand with accident prevention, and the designer should consider the part that finishes can play in this respect at the man/machine interface. This can comprise the surfacing of steps on a bulldozer cab, or the texture on the screw of a vernier gauge, and the area can be broadly divided into products for use in hostile environments and those that are more protected. The first category would include mining equipment, marine equipment, agricultural machinery and so on, while the second would cover domestic appliances, laboratory equipment, hand tools etc. In between comes an enormous variety of products.

Two extremes of friction are components of safety. On the one hand there is the need for secure grip, and on the other the need for deliberate slip. In many situations these must be combined, generally through the use of an appropriate finish. For example, the grained plastic of a car steering-wheel affords a secure grip, but also allows the hands to slip when required.

Ease of cleaning is also a safety consideration where hygiene is involved. Sterilisation by steam or chemical solutions is only effective if there are no traps or ledges on which foreign particles can lodge. Textures and patterns should therefore be restricted and plastics coatings or highly polished finishes may be preferred. Horizontal surfaces in particular, and those prone to hand contact will be more noticeable as dirt collectors. Remember that cleaning instructions, though important, are often disregarded. Finishes should be tested with harsh detergents, for example, because these will inevitably be used sooner or later by someone buying the product.

9.5 Cost

Inevitably, the question of finishes raises the issue of cost. An immaculate, high-gloss paint finish is more costly to produce than a textured one, but no volume car producer has dared to depart from tradition, which suggests that customers are prepared to pay more for what they want, even when a cheaper alternative would bring practical advantages. In the consumer goods market, comparative information on performance, reliability and safety is readily available, and it follows that a prospective purchaser who has done his homework may well make a final decision on the grounds of price or 'appeal', both of which are fundamentally affected by finish.

Commonsense, forethought and foresight will help the designer, but he must be sufficiently self-motivated to investigate field conditions, talk to users and service engineers and build up a store of information, rather than wait for details of accidents.

The final choice, however, may not be straightforward. For example, electrostatic painting techniques can be automated to eliminate the cost of skilled labour and reduce paint wastage by up to 40 per cent, but the capital investment will be relatively high and the plant may be less flexible in accommodating design changes.

Good design implies choosing the optimum finish for ambient usage, and cost should be linked with function, especially where a product is durable rather than disposable. In this case finishes should retain their good appearance for as long as the product can be expected to last. Sometimes the initial cost may appear high, especially where cheaper alternatives exist, but the justification comes in the long run. For example, we know of local authorities who bought steel columns for street lighting because their unit cost was far lower than for columns made from concrete, but the steel columns have a working life of 20 years and require regular painting, so their original cost may be more than doubled in the long term. The concrete columns do not require painting and last longer because they do not deteriorate below ground level. The 'lifetime cost' is the relevant factor.

The prudent designer of consumer durables knows that his customers' expectations extend beyond mechanical failure, and that expense, loss of goodwill and reputation can all result if finishes deteriorate in the course of 'reasonable' usage. Obvious counter-measures include the use of stainless steel instead of chromium plate in the case of bright finishes, and plastics coating instead of painting or galvanising may be used on outdoor products where abrasion resistance and weather resistance must be combined with good looks. The initial production cost may be higher, and it may be reflected in the higher cost to the customer, but there is little evidence to suggest that people are not prepared to pay for quality if the finish is allowed to demonstrate its presence.

9.6 Terminology

Before plunging into the technical aspects of different finishes, a brief explanation of the terms used is necessary. These are dealt with under three headings: colour, texture and pattern, and reflectivity.

9.6.1 Colour

The physiological aspects of colour are dealt with in Chapter 6. Briefly, the experience of colour is the result of varying wavelengths of visible light, and the human eye is sensitive between the extremes of 380 μm (blue) to 780 μm (red). Practically speaking, any colour can be specified according to the constituent pigments that produce it, and the designer need not feel bound to abide by the limited ranges published and coded as national standards, although there may be economic and practical advantages in doing so. Analysis of colour is usually carried out using formulae known as the ANLAB 40 system (widely used, especially in the textiles industry), or the more recent CIE 1976 system. Therefore if a designer wishes to 'invent' his own colour and use it as a standard for his company's products there are very few technical reasons to prevent him from having what he wants, especially for paint finishes, but practical problems do exist and will be dealt with later.

9.6.2 Texture and pattern

It would be very nearly correct to say that any surface less smooth than a mirror finish has a textured surface. Texture is a matter of degree, and a mirror has minimal texture and rough-cast concrete has a relatively high degree of texture.

Texture can be further divided into random texture and disciplined texture, the latter being more conveniently referred to as pattern. Random texture is a slight misnomer, since some degree of control is usually imposed.

A typical example of pattern would be the visual effect of different types of knurled finish on, say, a control knob. This in itself constitutes texture, but the effect can be varied by knurling on a matt or polished surface. Texture and pattern also combine in brickwork: the surfaces of individual bricks have a prescribed random texture, but the bricks are laid to a pattern. Textures can be varied to provide visual interest in the same way as the use of contrasting colours.

9.6.3 Reflectivity

In the field of paint finishes, reflectivity is often specified as a percentage, so a full gloss paint would approach 80 per cent reflectivity, a semi-gloss around 50 per cent, and a matt finish some 10 per cent. Reflectivity and texture are linked, because reflectivity depends on the way in which light is reflected from a surface. It is also linked with colour, because we perceive most objects as a result of the light they reflect. Different colours have different tone values and hence different degrees of reflectivity. The British Standard for testing gloss levels in paint finishes (BS 3900) generally relates to the American standard. Both adopt a 60° optimum (ie the reflectivity measuring head is positioned at an angle of 60° to the surface being measured) and the following table, relating descriptions to 60° measurements expressed as percentages, gives some definition to commonly used terms.

Dead Matt	0% (not measurable)
Matt	1–5%
Eggshell	15–20%
Semi-gloss	40–50%
Full gloss	80%+

9.6.4 Standardisation

The points made above really amount to an anatomy of finish. It is important to stress that, once a finish has been decided upon, the best way to ensure that it is applied properly is to establish reference samples for use by all those involved.

9.7 Painting of metals

Cast iron. Iron castings are widely used, since the material is relatively cheap and wastage through machining is not too serious. Modern casting techniques produce high-precision castings and cast iron has good natural resistance to oxidisation once its surface film has weathered. Vitreous enamel fused on to the cast iron surface has long been used as a finish. The rough-cast surface provides an excellent key and the high build properties of the enamel give a good finish. Other cast forms such as counterweights on forklift trucks or housings on machine tools are painted with air drying or nominal stoving finishes, probably more from the need for visual tidyness than for environmental protection. Machined surfaces may be coated with temporary protective greases or bitumen compounds, or cocooned in peel-off protective coatings if the surface must be clean prior to assembly.

Cast steel. Cast alloy steels are more sophisticated engineering materials, but both temporary and permanent finishes are the same as for cast iron. Many alloys are developed for special purposes, and these should be investigated specially with help from suppliers.

Sheet metals. Ferrous and non-ferrous piece parts, sub-assemblies and final assemblies almost without exception need pre-treatment and surface finishing. Painting and plating are generally used, and many techniques are available. Over the past decade industrial chemists have developed a wide range of different bases from which the specialist can choose. Consultation with experts is important if the designer is to get the best results in terms of the colour, texture and reflectivity he wants. In general, colour should not be an obstacle, provided that reference samples are used, and quality is usually very consistent. If colour mismatching occurs, the fault is likely to lie elsewhere, for example as the result of storage, stoving time and stoving temperature.

Paints are of course chemical formulations, generally made up of pigments and resin/solvents. The pigments provide the colour and the resin/solvents act as carriers and lock the pigment through evaporation or a more complex chemical change. Paints that are alkyd, acrylic or polyurethane based, for example, thus have specific solvents that have been developed for specific purposes. In industry, paint applied to sheet metal by spraying is generally cured by heat treatment, but where this is not practical for reasons of size or sensitivity to heat, paint may still be sprayed or brushed on but will often be formulated to cure rapidly at normal temperatures.

Although surface pre-treatment for paint finishing is outside the scope of this chapter, the designer should be aware of the advantages of textured paint finishes in disguising manufacturing blemishes. 'Wrinkle' or 'crackle' finishes have a long history, but trap dirt. 'Hammer' finishes, however, which are produced by a chemical reaction in the paint producing multiple 'sinking', do not have this disadvantage. Such finishes are rather out of fashion with designers who have found more versatility in terms of colour and surface texture in 'spattered' finishes which can be produced when spraying by altering paint viscosity and air pressure. Textured finishes do not lend themselves to screen printing, and self-adhesive labels can also be difficult to apply. All-over use of texture can appear monotonous, but transitions from smooth to textured (preferably not on the same component) can add interest even without a colour change.

At the other extreme, high-gloss paint finishes must be carefully specified, since on flat areas, and especially horizontal ones, it is easy to spot irregularities. We have already mentioned that the motorist does not seem to be unhappy about this, but if the product is anything other than a motor car the air becomes filled with complaints. Flat matt surfaces suffer from the same drawback, and in either case, the darker the colour, the more care is needed.

Surface hardness is reduced if the paint has a high content of metallic particles, but this can be improved if the finish is given a clear lacquer coating.

Non-ferrous castings, extrusions etc. Alloys based on aluminium, zinc, copper etc can generally be painted using the same materials and methods as sheet metals, and like sheet metal will also accept a variety of other types of finish. However, where castings, stampings, forgings and extrusions are concerned, their surfaces are likely to be less regular. This is where the use of pattern can be effective by breaking up the surface area through the use of low-relief ribs, bands, rectangles and so on. This can help to improve the overall appearance and eliminate some machining, but not give the paint shop an expensive task. Similarly, extruded or drawn sections can be visually improved by designing in some 'decorative' ribs to break up wide, flat surfaces which will tend to carry tool marks. Clearly, the introduction of such patterns must be considered carefully to avoid extra material costs, production difficulties or dirt traps on the finished product.

The pre-treatment of metal substrates is beyond the scope of this chapter, but it is worth pointing out that much useful information is to be gained by examining the processes that precede the finishing of a product. For example, aluminium sheet is often either caustic etched or shot blasted during pre-treatment, and the application of a clear lacquer on to the resulting surface provides an interesting finish. Another aspect which can only be mentioned briefly here is the question of suitability. Some paint finishes when combined with appropriate pre-treatment can offer quite remarkable durability and wear resistance. Such finishes may well be suitable when the item of equipment is necessarily expensive, although a domestic washing-machine, for example, will probably not qualify, even though it may well lead a rougher life. It is up to the designer to liaise with the chemist or specialist paint manufacturer to get the best possible result.

9.8 Plastics coating

This is a field that has expanded enormously in recent years. Plastics such as PVC, PTFE and members of the nylon family are in widespread use, but many other thermoplastics and thermosetting plastics, together with 'tailored' resins, are available for specific purposes. Specialist firms exist and should be consulted by the designer, but two main methods exist for applying coatings. The first and earlier method, involves dipping the component to be finished into molten finishing material. Powder coating is an alternative, particularly for components with a large surface area, and in this case heat is applied to fuse the plastics, which are sometimes applied using a fluidised bed technique.

Plastics coatings seem to have found their main application where components have to withstand severe use or environmental conditions. Nylon coating is often found on steel tube furniture, for example, and PTFE (polytetrafluoroethylene) is widely used as a non-stick finish on domestic cooking utensils and in food-processing equipment. There are grades of PTFE which can be used at working temperatures up to 300°C and down to $-200°C$. Polytrifluoromonochlorethylene is similar to PTFE but has better resistance to chemical corrosion. Polyvinylidene fluoride (PVDF) is widely used for protecting valves, pumps and similar components against corrosion in the chemical, food, pharmaceutical and nuclear industries. Thermosetting plastics, epoxy resins and so on should be considered where chemical resistance and exceptional durability are required.

Colours for plastics coatings tend to be standardised by manufacturers, and colour matching is only economic for very large quantities. It should also be pointed out that all finishes of this type increase the dimensions of components, and it is easy to overlook this at the design stage.

9.9 Electroplating

Electroplating is simply the process of depositing metals on to metallic or non-metallic substrates by the use of an electric current passing through a solution. The field is a very large one, and specialist advice should be sought where necessary, but the following outlines should be a starting point.

Nickel and chromium plating. Nickel coatings are applied as a decorative finish to many items, such as spanners, screwdriver blades, keys and can-openers. About three per cent of all nickel used in coatings is employed in engineering applications for new parts or for reclamation, where brightness is not normally necessary and deposits are generally thick. About five per cent is used in electroforming, and the remaining 92 per cent is used in the form of decorative coatings, 5 to 40 μm thick, usually under a top coat of chromium about 0.5 μm thick. The appearance of the final coating is largely determined by the finish of the nickel, and resistance to corrosion is very much dependent on the interaction between the nickel and chromium coats. Nickel-chromium coatings are applied to metal substrates such as steel, zinc, brass and others to increase resistance to abrasion and corrosion. The ISO has put forward recommendations for plating (including copper deposition) which set out four degrees of severity of service conditions (ISO R 1456—1970) and these are contained in a corresponding British Standard (BS 1224: 1970), approximately as follows:

Service condition number 4—Outdoors, exceptionally severe
                                  3—Outdoors, normal
                                  2—Indoors, with condensation
                                  1—Indoors, warm and dry

As an example, motor car components come under condition number 3 and office equipment under condition 1. While these standards are concerned with plating on metal substrates, procedures for plating on to plastics have been available for some years and this is an expanding field, with ABS (acrylonitrile-butadiene-styrene copolymer) as the most commonly used material. A problem is thermal shock, and thermal cycling is widely used as a test for adhesion of plating on plastics. Current practice also recommends copper undercoating to improve adhesion and resistance to cracking, peeling and discolouration.

Other plating finishes. Apart from nickel and chromium, metals such as zinc, cadmium, tin, copper and aluminium are widely used for plating purposes, either by electrochemical means or by hot dipping. They are generally used for protection rather than aesthetic reasons, but zinc plating provides an excellent basis for paint treatment, as do tin and cadmium. Aluminium hot-dip plating is particularly suited to the treatment of cast iron. Metal spraying is also worth a mention since, in addition to coating new parts and rebuilding worn ones, it can be used to provide an electrostatic shield on plastics components and, using Mumetal, it can provide anti-magnetic protection.

It should be stressed that the finish of the underlying substrate makes a large contribution to the effect of plated components, and finishing processes must therefore be considered as an integral part of the design. It is possible, and indeed often desirable, to combine satin and polished plated finishes on the same component, for practical or aesthetic reasons.

9.10 Aluminium

In the case of aluminium the practicability and variety of plated finishes is somewhat reduced compared with ferrous metals, but this is more than made up for by the large variety of anodic treatments that the metal and its alloys will accept.

Aluminium rapidly forms a fine oxide film on its surface, and for many light-duty uses this natural film will protect the metal from further corrosion. Anodising improves upon the natural oxide film by increasing its thickness and hardness considerably. As with plating, anodising should be viewed as a final treatment which has less effect on the finished appearance of a component than the process that precedes it.

Pure aluminium has relatively low mechanical strength, but is capable of taking a mirror finish which has nominally less colour than chromium plating finished to the same standard. Such a finish is, however, rather prone to damage and should be used accordingly. Aluminium is, however, generally used in alloy form, and such alloys contain proportions of copper, magnesium, manganese, nickel, silicon and zinc. It is essential to take specialist advice on anodising treatments for different alloys, especially when dealing with aluminium alloy castings, as the constituents in the alloy will react differently to the anodising process.

Aluminium alloys can be polished in a variety of ways to produce a brushed finish, textured by rolling, pressing or chemical etching, and shot, bead or vapour blasted. In addition to screen-printed treatments, a rather limited range of colour dyes can be introduced during anodising.

Aluminium sheet, extrusions and castings can be painted, but in most cases a base treatment such as an etching primer or chemical bonding agent is needed because of the oxide layer that is present. Aluminium sheet can also be obtained with a plastics film coating, eliminating the need for final finishing, and castings and extrusions can be dip-coated with plastics and can be coated in low-temperature vitreous enamel.

9.11 Stainless steel

The term stainless steel covers a large range of special steels, generally containing a proportion of chrome and nickel. The austenitic group of stainless steels are the most extensively used and have good corrosion and heat resistance and excellent weldability. Perhaps the most widely used alloy is 18/8, which denotes the chrome and nickel content respectively.

So far as sheet surfaces are concerned, it is seldom necessary to add any protective coating, but with a clean and 'keyed' surface the metal will take paint where this is necessary for nameplates, signs or control panels. Most of the available finishes are applied at the mill, from mirror-polished through brushed to very dull finishes, which will retain lubricants for press working. Textures can be introduced at the rolling stage to break up the finish and mask imperfections. More recently, techniques have become available that produce colours and also etched textures.

Electropolishing, which is a reverse electroplating process, is most effective in cleaning and polishing intricate parts and assemblies, but it will not remove scratches.

The high-temperature strength of stainless steel will permit vitreous enamelling, although this would only be necessary for extra corrosion resistance or decoration.

9.12 Plastics

A major advantage of plastics is, of course, that when used with appropriate production techniques they require very little in the way of finishing. Finishes can be determined in advance by using a combination of the moulding tool and the chosen material, and both thermoplastics and thermosetting plastics are widely produced using injection moulding techniques. In many cases, however, marketing requirements will not be satisfied with a single colour or single texture plastics part. In these cases it is possible to mould two colours in one tool, or make use of inserts or foils. Such techniques should meet practical requirements by producing, for example, keyboard buttons on which the legends cannot wear out. As mentioned earlier, variation from manufacturers' standard colours is only economical for very large quantities.

Many plastics can be given a metallic finish by vacuum deposition or plating (see 9.9), and plated plastics parts are increasingly widely used in the automotive industry to replace zinc-alloy die-castings. Transparent plastics mouldings can be masked before spraying to provide 'windows' for instruments, for example, and many other applications are possible. Vacuum deposition is principally used in industry to provide optical parts and mirrors or reflectors.

For relatively short production runs, the designer can use vacuum forming, GRP or similar compounds, or the newer realm of structural foam plastics. Vacuum forming allows the designer to specify a suitable thermoplastic sheet material of an appropriate thickness, but the choice of colour and texture is currently limited. The less chemical resistance the thermoplastic has, the more likely it is to accept a paint finish. ABS is a good all-round material for this application.

With GRP, the resin may be pigmented before production, but a popular alternative is to spray paint after curing. Paints having a compatible resin obviously give excellent adhesion, but most air-drying and low-temperature stoving formulations can be used. Texture finishes and variations in gloss are of course available.

Structural foam plastics or engineering foam plastics form a hard skin as they cure and will faithfully reproduce the mould surface, although too much texture can cause ejection problems. Visual interest can be provided by incorporating changes in levels and raised or sunk ribs. The materials are very tolerant of changes in wall thickness. Low-density foams will accept air-drying paint finishes, as will high-density ones, but the latter can also be formed integrally with plastic sheet or film, bonding to it during the foaming cycle. Plain or decorated foils can be incorporated during the moulding process or can be bonded on later. Post-form foiling permits surfaces to be given a wide range of finishes.

9.13 Conclusion

Finishes contribute to appearance, functional efficiency, reliability, durability and safety. Selecting finishes to achieve the appropriate level of quality in each of these areas may increase cost, but good finishes have a direct effect on market appeal and the extra money will in most cases be well spent. There is a relationship between the cost of finishing an article and its anticipated useful life. Certainly, as many manufacturing materials become scarcer and more expensive, and cost an increasing amount to extract, convert and process in terms of energy, the designer should consider how a small extra expenditure on finishes can preserve and enhance his product's usefulness and reduce lifetime costs.

10 Design for maintenance
Antony Gibbs and Charles H. Flurscheim

10.1 Philosophy of design for maintenance

The availability and performance of equipment in service is influenced by the amount of maintenance required and the efficiency with which this can be executed, and since maintenance spans the whole man/machine interface, design for it involves the use of integrated engineering and ergonomic techniques.

The specification (see Chapter 16) should give a guide to the nature of the maintenance required. Design can then contribute by introducing and developing procedures to reduce the need for maintenance; by providing monitoring to anticipate troubles; by ensuring simplicity and accessibility for essential maintenance operations; by making possible systematic fault analysis; and by providing logical instructions for fault-finding and maintenance.

It will usually be evident from inspection, and certainly from use, if design for maintenance has been carried out thoroughly, and when this is the case it will not only improve the facilities for maintenance but, since maintenance staff are appreciative of helpful design, it will also encourage high-quality workmanship.

Maintenance staff may be of three kinds. First, there are the people who own or operate the equipment. Although they may not be expert service engineers, they can handle basic day-to-day maintenance—for example, filling a car with petrol or putting a new battery in a calculator. Then there are professional service engineers, whose responsibility is to ensure the continuing efficient operation of the equipment. They would handle the 10 000 mile service on a car, using equipment and techniques beyond the resources of the owner, or change contacts on a circuit-breaker, or adjust instrumentation. Lastly, there are staff who handle major repair work and replace parts damaged in use—for example, repairing a dented car body, rewinding a burnt-out electric motor, or replacing eroded blades on a turbine.

10.2 Design to reduce maintenance

Following the principles mentioned above, the first and most fundamental aim is to attempt to design away the need for any maintenance at all wherever possible. After all, a need for maintenance implies that the product will eventually fail.

The most obvious approach to reducing the need for maintenance is through improving the quality of the technical aspects of design, which includes detailed attention to such matters as friction, lubrication, wear, corrosion, fatigue, creep, stress concentration, securing systems and even texture and colour.

A second approach is concerned with reducing maintenance through consideration of the man/machine interface, often using both engineering and industrial design techniques. This approach can be considered under a number of headings, as follows.

10.2.1 Prevention of troubles initiated before commissioning

Troubles in service sometimes originate from damage suffered by equipment prior to installation. This can arise for instance from inadequate packaging, or packaging which itself initiates incipient deterioration which might cause failure later. It is therefore important to consider the type of transport and storage and possible mishandling, and the class of personnel involved, so as to optimise the combination of packaging and product in terms of quality and cost. For example, it may prove cheaper to design fragile equipment, such as instruments, to withstand much higher accelerations than their duty in service might require in order to reduce the cost of packaging necessary for transit. Similarly, if machines running on ball bearings are likely to be transported or stored in circumstances that subject them to vibration, it may be economical to build in resilience in the packaging or in the machine in order to avoid brinelling of the bearings followed later by failure. High-cost materials have an attraction for vandals, and the machine and the packaging should therefore be designed so as to ensure that valuable components are not exposed or damaged.

Packaging can of course also contribute to marketing through favourable presentation, which is a separate design consideration, but the basic requirement for packaging in engineering is to provide protection. A further design requirement is for adequate and clear instructions. Instructions written in conjunction with the designers during the development of the product should give installers and operators all the information they require, in terms and language that will be understood locally, to enable them to carry out effective and accurate installation and to use the product without risk of damage.

10.2.2 Control system design

Good control system design will reduce incorrect operation and its consequences, such as overload, which may damage equipment. Simple controls which are clearly identified, easy to understand and non-critical in operation will help to reduce the need for maintenance.

10.2.3 Design for failure anticipation

Some types of fault develop slowly, and if their progress is monitored the need for preventive maintenance may be detected in advance. Developing faults can be identified by recording or noting measurements such as eccentricity of movement of shafts in bearings, temperature, pressure, lubricant and noise levels and their characteristics. The contribution of design in such cases may be to build in the measuring criteria, or to provide designs that themselves expose incipient troubles, for example by deliberate noise generation.

10.3 Design for maintenance

This function involves the identification of troubles, and maintenance procedures such as cleaning, measurement, adjustment, replacement, lubrication, testing and inspection, all of which involve design for the man/machine interface.

10.3.1 Fault identification

When incipient or actual troubles develop it is often difficult to identify the exact cause and location of the fault. For example, the failure of a system to respond correctly to automatic supervision may be in the controls, electronics or mechanical services. Poor operation of an internal combustion engine may be due to imbalance of one or more of the many air/fuel adjustments, mechanical errors in the engine itself, or problems with ignition. Hunting in the speed of a turbine can be caused by a number of problems in the hydraulic, mechanical or electronic systems associated with the governor gear or its supervision.

One way of aiding fault identification is then to build in, or to make provision for the temporary coupling up of, measuring systems that will analyse movements, gas composition, electrical faults, temperature etc, and so provide a logical basis for fault identification and correction. In control systems, built-in supervisory and testing facilities for maintenance are best segregated, as discussed in Chapter 3 (see also examples 18.8, 18.15, 18.18).

A second and very necessary approach is the provision of a logical analysis procedure in the maintenance instructions, identifying what is to be measured, the sequence of procedures and deductions from evidence obtained, written at a level that will be understood by maintenance staff, who may not possess the relevant specialised background.

10.3.2 Cleaning

Cleaning is a normal part of maintenance. Contamination from a dirty environment can be reduced by avoiding horizontal surfaces or patterns with horizontal ledges, pockets which can trap dirt, and materials that attract dirt electrostatically, as do some plastics. Contamination initiated by machines may be reduced by attention to oil seals, lubrication systems, and the efficient disposal of by-products and contaminants (see examples 18.14, 18.15).

10.3.3 Texture and colour

Cleaning includes the maintenance of appearance. Many devices have artificially textured surfaces, such as leather graining for the surface of moulded products to mask imperfections in the moulding and to disguise the effect of knocks and abrasions, both during assembly and afterwards. However, a pale-coloured product with a textured finish can get very grubby, dirt can work into the grain and be virtually impossible to remove, so both texture and colour have an effect on maintenance. But if the textured surface were coloured to match the local dirt, then however dirty it got, the dirt would not show. The lesson is that textured surfaces encourage the camouflage of imperfections and contamination in order to minimise maintenance, but are virtually impossible to clean properly.

Smooth finishes, on the other hand, can simply be wiped clean and polished, maintaining the surface finish and buffing out any blemishes that might appear through normal wear and tear. An added advantage is that the surface can be seen to be clean, a particularly relevant point in hygiene-conscious environments. An obvious example here is the telephone, which confounds the theories of the textured finish school by appearing clean and pristine after many years of use, simply through normal day-to-day dusting and polishing.

So we have two schools of thought: the first believes that camouflage is the best way to avoid maintenance, and the second believes that actual maintenance of a simple surface is best. Clearly, both have a case, and the designer must select the approach that is most appropriate to his needs.

We have touched on the use of colour to maintain appearance, but the maintenance of colour itself can be just as important. There is nothing more unsightly than a white plastics product that has yellowed with age. Colour deterioration can occur for a number of reasons. First, a self-coloured product may change simply through natural ageing. Then the effect of ultra violet rays, both man-made and in natural light, can have a drastic bleaching effect on many colouring agents, and this should be given careful consideration when selecting appropriate dyes or pigments. The third major cause of colour change is the staining caused by chemicals that react with pigments in finishes. A lathe finished in a pale colour (an admirable choice from the point of view of the maintenance department because it will quickly show up any contamination) may, because of the effect of cutting fluids and oil on the pigment, end up a dark grey colour. While most of this may be removed with emulsifying agents, the paint itself will eventually become indelibly stained by the oil. Provided that the designer bears in mind the fact that the colour of his product may be modified, he can usually avoid these effects by sensible selection of the colours themselves, the materials he uses, and the pigments or dyes used.

10.3.4 Lubrication

The need for lubrication can be reduced by the use of self-lubricating sintered or plastics bearings, or by building in automatic lubrication systems, taking care to ensure that the lubrication provided is adequate but not excessive. This may require the oil supply to sliding surfaces for example to be dependent on a combination of measurements such as time and movement, as in the horizontal boring machine (Chapter 18.15). When lubrication is required as part of maintenance, this can be simplified by providing obvious lubrication points, by ensuring that all lubrication requirements fall into a regular pattern, and by defining the frequency, the extent and nature of lubrication required for each position in the maintenance instruction.

10.3.5 Performance

When performance has to be measured by the temporary use of instrumentation, design should provide built-in positions to enable the necessary mechanical or electrical couplings to be made, ensuring accuracy of recording and simplicity of application.

10.3.6 Wear and leakage

Where measurement of wear or leakage is required, the design should provide built-in indication of the measurement points and a clear guide on acceptable limits and the corrective action needed if these are exceeded. Brake failure in a car is an example of a small failure which can have drastic knock-on effects, ultimately leading to complicated servicing and the replacement of expensive components, to say nothing of operator safety. In such important cases continuous visible indication of the state of wear and leakage may be desirable.

10.3.7 Accessibility

One of the major problems in design for maintenance is the provision of access for adjustment and replacement. It is not always appropriate to hire a 'small person' to carry out work in inaccessible positions, as was necessary for example inside the wings of Second World War bombers such as the Lancaster, and the contortions often required do not contribute to good workmanship. Adjustments should be easy to carry out and lock, non-critical in their character and easy to inspect. If parts need to be changed, the basis for this should be clarified in instructions, for example in terms of length of service, the integrated total duty carried out, or dimensional wear.

Design should anticipate the kind of detail maintenance that will be required, and aim to minimise the extent of dismantling necessary. For example, it is unreasonable to require complete dismantling to replace a simple washer which it is known will wear out, thus leading to maintenance costs out of all proportion to the cost of the part replaced.

Design should also take environmental factors into account. For example, extremely low temperatures can affect the type of maintenance work that can be carried out reliably, and may influence the nature of adjustments, or make it preferable to replace whole components rather than rectify them on site.

Design should permit use of tools such as ring or box spanners rather than open-ended ones, and cross-head Phillips-type screwdrivers, which contribute to ease of working.

Finally, if covers or housings need to be removed they should be dimensioned and positioned so as to provide adequate access, without exposing sharp edges, and they should be easy to remove and replace. Covers should be light enough to manhandle safely, or be so heavy that built-in or portable handling devices are clearly needed. It may sometimes be preferable to design equipment so that internal mechanisms can be withdrawn for maintenance, perhaps by swinging links or sliding rails, as in the chart recorder (Chapter 18.7), rather than to provide removable covers. Covers, whether fixed or removable, can contribute to tidyness, noise reduction, and protection from dirt and mechanical damage, but they may also hinder maintenance or interfere with accessibility. Unless such housings are required for safety reasons, these points should be taken into account when assessing their overall contribution [10.1] (see Chapter 18.21).

10.3.8 Prevention of assembly error

Design to prevent incorrect maintenance is also important. For example, if one-way valves or controls can be connected back to front, this will happen sooner or later (see Chapter 18.6). If O-ring seals are mounted in grooves in the upper face of an inverted flange joint, there is a risk that they will fall out, leading to incorrect positioning and subsequent leakage. The removal of covers and housings for maintenance purposes should not require the disconnection of wiring or control mechanisms, unless provision is made to prevent incorrect reassembly, for example by the use of coded wiring-harness plugs and sockets (see Chapter 18.10 and 18.11). The design approach should be to analyse what could be done wrong and take preventive action.

10.3.9 Safety

As mentioned above, maintenance operations are often carried out under unfavourable environmental conditions, in a hurry and with poor lighting, poor equipment or inexperienced operators, which increases the risk of accidents. Safety is fully discussed in Chapter 13, but specific aspects of design for maintenance associated with safety include: the weight and awkwardness of components to be handled, which require attention to built-in handles or handling equipment; machine guards, which when provided must be virtually foolproof against misuse; and sudden movements initiated in machinery during maintenance, which if guards are not practicable or have to be removed for adjustment, should be clearly described in the instructions, with warning notices on the mechanism itself and means of blocking such movements where necessary. Special care is needed with respect to guards on electrical circuits, for example to avoid any accidental contact with maintenance tools, and if guards have to be removed, warning notices on the machine or interlocks with circuit controls may be necessary.

Attention must also be given to the risks associated with mixed services, such as high-pressure hydraulics and electrics within a single enclosure. For example, switching in an atmosphere polluted by oil spray leakage can cause explosion.

Design should also take into account the possible consequences of maintenance procedures on control operations that must continue during the period when work is being carried out. Warning notices are essential to clarify prevailing limitations of the system, such as disconnection of instrumentation or valves locked in the closed or open position. Consideration of detail design is then needed to ensure clear titling and to prevent visual masking of controls and instrumentation by the warning system used, and this may require wider spacing of controls, the provision of mountings for warning labels in a safe manner, or the use of transparent notices. It is therefore important that design audits for safety should include consideration of the special conditions that apply during maintenance (see Chapter 18.1, 18.6).

10.3.10 Servicing skills

The level of labour available to carry out servicing is an important factor which must not be overlooked. Specialists in particular disciplines need little training and can adapt to the servicing requirements within their specialisation quite easily. However, a less skilled man will need more instruction and training to achieve the same level of competence. For this reason the designer must know what class of labour will be used from the outset, so that he can make allowances in his design and provide instructions for the degree of experience to be expected and the extent of special training that can be called for.

10.4 Checklist for design for maintenance

In considering design for maintenance it may therefore be helpful to consider the following basic questions.

1. In what environment will maintenance be carried out and is the design suitable?
2. What will be the expertise of maintenance staff relative to the work?
3. Will there be routine preventive maintenance and how often will it be needed?
4. If maintenance will be solely corrective, what might be the nature of the types of failure?
5. Will maintenance be by component replacement, by modular changes, or remedial adjustment?
6. Are all likely types of failure accessible, in terms of quality of the work required, and in terms of the cost of dismantling and assembly?
7. Would the time required for maintenance be reduced by providing better accessibility or greater breakdown and what would be the effect of these on initial cost?
8. Is design for adjustment satisfactory in terms of stability and precision needed?
9. Is built-in fault diagnosis equipment going to be used?
10. Does the equipment's packaging optimise costs versus troubles due to damage?
11. Is design for cleanliness on a satisfactory basis?
12. Is design for texture and colour on a satisfactory basis?
13. Is design for lubrication on a satisfactory basis?
14. Is design for measurement on a satisfactory basis?
15. Are covers easily removable and replaceable and are they likely to be damaged?
16. Does design prevent incorrect reassembly of vital components after maintenance?
17. Does design envisage use of tools which will avoid risk of damage to equipment and to staff?
18. What are the possible safety risks during maintenance arising from mechanical origins, electrical origins, compressed air origins, hydraulic origins, gravity, spring operation, high-pressure steam, leakage of gas or chemical fluids, fire, explosion, and lack of clarity in the visual state of the equipment, and has design taken these into account?
19. Are the operating and maintenance instructions adequate to ensure the quality of servicing desired with the personnel who will be carrying this out?

10.5 Reference

10.1 Kay, R. M. S. 'The application of industrial design to turbine generators', *Turbine Generator Engineer*, AEI Ltd, 1968.

11 Psychological factors in man/machine interface design

Peter Murdoch

11.1 Introduction

The relationship between man and machine is, in essence, a flow of information, and it is the effect of this flow on the individual operator with which this chapter is concerned.

It could be argued, in fact, that this very flow is the product itself. Information is given and received and the machine functions in accordance with the needs or wishes of the human user. What is important, in psychological terms, is that the design avoids as much as possible those elements which could inhibit, confuse, frighten or frustrate any of the people who might have a direct relationship with the machine.

The difficulty with respect to psychological factors in engineering design lies principally in defining what the problem is for a particular machine. Once this has been resolved, the design solutions, which will of course vary widely depending on the product, tend to follow fairly readily. This means that the designer must start out with at least some understanding of the psychological factors which affect people's relationships with machines.

11.2 Psychological factors

11.2.1 Misunderstanding

Information has to be transmitted accurately, but it also has to be transmitted in a form readily acceptable to the user. The design of instrumentation and control equipment, for both input and output of information, is therefore one area where particular attention has to be paid to the avoidance of misunderstanding. The ergonomic design of instrumentation and controls has been discussed in earlier chapters, and here it is necessary to emphasise only the need for clarity and logic in presentation of information and arrangement of controls.

For example, when a number of dial instruments define the state of the system, if their pointers can be arranged to align when the system is working correctly, error in interpretation is less likely than if they are all at different positions which have to be interpreted separately. Altimeters in aircraft can give dangerously misleading information if they are not re-set by the pilot to conform with local conditions. This implies a warning on the instrument panel, or perhaps very clear operating instructions, to avoid any chance of error.

A mirror image arrangement of adjacent and similar control panels can result in reversal of position of controls and instrumentation, and is a sure way of generating confusion, incorrect operation and accidents.

The steering system of a car can lead the driver to misinterpret stability and safety if it provides inadequate feedback information on the forces generated by turning which must be resisted by the tyres. The same holds true for aircraft if the flight controls present inadequate feedback of the stresses imposed on the airframe by change in attitude.

An example of information input is given by the use of keys in the cash dispenser (Chapter 18.4) where the logic of the instructions has been designed to reduce the probability of error by users who are unfamiliar with the system.

Glass doors are an everyday example of a false impression—they appear not to be there and people, visually misled, are in danger of walking into them. This can be avoided by providing a sign or a visible marking on the doors at eye level.

Machine operators can also misinterpret which push-button to press. This risk can be reduced in certain dangerous conditions by requiring two controls to be operated simultaneously. An example of this is the recording controls on cassette tape recorders, where devices, either on the recording mechanism itself or on tabs on the back of the cassettes, prevent inadvertent recording on pre-recorded tapes. Another example is the automatic lathe (Chapter 18.14).

Symbols themselves can be misleading. For example, do the F and S symbols on clocks and watches mean 'faster' and 'slower', so that an adjustment in the appropriate direction will make the mechanism speed up or slow down, or do they mean 'fast' and 'slow', in which case the adjustment should be reversed?

Whatever the attempted solution, complete protection against misuse or misunderstanding is impossible. But where information is being transferred, good design can often bring about an improved performance by the user and therefore less chance of accident or damage.

11.2.2 Pride and prestige

Both pride of ownership or control and the prestige derived from them can have important effects on the individual and must be considered during design. Pride of ownership will often result in a better communication between man and machine, greater care of it, and more attention to important detail and therefore less chance of accident or breakdown. A well engineered product, whether it is a car, drill, boiler, aeroplane or sewing machine, can create this pride, which can very often be combined with prior understanding or education of the user or owner.

However, it can be dangerous for the designer to assume such prior education. A young man may have great pride in his new sports car, but does this prevent him from driving too fast, or abusing it in other ways? Pride and prestige can lead to the design of superior quality products, but this relies not only upon the machine being well engineered, comprehensible and interesting, but also on the user knowing how to relate to it, with clear instructions provided and, possibly, educational courses.

The opposites of pride and prestige must also be considered in certain engineering situations. For example, aids to disability such as callipers or hearing aids are designed primarily to cope with a deficiency, but must also minimise any feelings of shame and dislike in the user—principally through cosmetic design. Heavy, unsightly or ugly apparatus of this kind will be unacceptable, however efficient it may be.

11.2.3 Fear, confidence and stress

There are two aspects of fear in relation to machines or their associated environments. One is subconscious fear, which may not always be immediately apparent to the person concerned but is based on past incidents, happenings or assumptions. The other is an immediate response to something which is perceived as frightening.

A minority of people suffer from medically recognised phobias, and these are dealt with in the next section, but the majority of people develop at some time certain fears in relation to machines or environments—fears which can be alleviated or lessened by careful design.

This chapter will cover later the important question of whether an 'audience' is captive or non-captive. However, fear is a more common reaction in a captive situation. Where a degree of choice can be exercised, as in the non-captive situation, most people are able to skirt around their fear. This means that the machine and the environment should be considered as an entity where a captive audience is concerned. Aircraft, trains, ships, submarines, cars and elevators all have captive audiences, and all are potential producers of fears resulting from enclosed places, flying, speed, being trapped, proximity to other people, malfunction, sickness—even unlucky colours.

In these products, which represent total environments, the engineer should be aware of the psychological factors involved in their design, including fear, and take steps to counter them at the design stage. For example, clearly visible earthing devices on high-voltage equipment, in support of elaborate safety procedures, will help to allay the fears of maintenance crews that the equipment may become live while they are working on it.

A classic solution to the overall problem is to present an *alternative* to the particular item or aspect which causes fear. Very often this can be in the form of a way out of a captive situation, through exits, emergency solutions, the provision of guides, experts, hostesses who look unperturbed, fail-safe systems or back-up or ancillary systems. Fear is alleviated by showing the possibility of an alternative, should anything go wrong. It is fear of the unknown that is most common and can be most effectively countered. The confidence of pilots in fighter aircraft, for example, is increased because they know they have an ejector seat as a last resort.

In circumstances where the product and the environment are not under the same design control, real problems do occur more frequently. A traditional operating theatre is such an unnerving environment that 'pre-med' sedatives are used to alleviate the panic induced by the equipment, the lights, the uniforms and surroundings. But much of this fear could also be alleviated by good design—by concealing parts of machines which are oppressive or overpowering and which do not have to be in the theatre area; by humanising components or parts which are essential to the operations, either by form or colour; or by designing them to generate patient interest in their application and use. An example is the linear accelerator (Chapter 18.12).

In contrast to fear, confidence in a product or system will contribute towards efficient operation by reducing psychological stress. This can be achieved by, for example, built-in testing and diagnostic features which enable a machine to be monitored and checked for incipient faults before they become serious; or by the introduction of fail-safe systems which will greatly reduce failure risks, (see the automatic flight control system, Chapter 18.8); or by providing warning instrumentation, such as oil-pressure or brake-fluid-level warning lights in cars.

It should be noted that, although a low level of mental stress may serve to improve human perception, and consequently response in control situations, excessive stress can result in premature or illogical decisions leading to errors. Design can be used to alleviate stress generated in demanding control situations by, for example, differentiating the character of warning signals, permitting time to think, from those that require an immediate response.

11.2.4 Phobias and sickness

It would be impossible for the designers of machines or products intended for a particular purpose to counteract all the phobias and sicknesses which affect a minority of potential users.

Medically defined illness, such as claustrophobia, acrophobia or agoraphobia (fears of confined spaces, heights or open areas) cannot always be totally countered in circumstances where they are likely to occur. However, motion sickness, which is caused by the disorientation of the inner-ear mechanism, can be alleviated either by specific drugs or by countering the motion and resolving some of the pressures which cause this illness. An example is the flight control system (Chapter 18.8), which reduces passenger discomfort through automatic stabilisation in all three axes.

What can certainly be achieved in most cases is the alleviation of such feelings by careful attention to the psychological effects of the machine and its environment. The general public are often exposed to bad design and engineering, which affect them psychologically. The effect is immediate among those few individuals already suffering from one of the psychological illnesses; with the rest, it is cumulative. No one is impervious to the environment in which they find themselves, and in subways, corridors, aircraft, elevators, cars and many other engineering products, the lack of attention to psychological factors shows itself all too commonly. Design isolated from the user and his environment can only be bad design.

11.2.5 Love, hate and indifference

Why do some objects or environments attract vandalism more than others? The vandalised object may be inanimate, but to the aggressor it represents something more than that—his envy, hatred, frustration, or an obsession with all that is new and thus often dislike for all that is old—even of society itself. The information the vandalised object conveys is what he is trying to destroy, in much the same way as we suggested earlier that the flow of information was in itself the product of the machine.

Good design is clearly not a panacea for all the ills of modern society, but it cannot be seen in isolation from them. A simple but effective example is the introduction of the parking meter, which aroused great hostility, notably in Paris, where many were ripped out of the ground and destroyed. In this country too, meters have been vandalised. A new design of meter, however, involves greater user participation, in that the timing mechanism is not automatic, but is actuated by the operation of a lever. This would seem to be a more acceptable design for the motorist because it requires a response rather than passively swallowing his money.

Shape is another important factor in design, as demonstrated by choice in cars. Men and women prefer different designs, often, it is said, based on sexual preference. Men opt for an aggressive extension of their own assertive personality, while women choose sleeker, more fashionable shapes.

Colour has an effect in design terms which is often very little understood. Research has shown that colour changes bring about entirely different reactions in the same environment, particularly in such places as intensive working areas and operating theatres. To paint a control room pink would be incorrect as it induces sleep more than most other colours. The interiors of aircraft should be in pale, pastel shades, inducing a sense of relaxation and comfort. In lifts a reassuring directness is needed, which means natural colours, wood and metal, because the user is entirely captive, and wholly dependent on the machine. For medical machines, apart from the design improvements mentioned earlier, the colours should be recessive, giving an impression of cleanliness. White and cream are used to convey this effect and to display the lack of dirt, and the machines blend into a white or cream background to be less threatening to the patient. With machine tools, however, colours should make the equipment stand out, so contrast colours are used such as yellow or orange.

These examples show different emotions—tribal, sexual and physiological—which must all be considered in the finished design of any machine or product. Naturally, someone who likes or loves a machine, as opposed to someone who hates it or is indifferent to it, will relate better and be more effective with it. The engineer who is aware of such psychological factors will perhaps produce a machine which avoids either hatred or indifference—it will not be vandalised or neglected but instead well used and prized.

The other approach to the problem of vandalism is to make abuse of the machine itself less likely or more difficult. Design of machinery which is unattended or sited in public places must clearly take this into account. In contrast to the obvious and vulnerable siting of parking meters, the cash dispenser (Chapter 18.4)

Figure 11.1 Vandal-resistant lift control panel with illuminated pushbuttons. (*Dewhurst & Partners*)

does not flaunt itself and has nothing particular to attract the vandal. The development of flush control surfaces, as in the lift push-button design (Figure 11.1), makes them much less attractive to the vandal, who will move on to some less complicated target.

11.2.6 Laziness or 'I know best'

Taking short cuts to reduce work is a very natural tendency, and this should be taken into account in design, more especially when it may affect operational or maintenance safety. For example, when dismantling for maintenance must be carried out in a particular sequence in order to prevent damage, and it is possible to omit some of the stages, mechanical interlocking or interference should be built in to prevent an incorrect sequence being adopted. If this is not practicable, clear instructions are essential, explaining why a specific sequence must be used and the consequences of not doing so. If it is possible to reduce effort in the supervision of a control operation by, for example, tying down a 'dead man's handle', then the design should be modified to make this impracticable.

11.2.7 Secondhand understanding

Present-day society relies heavily—indeed almost exclusively—on advertising, public relations and the communications media for its understanding of machines and environments. Much of this information is therefore garbled and inaccurate. This is not necessarily the fault of the information, but rather because any secondhand information is bound to suffer by being compressed, simplified or merely misunderstood.

Some car users, for example, do not react well to seat belts because of an unwarranted fear of being trapped in the vehicle after an accident and of being burned in a fire. Yet the statistics show all too clearly that this is a minor risk compared with injury due to impact.

Nor is it surprising that the approach of many people to machines within their environment is marred or incomplete, since it is hindered by many examples of irresponsible communication which, for their own purposes, paint only one side of the picture. Examples can be found in many types of advertising showing, for instance, kitchen aids such as food-mixers being 'simplicity itself'. The reality is that such equipment is far more complex than the advertisements show, and the housewife may be easily misled by this. Rotary lawnmowers are shown as light and easily handled, so users are sometimes unaware of the dangers resulting from the spinning blade.

All the psychological problems which we have so far discussed can and do arise from this type of secondhand information. The result is that many users of machines will approach them with prejudiced ideas of their possible malfunction, disaster or danger. They will, therefore, not relate well to the machine, and it is the job of the engineer or designer to appreciate this possibility and attempt to reassure them. Where the engineer realises that the objective is not just to design the 'perfect machine', suitable for use by the 'perfect user', but a machine which allows for the psychological factors already mentioned, then he is some way towards overcoming a major design hurdle. There are a number of methods of counteracting the problems inherent in providing a machine which can adapt to human variations, and it is to these methods that the second part of this chapter will be addressed.

11.3 Design guide

11.3.1 Communication

It was suggested at the start of this chapter that the very communication between man and machine in itself represents the product. In order to clarify this, and to explain why psychological factors are so important in the design of a machine and its environment, some guidelines on what is meant by communication will now be set out, followed by a method of analysis which can be used to determine the psychological factors relevant to a particular design. The diagram (Figure 11.2) shows the areas that should be investigated before embarking on a design project which involves communication between man and machine. One cannot isolate the machine (A) from these other parts, because its own activity helps to create the emotion (C) which may alter subsequent communication between itself and the man (B). Note too that the immediate environment (D) must be considered in relation to the activity taking place inside it. Again, this can have a bearing on the way in which the machine is designed.

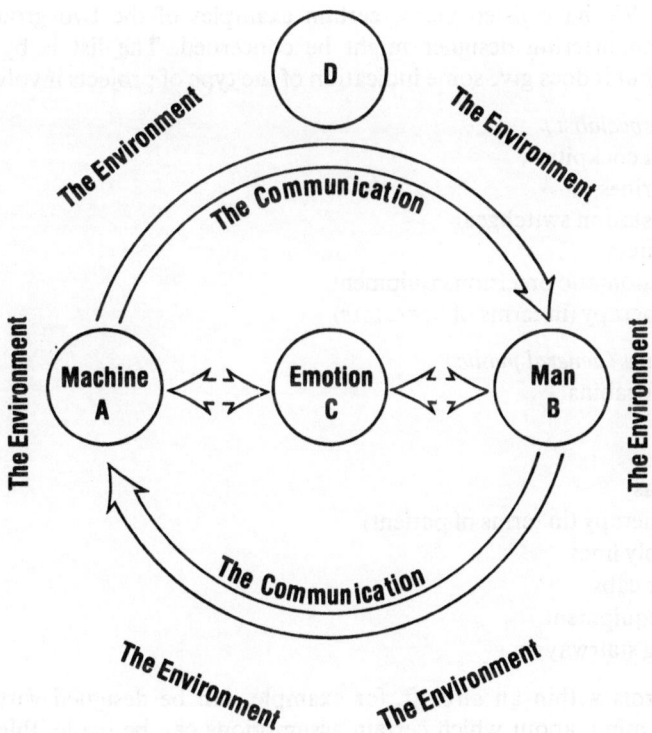

Figure 11.2 Communication considerations in a design project.

The majority of problems encountered in transactions between man and machine stem from inefficiencies in one or the other. There are, however, genuine communication handicaps which must be considered, and these fall into two categories—predictable and unpredictable.

Some *predictable* problems include:
　Inability to change eye focus in later life
　Need for sleep, tiredness
　Different languages in use
　Colour blindness, claustrophobia and other medical conditions
　Fear of certain equipment
　Weak hearing
　Slowness of mental reaction

Some *unpredictable* problems include:
　Breakdown of sight, misunderstanding
　Mental or physical breakdown
　Lapse of concentration
　Machine malfunction

It is important to establish the identity of the user or audience in considering psychological factors in design. The two categories commonly used are *captive* and *non-captive*. We have listed below certain examples of the two groupings with which the engineering designer might be concerned. The list is by no means exhaustive, but it does give some indication of the type of projects involved.

Captive (specialist)
　Aircraft cockpits
　Submarines
　Power station switchgear
　Computers
　Non-automatic precision equipment
　Deep therapy (in terms of operators)

Non-captive (general public)
　Aircraft cabins
　Lifts
　Cars
　Cameras
　Deep therapy (in terms of patient)
　Assembly lines
　Tractor cabs
　Office equipment
　Moving stairways

The controls within an aircraft, for example, can be designed with a captive audience in mind, about which certain assumptions can be made. Pilots are well educated and trained, medically tested and competent, so it is possible to establish at the outset performance criteria for the user. Psychological factors can then be studied within this controlled framework.

In the design of a car instrument panel, on the other hand, one is clearly dealing with a non-captive audience—the general public. One cannot assume intelligent understanding or a particular level of performance on the part of the user. The design must work effectively for all, rather than a specialist or an élite, educated group. This also means that one cannot design for individual psychological factors but must work to averages. There is a method of doing this effectively, but first let us look at the situation of a machine with a captive user.

11.3.2 Captive user

In general terms, when man and machine are in the captive category, the psychological problems will stem from the captive situation itself. The more controlled and specialised the environment of the man/machine interface is, the more important it is to keep the user from becoming either complacent or inefficient owing to pressure.

A monitoring system in which nothing happens will create lack of interest, wandering concentration, boredom and finally the operator may go to sleep. An alarm or something out of order may well escape notice until it is too late. The answer is to create an activity, even if it has no immediate purpose, in order to interest the operator and keep him alert. At a new major London brewery (Figure 11.3), psychological factors connected with the staffing and manning of the main control panel have been studied carefully. The control panel shows the influence of a design which takes account of the individual. Flat control panels for operator use, as opposed to simple display panels, are used to create a sense of involvement.

Figure 11.3 Brewery control panel.

The controls themselves are directly illuminated and the panel is sectioned for individual use. Control buttons are large and obtrusive, not flush and discreet. Although the plant is fully automated, the punched card control system is visible and active.

Employees were asked to undergo psychological tests before working in these isolated conditions. Regular visits by parties are actively encouraged, as is questioning of the operators. A series of brewing checks is carried out by the staff and several of the control operations continue to be done manually, although this is unnecessary from a technical point of view. It is well established that a man concentrates best during the first 20 minutes of any activity. After that time, concentration begins to wane. Where extended concentration is required, there must either be suitable breaks or an alternative task. For example, the high level of concentration required by airline pilots is supplemented by an autopilot and a co-pilot to provide suitable alternatives. And in the same way that a co-pilot needs to establish friendly relations with the pilot, so also must the automatic equipment be perceived as the operator's friend. A piece of equipment which looks badly designed is obviously going to receive less attention than one which appears designed to appeal to the senses, as well as functioning efficiently. An example of this consideration is amply demonstrated in the Chessel recorder (Chapter 18.7).

11.3.3 Non-captive user

Where there is a non-captive audience, it is the *freedom* of the user which creates the problem. Total freedom, although apparently appealing, is rarely the choice of the majority. Most people prefer to operate within certain concepts, philosophies, groups and organisations.

Faced with a product aimed at a non-captive audience, the designer must try to categorise the target group, since it is impossible to use design criteria applicable to the individual in such a situation. No groupings will, of course, contain a complete set of average people—the averaging comes when the variations within the groups have been resolved—but there are examples of well researched groupings which will provide considerable design data. Socio-economic groupings, for example, based on a social and, to some extent, financial scale, are often used for product design and development, especially by people involved in advertising and marketing. For the designer, however, it may be better to deal with more immediately apparent groupings.

Some examples of convenient groupings for the purpose of engineering design are given below.

Trained users: Control and operation can take account of instruction, rather than totally simplistic controls for first-time user.

Scientists: Assume well educated and therefore capable of understanding complicated operation.

Manual workers: Clear information, straightforward controls. Clear fail-safe provision. Repetitive work. Allow for interruption to process. Rugged.

Housewives: Bright, attractive, easy to clean. Clear assembly and dismantling.

Businessmen: Elegant. Sleek. Compact. Expensive-looking. Captive user controls. Can be complex but clear.

Secretaries: First-time operating instructions. Allow for carelessness or boredom. Design for environment.

Travellers: Fast operation. Clear warnings. Step by step activation. Considerations of fear, claustrophobia etc. Foreign users.

Disabled: Cosmetic, unfussy. Avoid confusion. Slow operation. Single operation at a time.

The particular status or opinions of the group concerned may play an important part in engineering design. For example, specialists who use a complicated piece of equipment, such as an electron microscope, tend to have very individualistic ideas of what is needed. They are also often in a position of authority to insist on what they want, so that flexibility in design to allow for individual ideas, perhaps by the use of modular construction, would be a valuable asset.

Another class of grouping can be identified with the habit and emotional characteristics associated with nationality or geographical climate. To take one example, the English, the French and the Italians all have different reactions to their cars and these are reflected in the design criteria. The English regard the car as an extension of their physical character—almost another limb. The French are often more detached and sanguine about machinery, not necessarily developing a love or obsession with the vehicle. The Italians like to put their cars in another class, with both elegance and performance the keynote. They like to be surrounded by elegance, and also first away from traffic lights! In domestic appliances, too, French design tends to be more utilitarian and practical, perhaps in the belief that no one loves machines, while English designers assume that the objects will be cared for, cleaned and polished.

Habits are another important source of group psychological pressure. Fear of the unknown or the unusual often plays an important part here. In Europe, for example, no single design of coffee-making machine has obtained universal approval because the way people prepare coffee varies greatly from place to place.

Similarly, in under-developed countries people's long exposure to simple ways of life to some extent make this a habit, and basic equipment will often be appreciated and provide a better performance-to-cost ratio in service than will more sophisticated designs. As an example, see the very practical, but basic, operating theatre table (Figure 11.4).

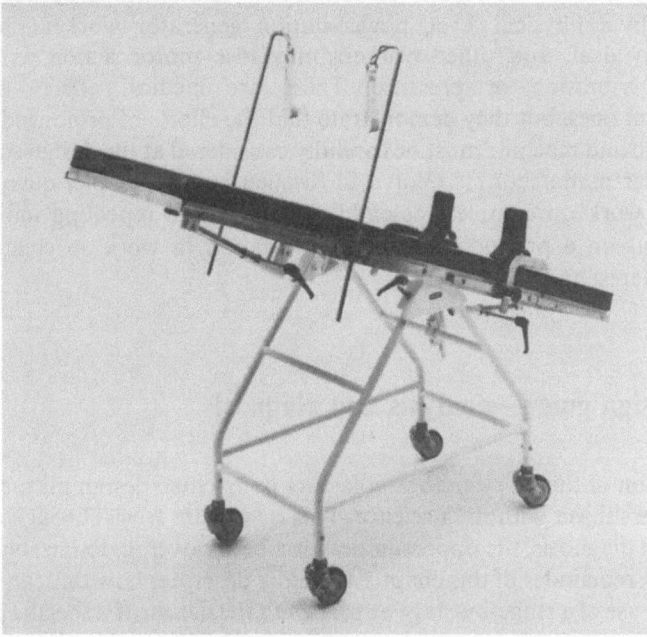

Figure 11.4 Seward Minor operating table. Design Council Medical Award 1975 (A. J. Seward & Co, designer David Reed).

By thinking about groups and observing them, therefore, the designer will begin to appreciate psychological factors and pay more attention to them. But one further problem arises when it becomes apparent that it is not solely the machine and the user who create the interface.

11.3.4 The environment

In certain situations, the machine alone is not going to be responsible for the emotions which are generated in the user. For example, a dentist's surgery containing a number of machines creates in the patient a pattern of fears, principal among which is the fear of pain and violent attack. It is made all the more frightening by the sense of being a victim, condemned to pain and suffering, without recourse to escape.

Machine design in itself will not directly affect this emotional reaction. But an advantage may be gained if the equipment is designed to look efficient and not too obtrusive (more fear might result if it were to be made altogether invisible). The colour scheme of the room, the acoustics and the lighting can all contribute to the emotional reaction of the patient, as indeed can the demeanour and dress of the dentist himself.

By placing the product to be designed in the context of its intended environment and by relating it to its surroundings, the engineer can build up a far better picture of its relationship with the user, patient or operator, and the psychological reactions which may be expected.

11.3.5 Duration

The way in which fears can be built up over a period of time was described earlier. The same is true of the results of continuous relationships between man and machine. On a physical level, power-station generator workers may become progressively deaf, and other workers may lose motor action as a result of continuous vibration or pressure. These are medical effects rather than psychological ones, but they demonstrate that the effects of prolonged interaction between man and machine must be carefully considered at the design stage.

Certain car manufacturers (Volvo in Sweden is an example) now arrange for workers to work on complete assemblies, rather than repeating individual part assemblies as on a production line. The variation in work so created helps to maintain interest and also efficiency.

11.4 Design guide—analysis and diagnosis

Consideration of these psychological factors in machine design may, at first sight, seem to be a subject without a science. This is far from true. Through a process of analysis and diagnosis, the opportunities for a better overall design solution can be created. The remainder of this chapter therefore describes how this can be achieved through the use of a stage-by-stage approach to the design of a specific product.

Step 1: Philosophy

Great as advances in engineering development may have been, there is still only a vague appreciation of the need to establish objectives which are clear, unambiguous and follow a logical pattern. There may well be objectives which are implied or assumed without being properly considered or explained. It is essential, therefore, to start any project by specifying the *objective*, in terms of an understanding of the exact nature of the man/machine interface and how it is intended this should operate; and the *machine*, in terms of how it should or does operate (see also Chapter 16). Often it is the second aspect which entirely controls the design, with the man/machine interface being considered of secondary importance, if at all (Figure 11.5).

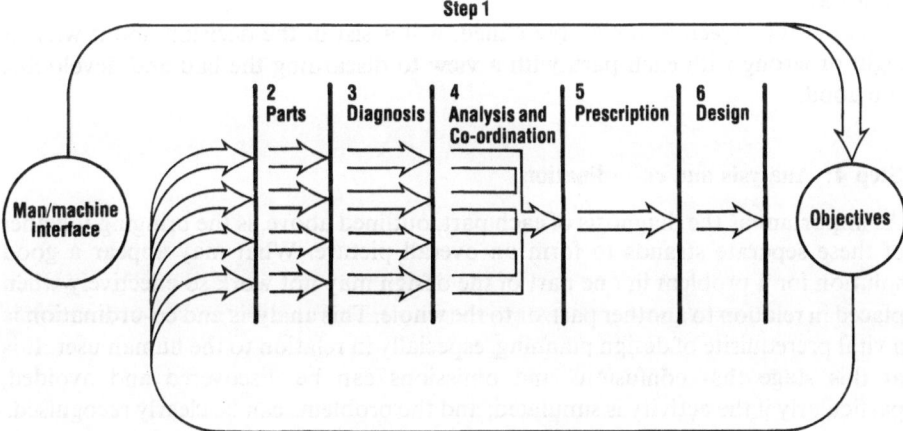

Figure 11.5 Stages in the design process.

Step 2: Parts

To study the objectives properly, it is necessary to break down the relationship between man and machine into parts, each of which can be examined separately. It is impossible to study the relationship as a whole (which may be why it is often not studied at all).

It is of overriding importance that these separate studies are in themselves controlled and directed by *the objective* which, as is shown in the diagram, is common to all the parts. By analysing, not in isolation, but with the objective constantly in view, the diagnosis and therefore the prescription will be kept as relevant as possible.

The ability to reduce the man/machine interface into parts is an important yet underrated design activity. As a guide, here are four distinct and different aspects which can be studied.

1. *Transfer of information from machine to man*: information can only be carried in a certain way—ie sight, touch, sound, smell—and will be carried to the brain through one of these senses or by a combination of them (the co-ordination of any combination is vital for the efficiency of the transfer).

2 *Response of the brain*: this can be anything from excitement, understanding, love or warm reaction to indifference, sleep, misunderstanding, fear, depression, anger, violence, phobia or confusion.
3 *Status of the person*: large, small, blind, deaf, stupid, intelligent, slow, quick, male, female, child, elderly etc.
4 *Status of the machine*: good, bad, indifferent, efficient, outdated, ugly, dangerous, safe, frightening, etc.

Step 3: Diagnosis

A diagnosis in this study will establish two characteristics in relation to the psychological factors involved in machine design. These could be described as: the *relative strengths* (the machine is clear in display, unambiguous etc) and the *relative weaknesses* (inhibiting factors for users, ambiguities, complexities, confusions etc).

Again, the *objective*, if clearly defined, will assist in the decision about what is right or wrong with each part, with a view to discarding the bad and developing the good.

Step 4: Analysis and co-ordination

As important as the diagnosis of each part, outlined above, is the bringing together of these separate strands to form an overall picture. What may appear a good solution for a problem in one part of the design may not work so effectively when placed in relation to another part, or to the whole. This analysis and co-ordination is a vital prerequisite of design planning, especially in relation to the human user. It is at this stage that confusions and omissions can be discovered and avoided, particularly if the activity is simulated, and the problems can be clearly recognised.

Step 5: Prescription

A prescription or brief for better man/machine interface is now possible, with the psychological factors affecting the different parts being brought together to ensure that the overall objectives have, so far as is possible, been complied with. This will create a correct basis on which the design can proceed.

Step 6: Design

The design itself can now be approached in relation both to the psychological objectives and to the technical considerations of the machine, as outlined above. By carrying out this diagnosis, rather as a doctor would do with a patient, there is a better chance of overall success with the relationship between man and machine.

12 Design of control components
Noël London and Howard Upjohn

12.1 Introduction

Virtually every machine we design, be it a simple cutting or levering tool or a complex device, requires a man/machine interface. In the simple case this interface takes the form of a handle and in complex devices the handle develops into a control panel. A further development is the fragmented interface, where the need for a variety of complicated and near-instantaneous inputs requires a number of controls to be operated by hands and feet. Hand-crafted implements which have evolved over centuries have handles which are both comfortable and elegant and in which the material is expressed in a form that recognises human contact. In more recent times 'ergonomic' handles have been developed in which the needs of the user have been allowed at least equal priority with the requirements of manufacturing processes and materials.

Control panels are produced by grouping the required display and control components together into an organised layout. The simplest group is a single lever plus one indicator, and this can grow to the complex array of an operating console or even the multiple control stations of a space-mission control centre.

These also have developed from the traditional, symmetrical approach, through centreline-dominated layouts, to the flow-line groupings which at least pay lip-service to an elementary operating sequence.

The usual task facing today's designer is to simplify the control interface, both in front of and behind the panel, and to produce a comfortable, easy-to-use design.

Control interface details can be evaluated as a part of the overall design process, both in purely aesthetic terms of form/contrast, pattern/rhythm, tone/texture, colour/hue balance and on the pragmatic grounds of material/process, ergonomics/cost-effectiveness, quality/ease of production, not forgetting of course the essential marketability of the product.

We can define the main categories of the control interface as tactile and visual (auditory alarms are discussed elsewhere) and these categories can be conveniently divided as follows.

Tactile/heavy—requiring reasonable physical effort. These are the levers, pedals and wheels used in operating heavy engineering products.

Tactile/light—requiring little physical effort. These are the toggles, knobs and push-buttons used to control electronic or electromechanical devices.

Visual/display—the displays giving information. These include analogue meters, digital displays, pilot lamps, CRTs etc.

Visual/graphics—relating to the design layout using pattern, tone, colour, line and typography to produce a distinctive panel presentation.

The tactile elements provide the means by which we instruct our machines. The visual elements give us information, either in response to the instruction or to enable monitoring of machine status. In general, all the tactile control elements (with the exception of recently introduced touch types) function as either pivoting or rotary levers. In basic mechanical terms a lever is a means of deriving a mechanical advantage where energy is being expended to carry out work. In their simplest forms, levers are classified as first order, second order and so on according to their mechanical advantage. The science of ergonomics has provided a great deal of information which must be studied if the designer is going to optimise the components described in this chapter, and these aspects of design have been discussed in Chapters 2 and 3. Here we are concerned with a wider philosophy, including the influence of ergonomics, which concerns the detail design and selection of controls.

12.2 Tactile/heavy

12.2.1 Hand levers

The category of tactile/heavy hand levers can be sub-divided as follows.

Hand-operated linear-action levers with a fixed pivot point operating through an arc in one plane.

'Joysticks' with a universal pivot operating through a conical field.

'Gearfield' action levers in which the operating cone is restricted by gates.

Although ergonomic aspects are dealt with in Chapters 2 and 3, the following points generally apply to these levers.

Maximum forces can be applied to levers gripped at shoulder level when standing, but at elbow level when seated.

Maximum applied force should not exceed 14 kg, although a momentary pull of up to 110 kg is permissible.

Push is greater than pull and both are greater from a sitting position than from the standing position. In both cases the fore and aft movement is referred to. Pulling across the body—ie side to side—allows for higher effort but less precision [12.1]).

Within anthropometric limits, the greater the length of travel the more precise is the control. As the length of travel is increased, the effort will be decreased proportionately.

Two further variants of both the pivotal and gear-lever hand levers can be listed as follows.

Hand levers (one-plane arc) which positively engage and disengage at extremes of travel, such as a dog-clutch control lever.

Hand levers (gate-restricted cone) which positively engage and disengage at extremes of travel, such as an automatic gearbox selector lever (Figure 12.1).

Hand levers (one-plane arc) which increase or diminish a control function incrementally over their arc of travel.

Hand levers (gate-restricted cone) increasing or decreasing function incrementally over their travel arc, such as fork-lift truck controls (Figure 12.2).

12.2 Tactile/heavy

Figure 12.1 Control for automatic gearbox. (*Vauxhall Motors Ltd*)

Figure 12.2 Fork-lift truck controls. (*Lansing Ltd*)

The following points relate to some aspects of these control levers.

Action and reaction to lever control may be predictable, such as a move from neutral to drive. The action is deliberate, the reaction is expected, and the machinery has carried out a command as programmed.

Alternatively the action may only put the machinery into a 'ready' state where its reaction can be varied by the operator moving a lever. Controls in this category require much more ergonomic study and must take environmental relationships into account.

There are few guidelines for the design of joystick controls other than the speed with which sweeping movements can be made, offset by the difficulty in achieving fine adjustments. As both left and right hands can be trained to give comparable performance, the joystick is better divided into two controls having x and y functions for two-dimensional operation, and combined with a foot control for three dimensions.

Data sheets are available which provide basic design guidance for a variety of controls, including levers. Reference to Chapters 8 and 9 will suggest appropriate materials and finishes. Other sources of detail design information are available [12.2, 12.3, 12.4]. Commonsense points to consider are generally those associated with comfort. Some levers, for example, may connect with a gearbox running at an elevated temperature, in which case some form of insulation is desirable.

12.2.2 On/off levers

In the simplest case of the operation of an on/off lever the limits of travel may be a sufficient indication. Alternatively the operation may have a visible or audible result. In some instances, however, lever operation must be interpreted through a display.

The display may be in the form of a signal lamp or pair of lamps, or any one of many systems for status indication.

Where levers are moving through an arc in any given plane, for example up and down, side to side, or fore and aft, the accompanying signals should be parallel with the line of operation.

Where the operation of a lever is more subtle and is attenuating an operation, the indicator, scale or other read-out should also follow the line. The convention is clockwise for increase, imagining the lever as a spoke in a wheel.

12.2.3 Feedback

It would be safe to predict that in the future the effort required for lever operation will tend to level out across the whole field of engineering, so that eventually the input force required is no more than that required to 'feel' that there is a response. This is an important point, and applies to all other controls in this chapter. It is part of the march of progress that machines will do more of our work for us, although a sense of deprivation may make us hesitant to welcome this. It is human nature to enjoy the feeling that our work responsibility is being transferred, but we are also reluctant to give up total control. Thus a lever, or any other control which may silently and effortlessly command a machine function, will actually be disliked if there is no resistance to its operation or feedback, as well as being more difficult to use as a control.

On a practical level, levers in frequent use, for example those controlling mechanical excavators in quarry conditions, must not become less efficient because of perspiration or grit. Abrasion from handling is remarkably rapid, so the handgrip must be able to be renewed easily. This also implies that integral legends should be given special consideration. The designer must also know if cold conditions are a possibility and the operators' hands will be gloved. Mock-ups should be evaluated using industrial gloves and handgrips modified accordingly.

12.2.4 Combination controls

Levers incorporating more than one function are well known, the motor car handbrake with its release catch being a widely used example. Less obvious perhaps is the motor-cycle handlebar with its twist-grip throttle. The handlebar is basically a lever and the throttle is a superimposed control. Combinations of this sort can serve a useful purpose where space is limited, where the operator's access is restricted or where there is a profusion of controls. Combined functions need not be given exaggerated forms to assist the newcomer; good design can suggest the function without this. Spacing levers correctly is most important. Be familiar with the clearance needed for the hand and make sure the lever travel is positively limited if it is operating between obstructions. Where several levers are in a line, as in the case of hydraulic controls for fork-lift trucks, excavators or cranes, care should be taken to arrange the control handles for convenient and safe operation based on ergonomic considerations. This may require cranked levers.

12.2.5 Space considerations

Levers by their very nature usually have to project into and move through space. Anticipate the extent to which they may be vulnerable. For example, they may catch clothing or they may be at a height which can contact the more sensitive parts of human anatomy or risk accidental movement from a vital setting. If the part to be actuated cannot be relocated and any directly attached lever would emerge in an awkward position, consider linkages or remote coupling. Many companies specialise in low-cost, reliable components. Look first before you design special piece parts!

12.2.6 Foot levers

Of the different categories of lever only the first, the linear-action lever with a fixed pivot moving through an arc in one plane, is really suitable for foot operation, and since the human foot does not have marked prehensile properties (particularly in an industrial safety boot) spring return is obligatory. In most cases pedal movement should depend on the operator's intentions and in these cases the residual spring force in the off position should be adequate to overcome any gravitational force applied to the pedal as the result of the attitude of the operator, ensuring that positive action is needed to cause pedal movement. The way in which pedals are used to control a motor vehicle shows that the feet are capable of sensitive control over prolonged periods if the direction of pedal travel obeys specific rules.

The pelvis can be considered as a stationary part of the human structure, the thigh having a moderately 'universal joint' with the pelvis. The lower part of the leg will work in one plane only from the knee joint, and the ankle, not being a ball-and-socket joint, is limited to some extent. Dexterity with feet is not biased in the same way as that of the hands and, although footballers may be more positively left or right footed, there is no conclusive evidence to show that pedal design need be biased as a result.

The complete use of the human transmission system from the hip joint to the toe is utilised in bicycling.

There are three types of pivoting pedals:

Pivot point at heel, thereby resting the weight of the foot on a solid surface;

Pivot point central, rests the weight of the foot but allows for precise incremental forward/reverse control with light bias spring (variations of this type place the pivot nearer the heel for some uses); and

Pivot point at toe, or top of pedal allowing the entire leg to exert a considerable force.

These pedal types would normally operate linkages or plungers to actuate the operating levers but a direct lever pedal is the *pendant pedal* with pivot point either above or below floor surface.

In each type the design of the pedal surface is dictated by general operating conditions. The traditional rubber pad is far from being suitable in all instances. Coarse open grating of a generous size would be more appropriate for operators of earth-moving equipment, who may be wearing heavy rubber footwear covered in mud. At the other extreme a control pedal for height or tilt adjustment of an operating table will need a small surface area incorporating a smooth but non-slip pad. Pedals of this type, when they are mounted above floor level, should always be limited in travel so that they cannot come to rest on the operator's foot. There are still some painful foot-operated guillotines with bar pedals and industrial footwear with reinforced toe-caps does not always provide immunity.

For an analytical study of pedal design a useful device is the Lauru platform, which measures vertical, frontal and transverse forces exerted by the body as the centre of gravity is shifted during work. It was employed for analysis of pedal design and indicated that a pedal with the fulcrum under the heel on the line of the axis of the tibia takes less time and force to operate than certain other designs [12.5].

12.2.7 Rotary levers (wheels)

The third category of lever defined in the tactile/heavy group is the control wheel.

The principle of operation derives from levers. Again, the wheel provides a mechanical advantage which is based on the ratio between the outer diameter and the inner diameter at which the load is applied. Where a control wheel has a projecting handle the similarity is obvious, but the comfort or ergonomics of actually handling one ends the comparison. A control wheel extends the degree of arc in a manner which is ergonomically appropriate. There are few applications where a lever is suitable beyond 180° travel. The wheel, however, permits uninterrupted transmission of control through 360°, multiplied by the number of rotations required.

So far as the motorist is concerned the steering-wheel is the most important control, since he must maintain sensitive manual contact with it throughout a journey. Superficially it might appear that a smaller diameter of wheel could be used on a car with power-assisted steering, but the actual requirements do not lead to such a simple solution. There are very many operational factors which influence the steering torque required by a car, the desirable degree of feedback from the servo system, and the extent of movement of the wheel rim that, in combination, will optimise control. The whole background of mechanical, ergonomic and physiological factors involved must be investigated by the designer before he can determine the most favourable size.

It is worth noting that, throughout many years of automobile design and development, the wheel has not been superseded by any other type of control. Certainly some unconventional shapes have appeared from time to time, but one suspects that the well-meaning stylist has been given too much liberty in such cases.

We might also wonder why the precise positioning of anything from a Mini to a multi-ton transporter is best served by a wheel. The answer seems to suggest that the action and reaction available to us due to the co-ordination between our eyes and both hands is of a very high order.

It therefore follows that where a control function requiring a high degree of co-ordination is known to be present, a control wheel should be considered. Before numerical control takes over completely in milling machines, lathes and other production machines it seems that scope for better design exists throughout many current capital goods. This is an area where convention has done much to stifle progress.

The best examples of man/machine relationship in the automotive field, for example, are found in the few examples where the steering-wheel is adjustable in at least two modes and the driving seat has additional adjustments. The increasing relaxation of limbs and trunk as mental adjustment develops while driving means that the initial 'optimum' setting of steering-wheel and seat should be able to be varied over the course of a normal journey. The designer should acknowledge that there is no such thing as an optimum solution to the problem.

12.3 Tactile/light

The second category, that of tactile/light controls, comprises toggles, push-buttons and knobs. With this change of scale we move into the infinitely more fashion-conscious world of consumer goods and rack/bench instruments. The hi-fi look controls many lo-fi devices.

12.3.1 Lever switches

The category of finger-operated linear-action levers with a fixed pivot point operating through an arc in one plane includes one, two or three-position switches with an actuator style ranging from a metal, ball-end toggle via a vaguely phallic-shaped plastic whimsy through tab levers, short and fat or tall and thin, to rocker actuators plain or fluted. And that just touches the surface. Among the host of proprietary components there are some good actuators which should be compatible with the character of most control panels.

The rocker switch is a less projecting adaptation of the toggle switch, though less positive in signalling its setting. Both types may be with us for some time for primary-function on/off controls because of their ability to handle large electric currents. While electronic engineers have an understandable preference for siting these as near as possible to their power input sources, the user may often be annoyed if this means he has to go searching for a button or other control before he can start up.

12.3.2 Rotary controls

The second group in the tactile/light category, that of rotary items operated by control knobs, divides as:

> Rotary control, finger-operated actuators with smooth (non-stepping) arc of movement (eg potentiometers); and
>
> Rotary control, finger-operated actuators, step-divided arc of movement (eg rotary switches).

Control knobs are used most widely in association with electrical or electronic equipment, and we are therefore in the range of finger operation. The direct sensory feedback of the human fingers is incredibly selective in the tactile aspects of hardness, temperature, vibration, plasticity, texture and movement in any direction, be it discrete or gross and at any speed.

Knobs generally imply a rotary action and the design of knobs should make this obvious. It is irksome to rotate a control knob only to find that you have unscrewed the top from a cleverly disguised lever control! Certainly knobs can be used in a push/pull mode, but the context should be sufficiently unambiguous to permit the distinction. Some clutch knobs depress to engage so that a setting cannot be changed accidentally.

Rotary switches with removable 'latch-keys' are sometimes used on microprocessor-controlled equipment. This is a safety measure to ensure that only an authorised key holder can break into the memory or other functions of a similarly critical nature.

A great deal of research was undertaken in the late 1940s on variation in the shape of control knobs for ease of tactual discrimination [12.6]. As a result, many unconventional shapes were in fact adopted as standards for aircraft cockpits. There has been ample evidence that the reduction in human operating error was significant, but the equipment designer is faced with a dilemma inasmuch as these 'unusual' shapes were not found to be acceptable in the wider market. Thus we still find hi-fi equipment with rows of identically shaped knobs, and scientific equipment and capital goods more often than not follow the same fashion. Manufacturers of control knobs depend on the market and it is worth studying some of their catalogues to see how conservative this has become. It is up to the designer to overcome the problem, if problem it is, of providing identification by other means when necessary.

The Photoplan microscope (Figure 12.3) is an example of design in which a large number of small controls have been given consistent integrated treatment.

Whatever the shape of the knob or material from which it is manufactured it usually has to be attached to a shaft or spindle. In electronics, diameters may be from 6 mm down to 3 mm. The main alternative methods of knob-to-spindle fixing are: clamping with a multi-jaw collet; pinching with one or two side grub screws; clipping with a spring device; and force fit—as with flexible plastic mouldings.

Collet fixing has a high degree of popularity at present because the method takes up reasonable tolerances and maintains concentricity. Settings can be

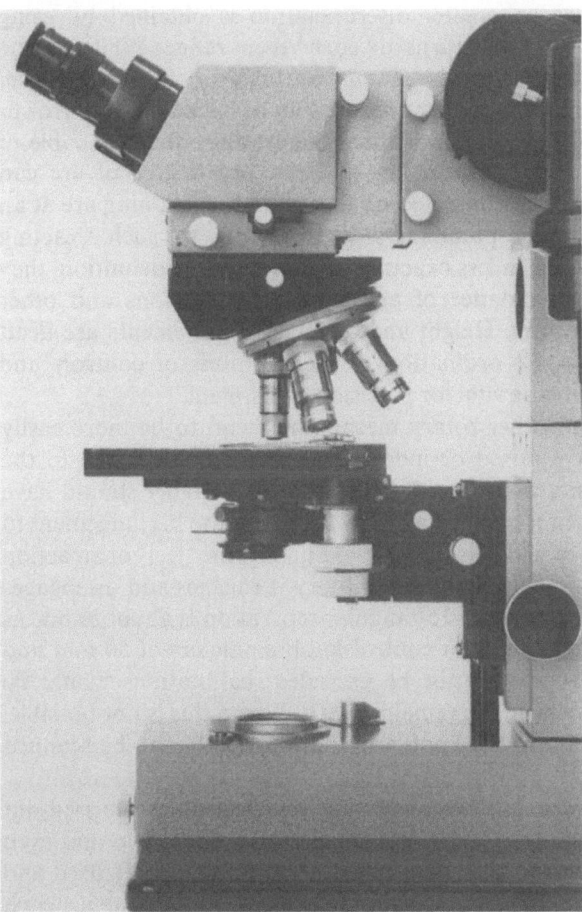

Figure 12.3 Vickers Photoplan microscope Type 41, showing co-ordinated design of control knobs, levers and wheels.

adjusted rotationally and axially, and when locked there is little risk of damaging the spindle. Considerable locking forces can be applied and, if these are overcome, the slipping can prevent damage to components. The collet mechanism uses as many as three components, but in quantity production automatic turning machines keep prices competitive.

Grub-screw fixing, particularly when two screws are used on a 90° spacing, has long been considered the engineers' favourite, but has disadvantages. Initial positioning provides adequate flexibility but spindles can be indexed by grub screws and removal and replacement can be affected. Vibration may also make screws come loose.

Patent fastenings are numerous and most efficient within their torque limits. Spring steel clips of various designs are used in conjunction with the natural engineering qualities of plastic to provide the necessary grip. Nearly always these fixings are pushed on to the spindle and removed simply by pulling. In some cases their effectiveness is improved by calling up a spindle of precise length with a flat or D section specified. Where control knobs are operating indexed switch functions this can lead to imprecise pointer location.

The two most commonly used components requiring the rotary knob are potentiometers and rotary switches. Potentiometers are generally smooth in operation, requiring little effort to operate and often functioning through an arc of

270°. In some instances, however, greater discrimination is obtained by using potentiometers which need up to ten turns to cover their range. While torque requirements are minimal, extremely precise manual setting may often be required. For example, on an analogue display, a meter needle may be backed by a mirror to give a parallax-free reading of 0.1 per cent (which the eye is perfectly capable of interpreting). The movement of a control knob within one degree of arc can position the needle accurately if the control knob design and positioning are at an ergonomic optimum. The design parameters which apply to such exacting operations should also be applied to less exacting operations if, by definition, they represent the optimum. The avoidance of projections, castellations and other stylised patterns is most important. Height and diameter requirements are dealt with in published data sheets. Co-ordination of the positions of controls and arm-rests, using non-slip textures, is vital for precision adjustment.

Indexed rotary switches and other rotary mechanisms tend to be more easily operated if control knobs have raised or indented grips running parallel to the shaft axis. Torque requirements vary considerably, and the designer should have the opportunity to evaluate components at an early stage and use his judgement to reject or modify any mechanism which is too heavy. Equally, the 'feel' of an action is also affected by the noise it makes or how positively it engages and disengages over its arc of travel. On this latter point 15° angular separation is about as fine as one would wish to go when dealing with control knob diameters of 50 mm and below. A total sweep of 270° should not be exceeded; calibrations would be difficult to read in most situations in the remaining 90°. Where this is not possible, it is better to add a skirt or dial to the knob and the whole 360° can be scanned against a fixed pointer.

It was stated in Section 12.2.4 that more than one function can be carried out by a given control. This is true of an array of control knobs where two and even three concentric tiers are employed. The dual concentric is more widely used and the lower control, that is the knob nearer to the panel, may often operate a switch action and is normally larger in diameter. The upper control knob will normally operate a potentiometer via a small spindle running through the main shaft. The purpose is very often to attenuate the preset values determined by the switch (see Figure 3.15).

12.3.1 Push-buttons

The final major category in this section is the push-button. This probably represents the element most likely to dominate the tactile interface of the future. The range of types and applications of push-buttons is large and continues to grow. This may well be because prodding with a finger involves the least physical or mental effort and there is some psychological satisfaction in knowing that so much can be achieved by so little.

It would be reasonable to suggest that the primary use of push-buttons will be in power supply on/off functions. By the same token the secondary usage will be on/off switching of functions other than power supplies. Third will be step-functions where, for example, a given range will be subdivided into increments selected by push-button. This was the role of the rotary switch for many years, but the technical and economic advantages of direct attachment to printed circuit boards has phased it out to a large extent.

12.3 Tactile/light

The most recent advances in printed-circuit development permit switching functions to be carried out merely by pressing a marked area on a panel (see Figure 8.9). The input effort is mainly mental because the physical effort is negligible. The output or feedback is, so far, only visible in terms of a change in a display or other piece of illumination. Since most of us when pushing buttons have become used to some form of movement plus some spring resistance and perhaps some audible confirmation that the task has been performed correctly, as for example in some pocket calculators, the 'mute' switch may take time to become acceptable.

Printed-circuit technology and component development have led inevitably to increased compactness in instrument design. An example of trends, evolving from the pocket-calculator keyboard, is the control panel with function group or groups, digital code bank and display read-out, already in use in the microwave oven (Figure 12.4), which foreshadows the form that will become standard practice for

Figure 12.4 Microwave oven, with controls derived from pocket-calculator practice. (*Toshiba Ltd*)

many products in the future. As more functions are packed into smaller areas and more controls are needed to utilise them, designers regret the fact that the hand of man, dimensionally speaking, remains the same. Push-buttons mounted in rows have been squeezed down to pitches of 12 mm and even much less, and many ingenious methods have been devised in order to make manual operation still possible. Study of pocket calculators and digital watches will illustrate this. Linked buttons, where a depressed button is released by depressing another button in the same bank, are gradually disappearing as microelectronics come into circuit design. The essential advantage of the mechanical switch, however, was its clear status indication obtained without the need for lights. The piano-key arrangement gave better contrast than most others in this respect.

Push-button travel can range from zero, with capacitance types for example, to a maximum of 7 mm with mechanical movements. Pressure requirements range also from zero, to a more usual level at 70 to 180 g where mechanisms are involved. Buttons may be typed as flat, sculptured, twin-shot moulded, engraved, illuminated, foil printed, and insert printed. The majority of control buttons are made from plastics and therefore the techniques available for moulding should be used as a design parameter.

The question of 'feel' with respect to the tactile and audible control response is dictated by mechanical operation. This has already been mentioned in respect of the 'mute' switch. However, there is evidence that strong preference exists for an action that offers definite resistance to finger pressure followed by an abrupt reduction in force as the contact is made. This preference is heightened by the addition of an abrupt sound which does not vary between buttons of a similar style. This can be a 'bleep' in series with all controls.

Before summarising and moving on to visual displays, we have to cover the area where the light/tactile control elements combine with the heavy. The use of push-buttons in heavy engineering seems to imply large components if one studies switch manufacturers' catalogues. There are of course logical reasons for this. Heavy engineering in general cannot maintain the clinical conditions found in the electronics industry. Suds, swarf, fumes, noise, vibration and extremes of temperature combine to produce environments which may be hostile and are always varying. Within such environments man/machine relationships set the designer extremely difficult tasks, particularly since man/man relationships are more often under stress in such situations. Push-buttons designed for hand operation may be shut off by a well aimed kick, with consequent damage to surrounding finishes when the kick is less well aimed. This is most often the case with smaller machine tools, where the switch-box is mounted low down and the statutory requirements of a large projection stop button contribute to the temptation.

Push-buttons are often required to be fully waterproof or oil proof and the use of rubber (or synthetic equivalents) for total enclosure is the conventional solution. As the flexible materials are produced by moulding processes it seems that opportunities for design improvement have been missed, and in many current bought-out examples the protection is simply a dome shape which gives no indication of the underlying switch head and thus the 'feel' is confused.

'On' switches should always be considered to be potentially more dangerous than 'off' switches. On heavy-duty equipment this is well recognised and a considerable number of regulations and recommendations cover their design. There is general agreement internationally and individual countries show only slight variations in the way the rules are set out. Legislation may not always be effective, however. For example, a test rig in continuous production use may require an operator to place a component on a jig and press a button in order that the machine will scan, test and measure to preset parameters. If, on this fictitious rig, the designer has provided two buttons sufficiently well spaced to need both left and right hand operation and, additionally, the system will stop immediately pressure is decreased on either button, then a potentially safe situation has been proposed, since the operator cannot put a hand into moving machinery. If, however, the positioning, size, surface texture, length of travel and effort required when pressing the buttons become irksome, and if a piece-work payment system is in force, the ingenious operator may well find that simply placing two weights on the buttons will give continuous operation so he or she has only to judge when it is 'safe' to change work pieces. The moral is that the designer must anticipate or imagine the most improbable ways in which his product may be operated and act accordingly.

12.4 Visual/display

The visual/display elements of the control interface provide the read-out and machine status information. The elements fall into three main categories.

Analogue displays, which include all types of meters and gauges. These are covered in some detail earlier in this book in the chapters on ergonomics and will not be treated again here.

Pilots, which are various types of a point light source, either self-coloured or capable of being colour coded by the use of a lens cap or filter. These lights are used to provide machine status information in both the 'on' and 'off' mode.

Digital displays, usually in the form of modular components which can provide numeric information. Alpha displays provide a screen upon which alpha information, from word symbols to complete text, can be shown. They range from miniature dot-matrix devices to large CRTs.

12.4.1 Pilots

Light sources not only provide status information but add life and sparkle to a control panel. Pilots can generally be subcategorised into filament, neon, and solid-state types.

The incandescent-filament low-voltage lamp in a lampholder, either miniature bayonet cap or Edison screw, with a range of coloured lens caps, represents the pilot type in longest use.

Neon indicators are generally used as mains pilots (with a series resistor) and overcome the size and limited-life problem of mains filament lamps. Both types, excluding the special requirement of the mains pilot, are being largely superseded by the solid-state light-emitting diode (LED). LEDs, available in red, yellow or green and with a light source buried somewhat deeply in the lens, are now available in several different shapes, and with both medium and high-intensity light sources.

12.4.2 Alpha-numeric

Liquid crystal and gas plasma type display panels are recent introductions which supersede the older and clumsier facilities which provided both alpha and numeric information. The older devices included both miniature back-projection component modules and light-source dot-matrix devices epitomised by the once-ubiquitous NIXI tube with its associated colour filter.

The recent display components are suitable for direct PCB mounting, as are, of course, the majority of today's control components.

12.4.3 Cathode ray tubes

Cathode ray tubes, both monochrome and colour, are widely used to provide flexible display information. Different sizes, from 100 mm up to 550 mm, all have their places in the display repertoire and the CRT offers the possibility of dispensing with light-source status display. The CRT display with input keyboard has developed as the standard computer interface and an example is illustrated in Figure 12.5.

Figure 12.5 CRT display with associated input keyboards. (*DATASAAB Ltd*)

Both green phosfors and amber phosfors have their champions, but there is little evidence as yet to suggest that either is superior.

A fundamental rule of physiologically correct vision is that we organise our visual field to place the highest luminance at the centre. Most display screens are darker than their surroundings and thus break this rule, but current studies suggest that good ergonomics and correct lighting of the work station will alleviate most of the problems identified.

12.5 Visual/graphics

The visual/graphics aspect of the control interface generally represents the area where the designer using proprietary components will have the greatest freedom for individual expression. Reference should be made to Chapter 7 for a comprehensive discussion of machine graphics.

12.5.1 Grouping

The quality of the panel design should evolve out of the techniques used to group controls and identify control groups. Grouping can be defined on the panel by outline boxes, tone-change patches, caption colour linking and line bracketing as well as by colour/tone grouping the tactile elements.

Pilot light sources can also be grouped, remembering that the lower-intensity modern light sources justify a dark ground around them. Modular pattern can be created, remembering that over-repetition reads as texture and not as pattern. The main parameters of control panel design problems will be the layout determined by PCBs and the overall aesthetic and ergonomic factors, and a certain amount of give and take is required during the design process. This will be better achieved if flexibility is possible on both sides of the panel.

The panel shown in Figure 12.6 is an example of the use of 'graphic patch' arrangement to aid clarity of presentation. The panel in Figure 12.7 is an example of arrangement to overcome the tactile problems produced by miniaturisation.

12.5 Visual/graphics

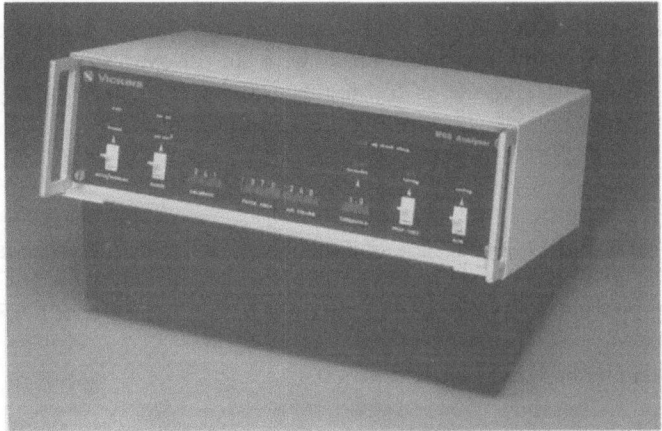

Figure 12.6 Clarity of presentation enhanced by the use of a 'graphic patch' technique on bench-top unit of M88 Analyser. (*Vickers Instruments Ltd*)

Figure 12.7 An example of the solution of problems posed by miniaturisation in the 2437 Universal counter timer. (*Marconi Instruments Ltd*)

12.5.2 Typeface

The typeface used to define the panel elements is also within the designer's choice, as discussed in Chapter 7. Although one can see examples where somewhat exaggerated letter forms have been used successfully, it is probably better to stick to the proven sansserif faces, Helvetica and Univers. Both of these faces offer a good range of choice from light through to bold and from condensed through to extended, and enable the designer to provide clarity and balance related to the specific problems of his product.

12.5.3 Production techniques

It is no longer necessary for control panel surfaces to be painted, engraved and filled, or even to look as though this process has been used. Photo-anodised aluminium plates with or without additional litho colour treatment, single and multi silk-screened first-surface metal or plastic, single and multi silk-screened second-surface clear plastic, melamine laminates and the more recent introduction of multi-layer display panels with integral light-source colour filters, touch-type push-buttons and display window apertures faced with an all-over smooth or textured surface skin, offer a wide choice of techniques which can express quality and achieve substantial cost savings over the old knife-and-fork methods.

12.5.4 Component colour

When deciding on panel colours a knowledge of appropriate standards practice is required. At the time of writing the standard most likely to affect thinking is IEC Publication 73 to which parts of BS 4099 (and German VDE 0199 and French NF 64410) relate.

A very general abridgement of IEC 73 is as follows:

Indicator lamps

RED	DANGER	—Potential danger or action
YELLOW	CAUTION	—Change of condition
GREEN	SAFETY	—Safe situation—proceed
BLUE	SPECIFIC	—Not covered by above
WHITE	NON-SPECIFIC	—Use if in doubt

Push-buttons

RED	STOP/OFF	—Stop all (or part)
YELLOW	INTERVENE	—Suppress abnormal
GREEN	START/GO	—Start all (or part)
BLUE	SPECIFIC	—Not covered by above
GREY, WHITE, BLACK	NON-SPECIFIC	—Any general function

Illuminated push-buttons

INDICATION	Light 'on' shows action required Depression will extinguish YELLOW GREEN BLUE
CONFIRMATION	Light when depressed, remains on May be flashed during run-up period WHITE

12.5.5 Name and rating plates

Corporate graphic features divide into two main areas: first are badges or emblems which express corporate identity through the product line; and second are technical labels which include operating instructions, mains supply information, serial numbering, ownership, servicing status etc.

The 'badge' treatment can range from an embossed, three-dimensional element set into a plated die-cast frame to a restrained two-dimensional silk screening or transfer direct on to a housing surface. It is always an important visual design element and can relate to a range as well as to an individual product identity.

The placing of this element is important in the same way as are the control elements discussed in this section, but there are here the additional requirements of range identity and corporate integrity. The designer must make the difficult decision of balancing aggression against timidity and vulgarity against self-effacement.

The second grouping, technical labels, are often ignored and consequently contain antique qualities of process, fixing technique and graphics. They should be considered as potential linking elements across a product range and a modular label breakdown and presentation should be the basis of the design approach.

A variety of useful processes are listed below, for both badges and technical labels.

Metal—polished chrome-plated die-casting with colour enamel infilling.

Metal—polished chrome-plated picture frame housing a variety of graphics inserts.

Metal—die-cast bronze or brass with colour enamel filling.

Metal—photo-anodised aluminium self-adhesive labels with or without additional colour printing processes. (This process also offers one of the insert processes for frames produced by the above methods).

Metal or plastics—silk-screened colour and copy on to die-cast, blanked or moulded carrier 'plates'.

Plastics—injection-moulded acrylics etc with paint and/or metallic finish processes.

Plastics—injection-moulded plastics wholly or partly metallised.

Plastics—photoprinting-produced plastic film labels with optional metallic-style finishes.

A large number of subcontractors offer a very wide range of types and processes related to quality and quantity requirements.

12.6 References

12.1   Hunsicker, P. A. 'Arm strength at selected degrees of elbow flexion.'
12.2   Dreyfuss, H. *The Measurement of Man.* Whitney Library of Design, New York, 1960.
12.3   McCormick, E. J. *Human Factors Engineering.* McGraw-Hill Book Co, New York, 1976 (fourth edition).
12.4   Van Cott, H. P. and Kinkade, R. G. (eds) *Human Engineering Guide to Equipment Design.* US Government Printing Office, Washington, 1972 (revised edition).
12.5   Lauru, L. 'The measurement of fatigue.' *The Manager*, vol 22, 1954.
12.6   Jenkins, W. O. 'The tactual discrimination of shapes for coding aircraft type controls' in Fitts, P. M. (ed) *Psychological Research Equipment Design.* USAAF Aviation Psychology Report 19, 1947.

13 Industrial design and safety

Professor Richard T. Booth

13.1 Introduction

A vital consideration in the design of engineering products is that the product should not endanger people using it. The need for safety is therefore incontrovertible, but there are disagreements between designers, purchasers, and government inspectors concerning the scope of the designer's responsibilities for safety, and the criteria that determine the acceptability of individual designs. In general, the standard of safety of many engineering products is well below the ideal. Two factors have led to this situation: until recently the law relating to machinery safety imposed duties primarily on the users of equipment; secondly, the law was (and to a large extent still is) concerned with one aspect of machinery safety, namely the provision of guards. In fact machinery accident prevention demands much more than the fitment of physical guards designed with the primary objective of meeting stringent but narrow legal standards.

The objective of this chapter is to review the scope of the designer's duties in legal, technical, and procedural terms for safe design, and to describe in detail and with the aid of case studies the application of these general obligations to real situations.

The dangers considered in the chapter are the risks of injury from moving parts (or otherwise dangerous parts) of machinery. The chapter considers the following areas.

> Legal duties which relate to designers of engineering products, and related obligations on users of such products.
>
> Identification of dangers and basic safeguard design.
>
> Risk assessment and approaches to control of danger.
>
> Design procedures.
>
> Review and summary checklist for safe design.

Certain aspects of safety are not covered in this chapter. No consideration (except in passing) is given to stress analysis and related studies to prevent structural and component failure. No consideration is given to dangers immediately associated with manufacture or demolition. Nor is reference made to measures to control of the emission of dust, fumes, and noise from engineering products since these are referred to elsewhere in this book in Chapter 4. In practice, however, a number of the procedural measures discussed below are just as applicable to these technical problems as to the matters discussed in detail here.

13.2 Legal requirements

The principal duties and liabilities in law which relate to the safety of engineering products are shown in tabular form in Figure 13.1. The law relating to product liability in general is vast and also currently in a state of flux, and no reference is made here to consumer protection legislation and the law of contract. The following points need to be emphasised when considering the duties imposed by the Factories Act 1961 and subordinate legislation, and the Health and Safety at Work Etc Act 1974.

Act or directive	Section of Act or subordinate regulation	Person on whom duty is imposed	Details of principal duties or liabilities	Character of duty
Health and Safety at Work Act 1974	6	Designers, manufacturers, importers, suppliers of articles	(i) To ensure that any article for use at work is safe when properly used (ii) To carry out any necessary testing and examination to fulfil item (i) (iii) To make available adequate information and conditions relating to the safe use of the article	'As far as is reasonably practicable'
		Designers, manufacturers of articles	To carry out any necessary research to discover risks to which the article may give rise	
Common Law (Liability in Tort)	–	Person responsible for defective goods	Plaintiff must prove (i) negligence (ii) injury caused by defect and was foreseeable consequence of defect	'Reasonable care'
EEC Directive (draft)	–	'Producer' of article	Liability for injury etc caused by a defective product (a product has a defect when it does not provide for the safety which a person is entitled to expect)	Strict (absolute) liability
Factories Act 1961	12 - 13	Occupiers of factories	Requirement to guard prime movers, electric generators, flywheels and transmission machinery	Strict (absolute) liability
	14		Every dangerous part of any machinery	
	17	Any person who sells, lets, or hires powered machinery	Sink, encase or guard bolts etc on revolving shafts Encase gears including friction gearing	

Figure 13.1 Table summarising legal and other provisions related to the safe design of engineering products. Certain details of safeguards under the 1961 Factories Act concerning particular types of machinery are excluded.

First of all, the duties in the Factories Act 1961 are absolute duties. This means that if a machine is not safe then it cannot be used, even if it is impracticable to make it safe. However, the duties of the 1974 Act are stated as being so far as is 'reasonably practicable', which means that the standard of safety required is limited by the cost of the safety precautions in relation to the risk.

The second point is that, although the principal duties in the Factories Act 1961, for example the duty to fence securely every dangerous part of any machinery, relate to occupiers of the factories, the designer should recognise that the purchaser of his equipment will have to comply with these requirements at least in principle.

Finally, it will be noted that before the 1974 Act the duties on the seller of any machine, and hence its designers, were confined to certain specific duties—for example to sink, encase or guard bolts on revolving shafts. However, the 1974 Act, Section 6, while not an absolute duty, is in fact much more stringent than before. These new obligations to ensure that any article is safe when properly used are very demanding—not because of their stringency in detail but because, in contrast to the old fixed rules, they require designers to analyse in detail the dangers associated with their designs, and to devise for themselves the most appropriate control measures. Hitherto the designer's work was largely done for him by the legislators. The obligation to carry out tests, examinations, and research, coupled with the obligation to make available adequate information about the safe use of the product, are clearly far more open-ended than the straightforward duty simply to fit guards over specified parts of machinery. Guidance on the extent of the new duties is available from the Health and Safety Commission [13.1].

13.3 Identification of dangers

The best way of deciding which parts of machinery are dangerous is the obvious one. It is to visualise the ways in which a person can be injured by a component; the immediate sequence of events leading to injury; and the location and potential severity of the injury. This approach is helpful in the correct choice of safeguard.

People can be injured by machine components in five different ways. Certain machines may give rise to sequential injuries from a combination of the forms of harm. The designer should examine the potential for harm of every component/agent of danger associated with each form of harm listed here, and summarised in Figure 13.2.

Traps. The limbs (in some cases the whole body), which need not themselves be moving, may be trapped between closing or passing motions of machines. In other cases a trap occurs when the limbs are drawn into a closing motion of a machine. These are often called in-running nips.

Impact. Injuries can result from being struck by moving parts of machinery.

Contact. Sharp or abrasive surfaces, or energies (hot, or electrically live components) can cause injury on contact. This category includes contact with fast-moving circular saws.

Entanglement. Injuries involve the entanglement of hair, rings, gloves, cuffs or ties in moving (usually rotating) machinery.

Ejection. The previous categories involve the need for safety devices to prevent people reaching into danger areas. In contrast, ejection involves the throwing out

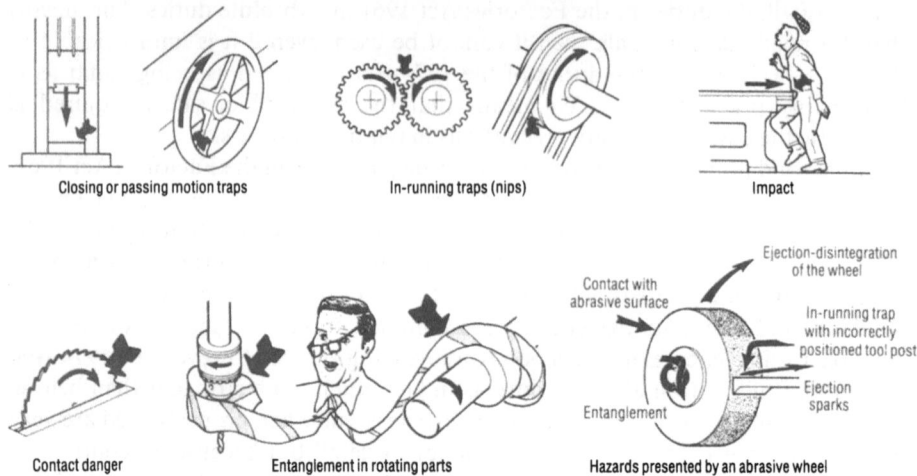

Figure 13.2 Graphical representation of six major sources of danger.

of materials, such as particles, swarf, chips, sparks or molten metal, or broken machine components, such as a burst abrasive wheel or a spindle moulder cutter.

The list of dangers describes the mechanical (and other) sources of danger from engineering designs. It is important to identify the dangers themselves, and also the prerequisites for these dangers in terms of design and other errors. (This topic is developed further in subsequent sections). At this stage it is necessary to distinguish between *continuing dangers*, which are dangers associated with normal working conditions and which may occur even when the equipment is performing its designed duties without malfunction; and *contingent dangers*, which are either dangers that arise directly and immediately from the failure of structures, components, and machinery safety devices to perform their designed function, or are dangers arising from the incorrect assembly of vital parts, during either initial assembly or post-maintenance reassembly procedures. Clearly the way to eliminate the possibility of the latter danger is to ensure that it is only possible to assemble parts in the correct way.

The dangers identified here, and the solutions proposed, relate primarily to mechanical dangers. The detailed study of fire and explosion dangers from the build-up of flammable gases or vapours in or around machinery is outside the scope of the present chapter. Clearly designers who have to incorporate flammable substances in their designs should analyse in detail the contingencies and human errors that may lead to the formation of an explosive mixture, and the potential sources of ignition. Where electrical installations are employed in potentially flammable atmospheres, certified 'flameproof' or 'intrinsically safe' electrical equipment should be used.

13.4 Basic safeguard design

It should be remembered that guards are only required where the designer has failed to achieve intrinsic safety—the avoidance of trapping and similar dangers.

13.4 Basic safeguard design

The considerations for choosing guard materials are: strength, stiffness, durability (to cope with both continuing and contingent dangers), and the possible effect of the guard material on machine reliability. A solid guard may create problems of cooling and visibility; there are both operational and safety reasons for a clear view of the danger area.

Guard types are listed below in a generally accepted decreasing order of preference. However it is questionable whether a fixed guard is superior to a well-designed interlocked type; the reverse is probably true.

13.4.1 Fixed guards

A fixed guard should, when fitted, prevent all access to the danger. Clearly, two factors determine the protection afforded: the method of fixing and the size of any openings in relation to the distance from the opening to the danger.

BS 5304:1975 [13.2] states that these guards should be fixed in such a way that they can only be opened with the aid of a tool. This is insufficient guidance—some kinds of tool are more readily available for improper use than others. The table (Figure 13.3) gives a hierarchy of fixing methods.

Quick-release catches Wing nuts	Not acceptable
Secured by fastening requiring the use of a tool	
Cheese-head screws	
Phillips/Pozidrive screws (self-tapping screws are unsatisfactory)	
Hexagonal nuts/bolts socket (Allen) screws	Increasing acceptability
Shrouded or countersunk socket screws	
Padlock (keys under system-of-work control) coupled with shrouded or countersunk socket screws	
Riveted or welded	Official access difficult

Figure 13.3 Methods for fixing fixed safety guards.

An ideal fastening for a fixed guard should be very difficult to undo without the authorised special tool; quick and simple to undo with the appropriate tool and authority; quick and simple to replace with the appropriate tool; and quick and simple to replace even when the fastening has been undone without the authorised tool. Note that hexagonal bolts can be defeated with an adjustable spanner and unshrouded socket screws can be defeated with a 'Mole' wrench. Careful design of fixed guards is required to prevent traps being created between the guard and moving (but otherwise safe) parts of the machine.

13.4.2 Interlocked guards

Two basic criteria must be observed: until the guard is closed the machine should not be capable of being started; and in reverse, the guard should not be capable of being removed or opened until the dangerous parts have come fully to rest.

Interlocked guards can be hinged, sliding, or removable. The crucial feature is the detail design of the interlocking mechanisms. The mechanisms must be reliable and the system should fail to safety in the event of all foreseeable contingencies (including machine failure or failure of the safety devices).

Provision should also be made for overriding the guard officially as part of a permit-to-work system if this is required. It creates an unfortunate impression if supervisors defeat the device by methods potentially available to other workers. Separate devices, such as locks, are required.

The choice of interlocking method depends on the motive power of the machine, the risk of continuing dangers, and the consequences of failure of the machine or safety device. The system chosen should be direct and simple. Complex systems with potentially unreliable, fail-to-danger elements which are difficult to understand, inspect and maintain are unsatisfactory.

Interlocking mechanisms can be divided into those which lock the guard with the source of motion or danger, and those which lock with respect to the motion itself. It is vital to consult BS 5304:1975 for a review of the acceptability of different power and power-control interlocks. The methods described may be used in combination to achieve a complete interlocking system.

Power-source interlocks include:

Direct manual switch (or valve) interlock (Figure 13.4);

Cam-activated limit switch (or valve) interlocks (Figure 13.5) which are versatile and highly effective. Two features are crucial: the mechanical arrangement of the switches or valves and the location of the switch or valve in the circuit. A hydraulic valve is a less positive interlocking element when it forms part of a pilot circuit;

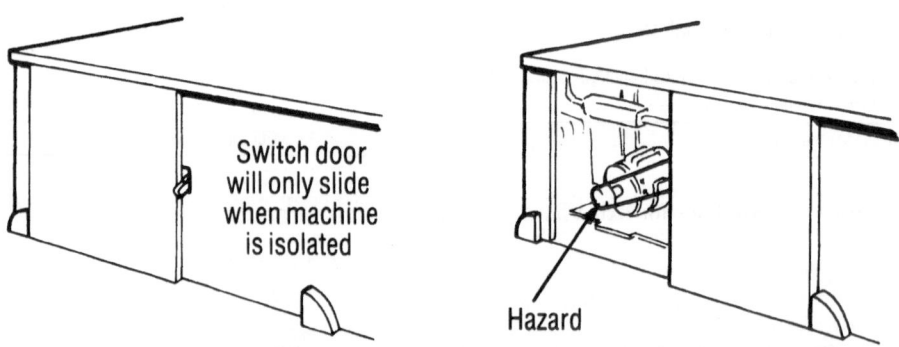

Figure 13.4 Direct manual switch interlocked guard.

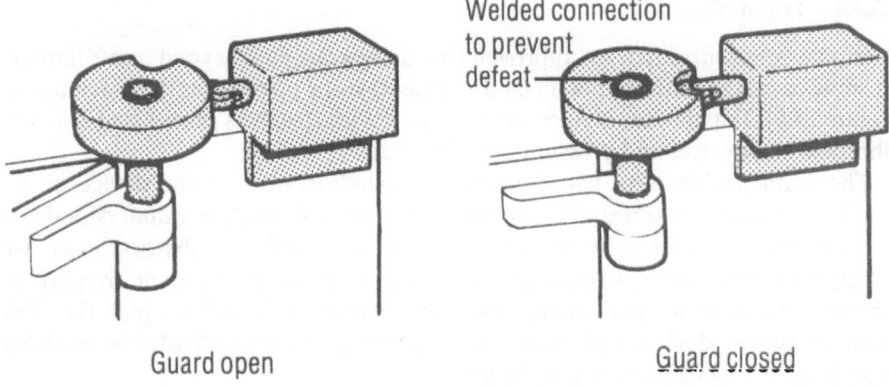

Figure 13.5 Cam-activated limit switch interlocked guard.

Trapped key interlocks (key exchange system) require the same key to switch the supply and open the guard. When the supply is on, or the guard is open, the key is trapped in the relevant keyhole. A time-delay feature can be incorporated in proprietary systems. The success of this method largely depends on the ease of improper fabrication of a master key;

Captive key interlocks involve the combination of a latch and an electrical switch within a single assembly; and

Magnetic switch interlocks. Proprietary magnetic switches are available and can be used for interlocks. They have the advantage that the guard does not have to be hinged to the machine. They are potentially fail-to-danger devices but they are remarkably reliable, although they can be defeated relatively easily with a spare magnet;

Motion interlocks. Mechanical interlocks which prevent the opening of a guard until the machine is at rest; and *Mechanical scotches* which prevent motion when the guard is open. These are combined in Figure 13.5.

Motion interlocks are often required as a back-up to power-source interlocks to cope with contingencies.

13.4.3 Automatic guards

The guard is designed to force the operator's limbs or body out of the trapping area when a trap is created. Applications are limited to slow-moving, long-stroke presses and guillotines. Automatic guards readily create their own trapping and impact dangers. Ergonomic considerations are important.

13.4.4 Trip devices

A trip device ensures that an approach to a dangerous part beyond a safe limit (or in some cases the initiation of an entanglement) stops or reverses the machine (see Figure 13.6). Trip devices include mechanical devices (such as probes and barriers) photoelectric devices and pressure-sensitive mats.

The main problems of trip devices are maladjustment of the device or the machine's brake, or failure of these elements. It should be clearly understood that trip devices should be activated by the involuntary action of the person at risk. Emergency stop wires, suspended for example above a conveyor, require an operator to have a spare hand and the presence of mind to pull the wire. Emergency stop devices have their value, but the protection afforded is much less than from a well-maintained trip device.

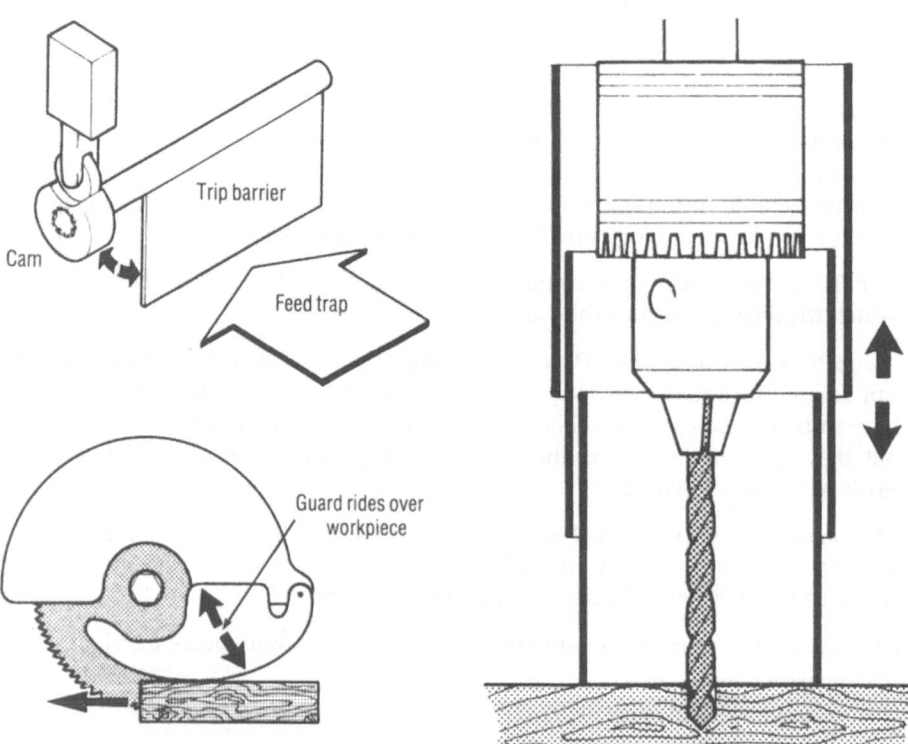

Figure 13.6 (top) Trip device.
Figure 13.7 (right) Adjustable guard.
Figure 13.8 (above) Self-adjusting guard.

13.4.5 Adjustable guards

Adjustable guards comprise a fixed guard with an adjustable element. They are widely used for woodworking and toolroom machines. The protection afforded by adjustable guards is usually poor. Residual access remains when the guard is correctly adjusted, and adjustment is often inconvenient and time-consuming in relation to the time the job takes to do. Visibility is also often badly impaired, thus creating other dangers. Adjustable guards, particularly in machine toolrooms, are regularly defeated, and it is only a slight exaggeration to say that in most cases the adjustable element of the guard does not protect people from accidents—it merely shifts the blame. Figure 13.7 shows a typical application of adjustable guards.

13.4.6 Self-adjusting guards

Self-adjusting guards prevent access to the workpiece except when the guard is forced open by the passage of the work. They usually incorporate a spring-loaded pivoted element, for example Figure 13.8.

13.4.7 Two-hand control devices

Two-hand controls protect (or may protect) the operator, but not third parties. This is their main snag. However, the poor reputation of two-hand controls has been created largely by badly designed controls which can be defeated easily and which do not cope with contingencies. The controls should be spaced well apart and shrouded. The machine should only operate when both controls are activated virtually simultaneously and the controls should require resetting between each cycle of the machine.

13.4.8 Ergonomics of machinery guards

A crucial requirement of all types of machinery guard is that the people at risk should not be able to get their limbs into the danger area. A number of standards exist which give guidance on this matter, but unfortunately they are somewhat contradictory, and research is required to establish the appropriate safe openings and anthropometric measurements. Essentially, two types of openings exist: first, narrow openings, for example slot-type openings, in machinery guards; and second the distance of a danger from a barrier rail of a given height. For narrow openings in slot-type guards, the Chief Inspector of Factories in his 1975 Report [13.3] recommends the following formula:

$$Y = \frac{X}{12} + 6$$

where Y is the size of the opening and X is the safe distance from the opening to the point of danger, both in millimetres. This recommendation appears to supersede the safe opening standard given in British Standard 5304:1975.

Very limited information exists to specify the distance from barrier rails to danger points. The most commonly used figures are those provided in the German Standard DIN 31001:1976 [13.4]. In fact this recommendation, which is designed to protect 95 per cent of the workforce, appears to allow a substantial amount of unsafe access to dangerous parts of machinery. Booth and Thompson [13.5] found that the German standard did not protect an equivalent percentage of the British male workforce at most barrier heights (see Figure 13.9). In the absence of an agreed standard, barrier rails should be located with reference to the maximum reach figures given by Booth and Thompson.

When considering guards for live electrical circuits the anthropometric issue just described is not the only, or indeed the most important, consideration. Risks occur

Distance of hazard point from floor (mm)	Height of edge of safety feature (barrier) (mm)						
	2200	2000	1800	1600	1400	1200	1000
2400	100* 308 +208	100 425 +325	100 534 +434	100 486 +386	100 591 +491	100 701 +601	100 761 +661
2200	250 321 +71	350 528 +178	400 699 +299	500 718 +218	500 794 +294	600 920 +320	600 994 +394
2000		350 537 +187	500 785 +285	600 824 +224	700 920 +220	900 1071 +171	1100 1171 +70
1800		— 477 +477	600 826 +226	900 890 −10	900 993 +93	1000 1189 +189	1100 1295 +195
1600			— 271 +271	500 835 +335	900 898 −2	900 1026 +125	1000 1267 +267	1300 1397 +97
1400				100 764 +664	800 869 +69	900 1026 +125	1000 1318 +318	1300 1456 +156
1200					500 788 +288	900 976 +76	1000 1322 +322	1400 1484 +84
1000					q 300 637 +337	900 889 −11	1000 1310 +310	1400 1482 +82
800						600 740 +140	900 1273 +373	1300 1449 +149
600						— 422 +422	500 1160 +660	1200 1373 +173
400							300 1116 +816	1200 1313 +113

*100 = Safety distance (each distance + safety allowance) recommended by DIN 31 001.
308 = Reach distance (+ 3 S.D.) found by experiment.
+ 208 = Distance by which experimental reach exceeded (+) or failed to reach (−) DIN standard.

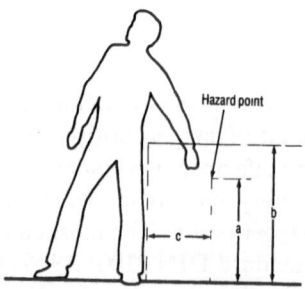

Figure 13.9 Table comparing maximum reach values obtained by May [13.5] and the German DIN 31001 standard.

as a result, for example, of maintenance men using tools such as steel tapes which can be pushed through openings that prevent finger or hand access. It might therefore be argued that there is no such thing as a 'safe' opening in an electrical circuit guard.

13.5 Risk assessment and approaches to control

The Factory Inspectorate (now part of the Health and Safety Executive) states [13.6] that about three-quarters of all moving machinery accidents are preventable with reasonably practicable precautions. Broadly, half the preventable accidents are caused by the failure of employers to provide appropriate safeguards. The other half are associated with the removal or misuse of the safety devices by the injured worker or his workmates. Many safeguards invite defeat and misuse.

Essentially the risk (that is, the probability of an accident) depends upon the actual situation (the 'objective danger') and the attitude and skills of the people at risk (the 'subjective risk'). Accidents are most likely to occur either if there has been a failure to identify the danger (that is, a failure to recognise that harm can, in fact, occur) or if the control measures adopted are insufficient to deal with it. Three control strategies have been adopted historically.

The first of these comprises efforts to make the personnel at risk more able, and more motivated, to cope with dangers (dealing with a problem by trying to improve the subjective appreciation of risk). The second strategy has been to provide physical safeguards (or protective clothing) which have complied, perhaps, with legal requirements but which have failed to cater for machine operation and the vagaries of behaviour (a half-hearted attack on the 'objective danger'). Finally there has been the provision and maintenance of physical safeguards coupled with appropriate and monitored 'safe systems of work' designed to match the need of the people at risk, and to cope with machine failures and irrational behaviour.

The first two methods depend for their success upon changing people's attitudes and skills in relation to safety, or selecting people with the appropriate attitudes and skills. Recent research [13.7, 13.8] suggest that only small (and usually short-term) benefits are possible with these methods. Selection, training, propaganda, and supervision simply cannot be relied upon to persuade someone to use a safety device or act safely when the benefits of acting safely are remote and nebulous (although the 'costs' are paradoxically immediate and painful). In contrast, the third strategy, particularly if underpinned with law, has proved remarkably successful in this country.

The table (Figure 13.10) summarises the criteria that determine the magnitude of the objective danger and the technical, procedural and behavioural measures used to reduce the probability of an accident or to reduce the severity of injuries. It is important to note the distinction between the build-up of danger, which depends on factors such as the ease of access to the dangerous parts of the machinery and the need to approach the dangerous parts, and the stage of imminent danger. From the table and from earlier comments it should be appreciated that it is better to prevent a build-up of danger than to try to prevent injuries in circumstances of imminent danger. The approach whereby a distinction is drawn between the build-up of danger and imminent danger is due to Surry [13.8].

Objective danger and risk reduction criteria		Methods of reducing danger		
Danger build-up		Design (technical) measures	Procedural measures	Behavioural measures – to improve subjective risk ('instructions for safe use')
The agent of danger (continuing and machine contingent dangers)	Potential of harm The form of harm (traps, contact etc) Probability of failure; probability of incorrect assembly (contingent dangers)	Intrinsic safety, or reduce injury potential Improvements in component and system reliability and fail-to-safety (contingent dangers) Design for assembly where the correct method of assembly is the only feasible method	Planned maintenance schedule: – to prevent machine unreliability – detailed instruction for correct assembly	Information about dangers should be made available to the users of the equipment
Ease of approach leading to imminent danger	Extent to which access restricted by physical safeguards Extent to which access made possible by misusing/defeating/removing safeguard	Make access more difficult: – rigorous enclosure – safeguard fastenings and mechanisms more difficult to defeat (but capable of being officially overridden) – design to improve safeguard reliability and fail-to-safety	Planned maintenance schedule: – to check guards in position – to prevent safeguard unreliability	Warning notices and information about safe systems of work Motivational training to reduce risk taking and alteration of eg incentive schemes to reduce benefits which accrue Supervision
Need for approach leading to imminent danger	Frequency of access Proximity of access Circumstances of access: – machine on/off – personnel involved – official need – unofficial need	Design to minimise need for access (eg remote lubrication of gears)Safeguards (eg interlocked guards, trip devices) which permit necessary access without danger, (but prevent access leading to imminent danger)	Safe system of work (to minimise need for access) Permit-to-work system to formalise precautions to be observed with clear allocation of responsibilities	*Note: safe behaviour is attainable only where the inconvenience of defeating the safeguard exceeds the benefits which accrue from defeating it*
Imminent danger				
Human errors	Criteria include: – complexity of task – repetitiveness of task – level of arousal – perception of imminent danger – skills to avoid injury	Ergonomic layout of controls to minimise errors Design to make dangers as obvious as possible	Specification of permit-to-work should identify potential errors Provision of protective clothing and equipment	Training to improve perception and skills to avoid injury

Figure 13.10 Table of risk assessment and reduction.

The main feature of the table is that it shows the technical measures for control linked with the procedural and other measures which may be necessary. The distinction is between technical measures, which the designer should adopt at the drawing-board, and procedural and behavioural measures that he should generally be specifying as 'instructions for safe use'.

In order to explain the use of the table, studies of accidents follow to show the range of options available to the designer and to the user of the equipment to prevent the injury which occurred.

13.6 Case studies

13.6.1 Acrylic fibre line (Figure 13.11)

Figure 13.11 Acrylic fibre line.

The machine operator was fatally injured when he was drawn into the trap between (in the first instance) roller and fibre.

The arrow indicates the in-running trap. At the time the injured person was, as a routine operation, cutting a broken fibre from the roller. Nobody was at the time available to stop the machine.

The crucial point is that the components concerned were not perceived to involve a risk of serious injury. Referring to the table there was a failure to identify the potential for harm of the agent of danger; access was unrestricted; and there was a substantial need for access for operational reasons. Prevention of repetitions of this accident demand an improved standard of machinery guarding coupled with methods for cutting broken fibres without the need for close access.

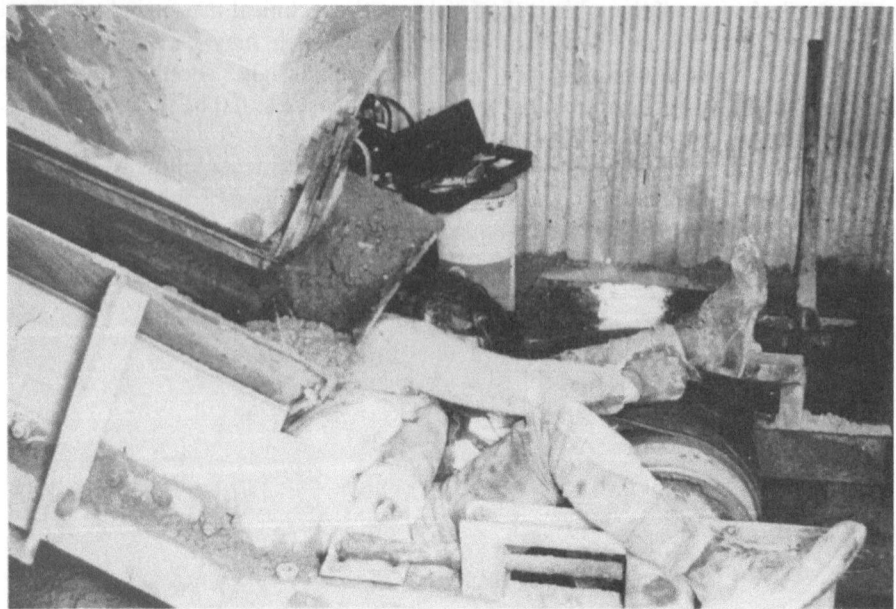

Figure 13.12 Quarry conveyor.

13.6.2 Quarry conveyor (Figure 13.12)

The guard provided was easy to remove. Both maintenance and labouring personnel had reason to remove the guard. The dead man, who had been employed for two days, banged his head on a roof beam and fell into the conveyor. Prevention could be achieved by a combination of improved guard design coupled with measures to ensure that the machine could be maintained and cleaned without the need for guard removal.

If interlocked guards were not fitted then a permit-to-work system would probably be required in order to formalise precautions to be observed. If an interlocked guard is installed in circumstances such as this then a planned maintenance schedule is required to prevent safeguard unreliability.

13.6.3 Dough divider (Figures 13.13a and b)

On these machines dough tends to bypass the seals and clogs the machine. Frequent cleaning is required. The interlocked guard, which incorporates a 'normally open' limit switch invites defeat, and can be easily defeated. The machine operator tends to be under pressure from his workmates to defeat the guard.

Problems of this kind can be overcome by reducing the need for unofficial access (for example by providing two dividers in the bread manufacturing line), and by an improved design of the interlocking system.

13.6 Case studies

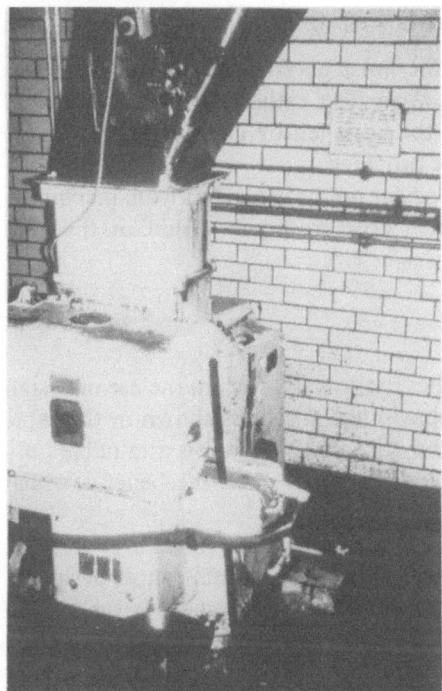
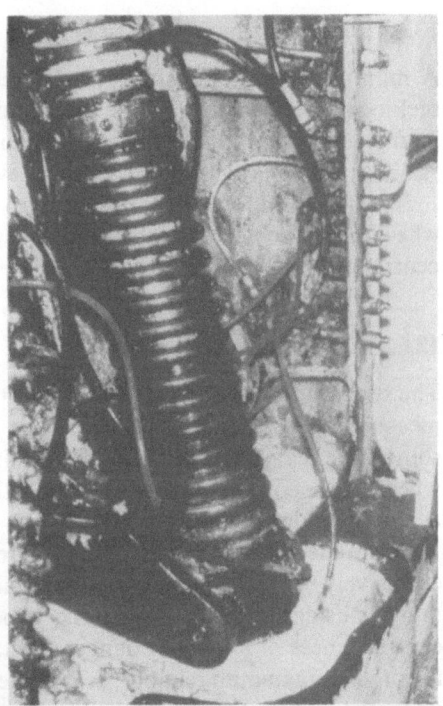

Figures 13.13a and b
Dough divider, external view (above) and internal view (top right).

Figure 13.14 (right)
Furnace charging hoist.

13.6.4 Furnace charging hoist (Figure 13.14)

A man was very seriously injured when the hoist platform fell on him. He was seeking to clear a jam in the hoist guideways. The rather notional interlocked guard was irrelevant in this case because the hoist fell by gravity. There was no 'slack-rope' or gravity-fall scotching device provided.

It is clearly important for designers to consider not only dangers from machinery when under power, but also dangers which arise from the weight or the high centre of gravity of their products.

13.6.5 Discussion

The suggestions for control in the case studies exclusively concern the use of design measures or procedural measures. The behavioural measures shown in the table stand little real prospect of success on their own. Safe behaviour is attainable only where the inconvenience of defeating the safeguards exceeds the benefits which accrue from defeating it.

Once the situation has become one of imminent danger, safety depends exclusively upon the avoidance of human errors. In many circumstances designers must consider in their design the expected frequency of human errors in the circumstances of imminent danger. Kletz [13.9] gives some rule-of-thumb criteria for predicting human reliability. The essential point of his figures is that, depending upon the circumstances, human errors can be expected with frequencies between one error in four events to one error in ten thousand. Even the latter figure may represent an unacceptable risk for employees who are working repetitively and continuously on the same task for many years, and where the consequence of an error is a serious injury.

13.7 Design procedures

It is essential that the designer adopts a formal design procedure to ensure that all relevant safety aspects are considered adequately at the design stage. The procedure should consider not only the matters shown in the table (Figure 13.10); it should also take account of stress-analysis aspects of safety and reliability, explosion and fire risks, the question of foreseeable errors in fabrication, and prevention of health risks from noise or dust emission. Certain systematic methods are available [13.10] to ensure that all dangers are identified, including hazard and operability studies and fault tree analysis. It is an open question whether these sophisticated design procedures are in fact necessary in the design of individual items of equipment, and they are probably more relevant to the problems of process design. Therefore for the purpose of the present chapter the procedure for safe design should involve a systematic evaluation of the dangers and measures for assessing risk presented in this chapter, coupled with an appropriate, balanced, choice of methods for reducing danger, also shown in the table. When the designer has adopted the best reasonably practicable design measures to minimise danger in his design, he is also obliged to make available 'instructions for safe use'. The procedural and behavioural measures shown in the table give a summary of the scope of the instructions which may be required in order to comply with the duties under the Health and Safety at Work Act (Figure 13.1). It is not the purpose of this chapter to describe permit-to-work systems in detail. A satisfactory description of the requirements is available elsewhere [13.2].

13.8 Conclusion

There is substantial evidence that the provision of a high standard of safeguarding when linked with appropriate systems of work, permit-to-work systems and safeguard maintenance programmes is the most effective method of preventing machinery accidents. It goes without saying of course that intrinsic safety is the ideal.

The following seven questions represent a useful checklist for a designer in considering whether the safeguards are appropriate for his equipment.

1. Does the safeguard totally prevent approach leading to imminent danger when in its correct position and when working properly?
2. Is the safeguard reasonably convenient to use (ie does it interfere with either the speed or quality of the work); are there foreseeable reasons why it is necessary to override the safeguard, or to defeat it?
3. How easy is it to defeat or misuse the safeguard?
4. Does the safeguard cope with foreseeable machine failures?
5. Are the components of the safeguard reliable and fail-safe?
6. Is the safeguard straightforward to inspect and maintain?
7. Are the instructions for safe use of a machine and safeguard adequate to cope with all foreseeable dangers in use?

Other, intrinsic, aspects of design that are relevant to safety, especially as it relates to maintenance, are covered by questions such as:

8. Is the design safe in terms of gravity and balance?
9. Are the design and operating procedures safe for maintenance of electric and high-pressure fluid services?
10. Are the design and operating procedures safe for maintenance of high-energy springs and heavy components?
11. Does design ensure safety by preventing incorrect reassembly of components that could cause accidents?
12. Does design contribute to safety by offering good visibility and accessibility for routine inspection of vital components such as fasteners that may require tightening, or areas where corrosion could lead to failure?

13.9 References

13.1　Health and Safety Commission, Great Britain. *Articles and Substances for Use at Work: guidance for designers, manufacturers, importers, suppliers, erectors and installers.* Guidance Nots GS8. HMSO, London, 1977.

13.2　British Standards Institution. *Safeguarding of Machinery.* Code of Practice 5304. BSI, London, 1975.

13.3　Health and Safety Executive, Great Britain. *Health and Safety Industry and Services.* HMSO, London, 1975.

13.4　German Standard. *Safety Requirements for the Design of Technical Equipment: protecting devices; definitions; safety distances for adults and children.* DIN 31001, 1976.

13.5　May, S. *An investigation into the forced reach of male workers over safety barriers and through large openings in machinery (a pilot study).* Unpublished MSc Thesis, University of Aston in Birmingham, 1978.

13.6　Booth, R. T. and Thompson, D. T. 'Anthropometry and machinery safeguarding: an investigation into forced reach over barriers and through openings. Part 1: reach over barriers.' *Journal of Occupational Accidents*, vol 3, 1980.

13.7　Department of Employment, Great Britain. *Accidents in Factories; the pattern of causation and the scope of prevention.* HMSO, London, 1974.

13.8　Hale, A. and Hale, M. *A Review of the Industrial Accident Research Literature.* HMSO, London, 1972.

13.9　Surry, J. *Industrial Accident Research.* University of Toronto, Department of Industrial Engineering, 1974.

13.10　Kletz, T. A. 'Human error: some estimates of its frequency.' February 1974.

13.11　Lawley, H. G. 'Operability studies and hazard analysis.' *Chemical Engineering Progress*, vol 70, April 1974.

14 Models as an aid to design

Jack Hilton

14.1 Introduction

Three-dimensional modelling techniques offer valuable support to the design process because of the wide range of information they provide in a single entity, compared with a number of drawings or sketches. Models can be used to investigate ergonomic and visual factors in design and can be adjusted easily to demonstrate the effects of changes. They are also an excellent medium for communicating ideas within a design department and externally to users of a product. Models contribute to the design process especially through:

Design to achieve aesthetically acceptable form and colour;

Design of detail and control layout for ergonomic effectiveness;

Design for safety, for example as affected by positioning of components;

Design of systems in terms of interference between components, accessibility for erection and maintenance, and provision of photographic guides for erection purposes;

Design of detail to aid in stress analysis; and

Commercial objectives during development and in subsequent exploitation.

Models are used today more extensively than in the past because of the growing complication of machines, systems and their control, and more exacting demands for high ergonomic and aesthetic standards; and because of reduced relative costs, as the result of the availability of more effective materials, scale details and adhesives for model-making. Nevertheless, although models comprise a very small part of development budgets, their cost can be a disincentive and it is therefore important to restrict their quality, accuracy and extent of detail to the level that will satisfy the needs of the project, more especially as the requirements from models can often be met by fairly rough construction.

Models can be made either in-house, or by external specialists, or collaboratively. The choice will depend on such factors as the quality of models required and the skills that can be offered by the in-house organisation, including the availability of model-making tools and familiarity with model-making materials and experience of their use, relative to the type of model needed, the time available and cost. When model-making is a new venture, it could well prove worthwhile to acquire some skilled help.

There are therefore a number of different classes of models, the application of which is discussed below, together with examples.

14.2 Full-scale models and mock-ups

Full-scale models are almost essential when the problems to be resolved are ergonomic, and they can be simple or elaborate depending on requirements, the method of building and materials used. They can be valuable to designers, users and those responsible for manufacturing processes, they can highlight early in the design process anomalies that might be difficult and expensive to rectify at a later stage, and they contribute to decision making.

Conventional tools will suffice. For materials, rolled steel sections and fabricated parts can be simulated in wood or blockboard; sheet metal by hardboard, cardboard or even linoleum. Actual details such as controls, switches or instruments can be used if available, or three-dimensional models of these may be necessary. In some cases, as when designing large control consoles, it may be adequate to glue full-scale photoprints or drawings of items such as dials on their appropriate panels, which affords a quick way of determining their best visual and ergonomic arrangement. Straight pipework can be represented by wood dowelling, bends by plastic dowelling or tube, and large diameter pipes by cardboard tubing, all of which are available in a wide range of diameters.

Suitable finishes include vinyl emulsion, which is easy to apply and quick drying, and oil or synthetic paint applied by spray equipment or by aerosol pack for small areas. An example of full-scale procedure for a textile laboratory machine is shown in Figure 14.1, in which the designers are using cardboard cutouts of components, and later models of controls and gauges, placed on a full-scale blackboard sketch to optimise ergonomic positioning. This was followed with a full-scale three-dimensional model (Figure 14.2) to demonstrate configuration, ergonomic features and colour treatment.

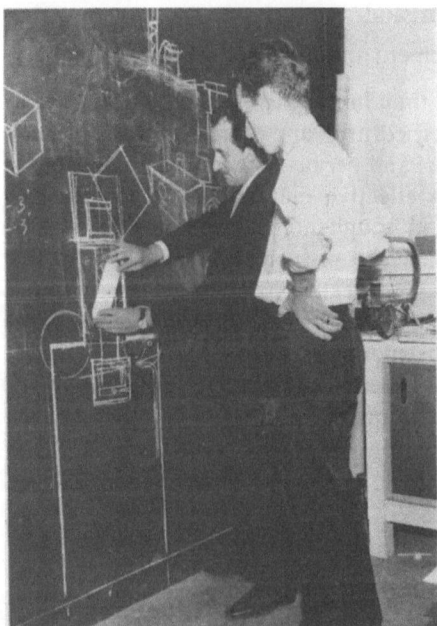

Figure 14.1 Outline sketch of a textile-finishing laboratory machine drawn full size on a wall blackboard. (*Mather & Platt Ltd*)

Figure 14.2 Full-scale three-dimensional model to demonstrate overall configuration, ergonomic features and colour treatment. (*Mather & Platt Ltd*)

Figure 14.3 Full-scale mock-up of a self-propelled vegetable harvester. (*Mather & Platt Ltd*)

A mock-up of the front end of a self-propelled vegetable harvester is shown in Figure 14.3. The driver's cab and large crop pick-up equipment had to be arranged to ensure the maximum unobstructed field of view to the driver during harvesting. The controls were represented in mock-up form, together with an actual driver's seat, and the model was used as a design laboratory during the whole of the development procedure, involving collaboration between interested customers and the designers.

The objective of the full-scale model of a flameproof motor (Figure 14.4) was to demonstrate appearance and a new method of manufacture. Materials used were blockboard, hardboard, actual clamps withdrawn from the part stores, and a steel tube blanked off at the end to represent the shaft. This model took an industrial designer and a model-maker 10 working days to complete.

Full-scale models can also be used for exhibitions, and Figures 14.5 and 14.6 show an exhibition model of a centrifugal pump during construction in a pattern shop, and completed. The quality of finish required can make the costs of such models relatively high, but the cost of preparing, transporting and financing the capital for an actual and very heavy product could be prohibitive.

Figure 14.4 Full-scale model of flameproof electric motor. (*Mather & Platt Ltd*)

Figure 14.5 Full-scale model of a large centrifugal pump during construction. (*Mather & Platt Ltd*)

Figure 14.6 Completed model of centrifugal pump ready for despatch to an exhibition site in the Middle East. (*Mather & Platt Ltd*)

14.3 Scale models

Scale models are a convenient method for displaying three-dimensional information such as overall configuration and structural details, and for exhibitions when actual equipment is too large, but they have severe limitations in terms of ergonomic factors. The most suitable scale for each project may be determined by the detail to be modelled and often by portability for exhibitions.

Plastic sheet, rod and extrusion offer the best finish with the least labour; wood should be medium-hard to hard, close grained and as knot free as possible, although soft woods can be used where high finish is not critical. Thin gauge aluminium is usually preferred to brass or tin-plate as quick set adhesives usually make soldering unnecessary. If a number of identical components are required, such as electric motors or gearboxes, these can be moulded in epoxy-resin or similar material, using a flexible mould of the part required from which any number of mouldings can be produced. These will need final touching up, but the process saves time and cost. Painting can be by hand or by aerosol pack, but as many parts as possible should be painted prior to final assembly. A model-maker's tool kit will be a useful addition to the conventional tools required to work the selected materials.

The scale model of the self-propelled vegetable harvester (Figure 14.7), the front end of which was reproduced full-scale (Figure 14.3), was used to demonstrate overall configuration and for sales purposes, and was built from wood, aluminium sheet and brazing rod.

The model of the frame of a bus (Figure 14.8), made largely from wood and cardboard, was used to help in evaluation of structural design and manufacturing processing.

Scale models of large equipment can be an ideal solution for exhibitions, and Figure 14.9 shows a model of an automatic packaged boiler. The cost of such highly finished models can be relatively high, but may be justified by their use for commercial purposes and for lectures.

Figure 14.7 (left) Scale model of self-propelled vegetable harvester. (*Mather & Platt Ltd*)

Figure 14.8 (below) Design model for structural evaluation of a Leyland National bus chassis. (*British Leyland*)

Figure 14.9 (bottom) Automatic boiler model. (*Davey, Paxman Ltd, model by Thorp Models*)

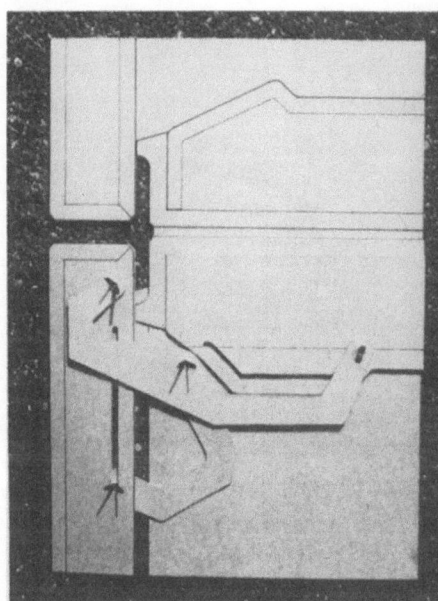
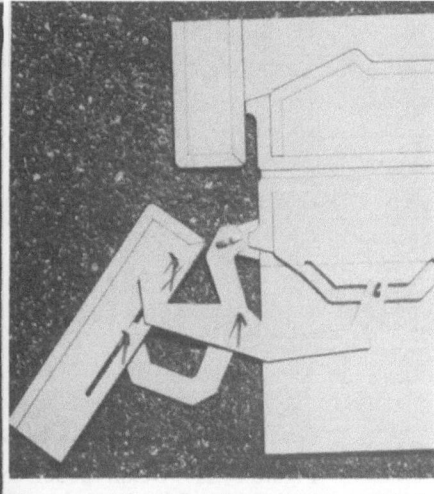

Figures 14.10 Two-dimensional model of hinge mechanism. (*Bill Moggridge Associates*)

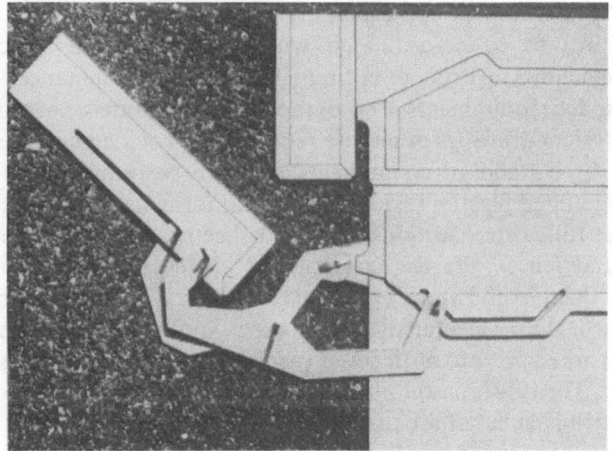

14.4 Models of details

A wide range of catalogue details are available but when new components such as handles, controls and other mechanisms are necessary, models, which should normally be full-scale, can help with their aesthetic, ergonomic and functional design. If functional testing is required, these must of course be built from materials providing adequate strength, but in other cases modelling compounds or plaster of Paris may be suitable.

A simple model of a hinged linkage developed for an unusual transfer movement is shown in Figure 14.10. The model of a compact computer keyboard (Figure 14.11) illustrates the effectiveness of this approach for design communication, and such models can also help in production processing.

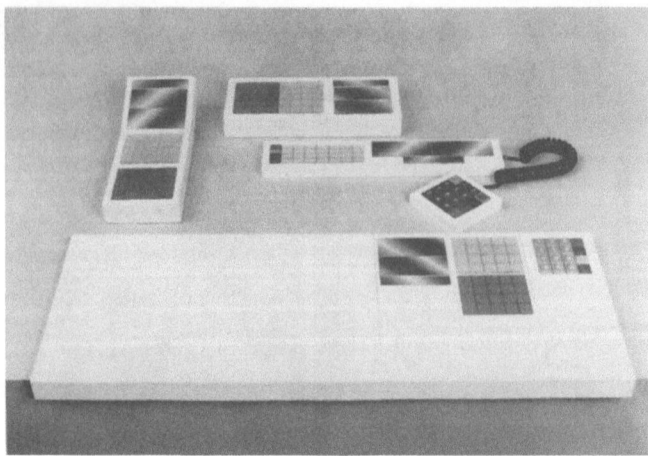

Figure 14.11 Detail models for design of computer keyboards. (*Bill Moggridge Associates*)

14.5 System models

System models are especially valuable when complex pipework and cable runs have to be co-ordinated with equipment and structures, and for elaborate machines, in order to optimise functional, maintenance, safety, aesthetic design and space requirements. Photographs of such models can reinforce, or even replace, the system drawings otherwise required to provide guidance for erection on site. Most of the standard components likely to be required are obtainable as die-cast or moulded plastics scale models in kit form, as also are extruded plastic simulations of rolled-steel sections, rods and sheets, and environmental features such as trees and fences. Plastics components are easy to work, offer a smooth finish, are available in a variety of colours, and, can be bent or formed with application of a little heat. Dental plaster can be coloured with pigment, after mixing and be worked prior to setting and varnished to provide effects such as sea water.

The off-shore oil platform model (Figure 14.12) was built to demonstrate the arrangement of fire protection and detection equipment installed in the processing area and living quarters, using materials as above and perforated aluminium sheet, plastic tube, nylon thread and brazing rod.

In-line production machines can involve space problems, and the model of a new shrink-wrap packaging system (Figure 14.13), fabricated entirely out of wood, was made to demonstrate to clients the economy in space achieved, the appearance and colour scheme.

A model of a large turbine generator with associated steel foundation and equipment is described in Chapter 18.21.

14.6 Models for work station design

The design of work stations for complicated duties can present a wide range of operational and maintenance problems, the optimisation of which can require a series of modelling techniques.

Figure 14.12 (right) Systems model of an offshore oil platform to show arrangement of fire protection and detection equipment. (*Mather & Platt Ltd*)

Figure 14.13 (below) Scale model of a shrink-wrap packaging system. (*Mather & Platt Ltd*)

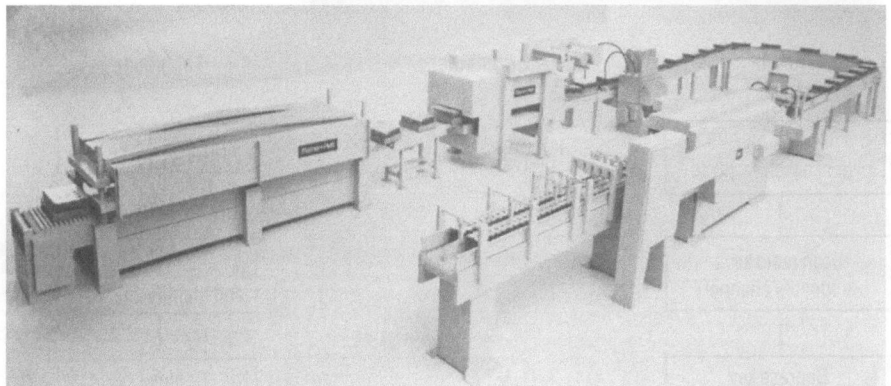

An example of such procedure [14.1] is the design of a work station including a computer, radio and telephone systems, designed for emergency public services, the sequence of development for which was as follows.

(a) Formulation of an ergonomic specification summarising the operator requirements in relation to the individual components of apparatus, in service and for maintenance; a typical summary of 'reach-to' and 'vision-of' equipment is shown in Figure 14.14a.

(b) Flow charting of most frequently occurring sequences of activity in relation to various items of equipment (Figure 14.14b).

(c) Production of polyurethane foam models, $\frac{1}{8}$ inch scale, for generation of alternative layouts (Figure 14.14c).

(d) Production of full-scale ergonomic rig model permitting alternatives suggested in (c) to be evaluated by users. The detail layout of radio and telephone controls received special attention at this stage, with functional grouping arranged to produce logical sequence of movements complying with the requirements in (a) and (b) (Figure 14.14d).

(e) Production of full-scale prototype console checking tolerance, manufacturing problems and finishes.

(f) Final design incorporating minor changes and manufacture of console (Figure 14.14e).

Item		Visual			Reach			Frequency of use		
		Not important	Peripheral	Optimal	Not important	Peripheral	Optimal	Low	Medium	High
VDU: (WDM 2020)	screen			×	×					×
	status display	×			×				×	
	keyboard			×			×			×
	brightness control		×			×		×		
	on/off switch		×		×			×		
VDU: (WDM 2012N)	screen			×	×					×
	status display	×			×				×	
	brightness control		×			×		×		
	on/off switch		×		×			×		

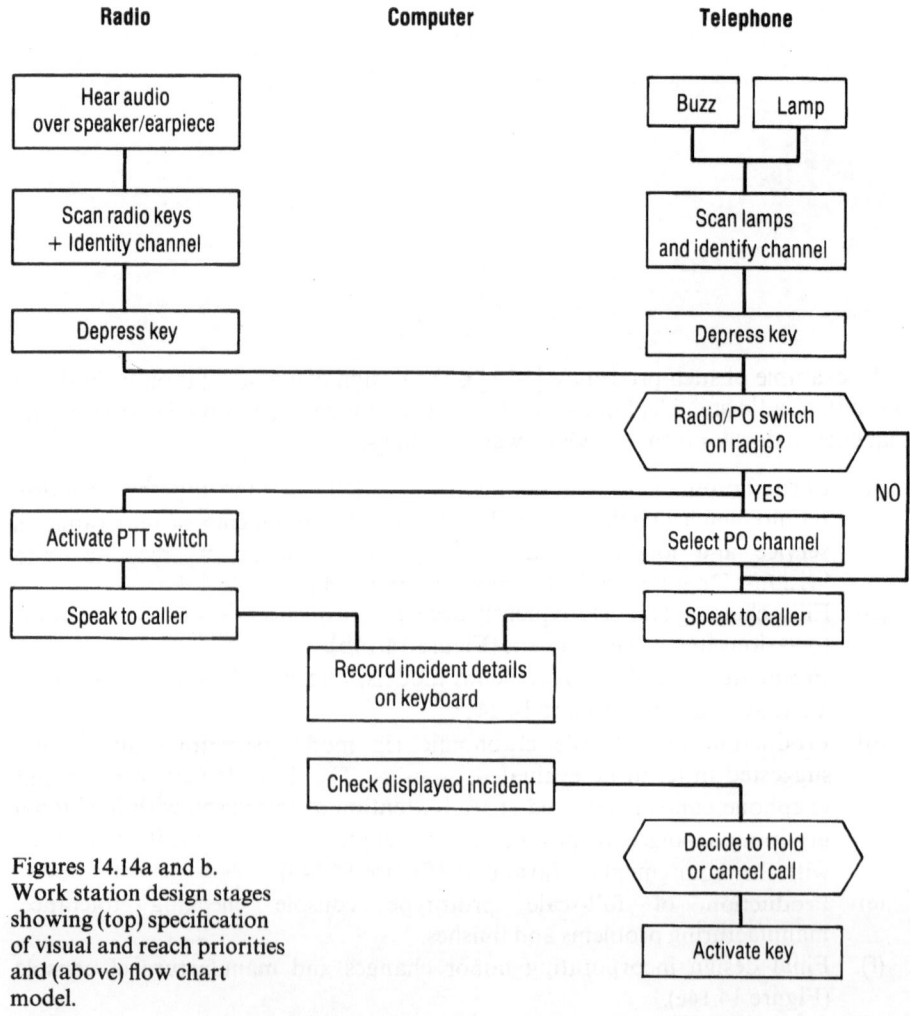

Figures 14.14a and b. Work station design stages showing (top) specification of visual and reach priorities and (above) flow chart model.

14.6 Models for work station design

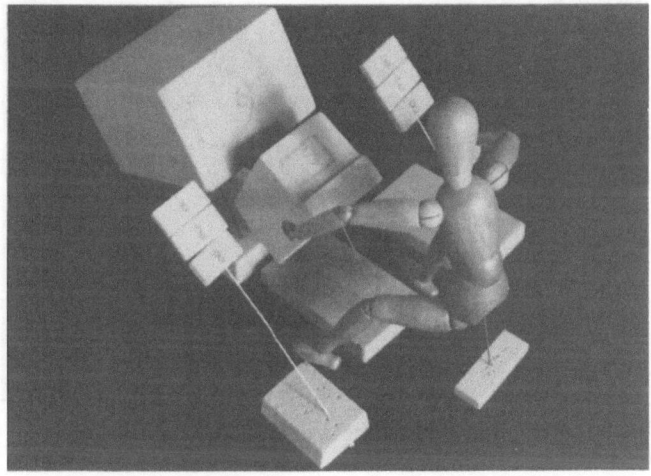

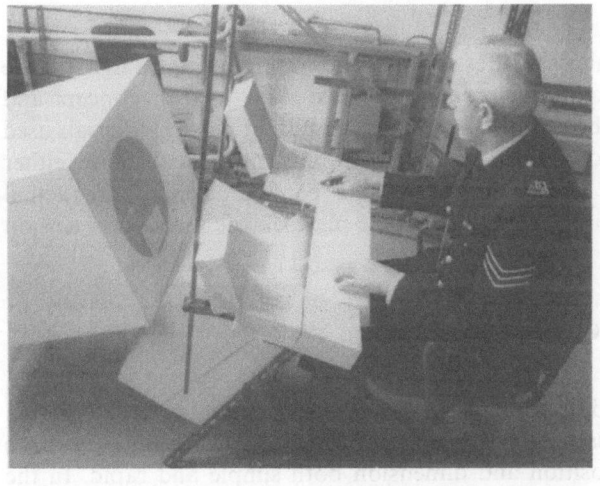

Figures 14.14c, d and e. Work station design stages showing (top) polystyrene foam model, (middle) full-scale mock up and (right) final equipment design. (*All pictures John Wood, Communications Complex Design Ltd*)

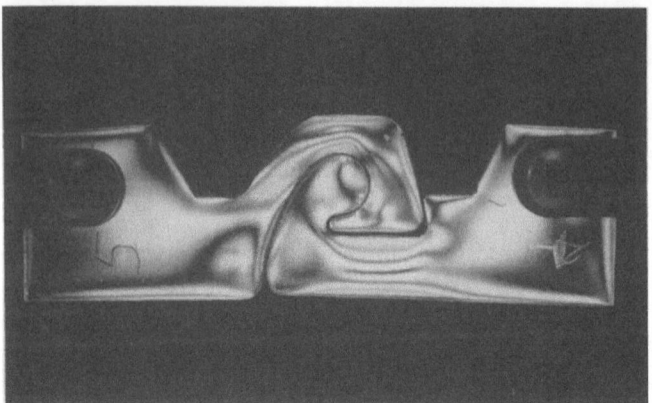

Figure 14.15 Photoelastic stress model under polarised light.
(*ICI Ltd*)

14.7 Models for stress analysis

Photoelastic models for analysis of stress or distortion using polarised light can be valuable in the design process, but require specialist skills in their concept and manufacture. Models must be very accurate, the transparent plastic materials used must be treated with respect and the integrity of joints must be absolute, either using photoelastic cement or by simulating the configuration of the actual fastening to be used. Figure 14.15 shows a simple photoelastic model of a tension joint under stress.

14.8 Computer-aided ergonomic models

Advances in computer-aided design (CAD) have brought computer representation of three-dimensional models within the range of economic feasibility and make evaluation of changes in position and dimension both simple and rapid. In the present state of the art, this is most likely to prove cost effective when prior computer studies can be expected to reduce significantly the extent of subsequent modifications and testing on full-scale three-dimensional models, or when the same programme can be used for a range of designs, or for periodic design reviews, with the assistance of specialised expertise in programming.

CAD systems can provide a three-dimensional picture of a man-model working in a machine and the surrounding visual environment, displayed on a graphic screen or plotted out as a drawing. This permits a variety of evaluations to be carried out including a check on the ergonomic capability of the man-model to operate controls, or adjust to the working space and postures required, and an evaluation of the field of view of the man-model in terms of instrumentation and the environment.

An example [14.2] of a man-model looking rearwards to monitor the action of an implement being towed, while at the same time carrying out his normal duty functions, is shown in Figure 14.16. A typical view from the driving seat of a straddle carrier is shown in Figure 14.17, the view including another straddle carrier seen in the distance.

14.8 Computer-aided ergonomic models

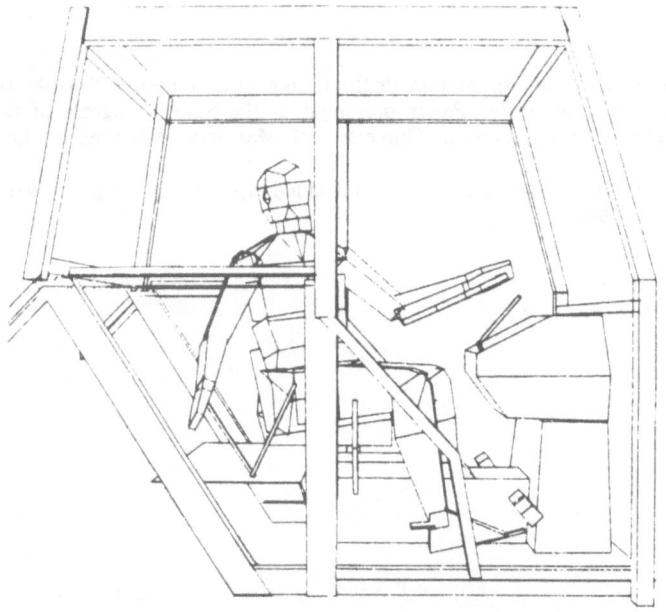

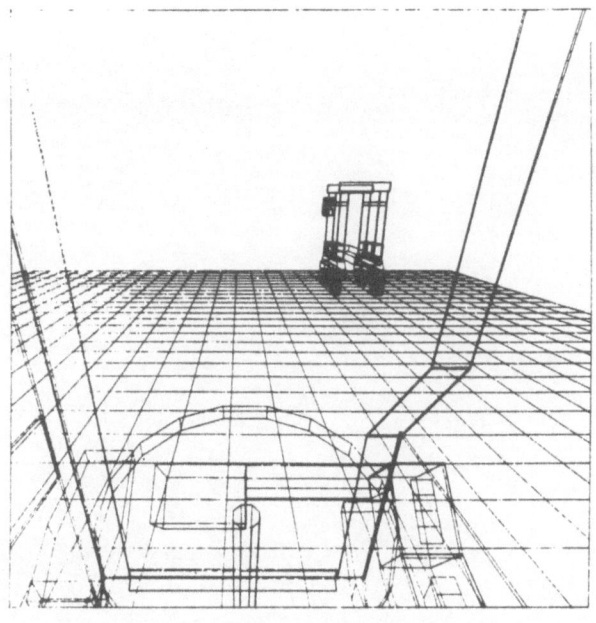

Figures 14.16 and 14.17 Computer-aided model SAMMIE showing (top) typical tractor-driving posture and (right) driver's view from straddle carrier. (*Compuda Ltd, subsidiary of National Research and Development Corporation.* Studies made by Production Engineering and Production Department of Nottingham University)

Computer-aided bio-mechanical analysis using models is another new approach to optimisation of ergonomic design when high-speed changes in position occur between the human body and associated mechanisms, as for example with ski bindings or aircraft ejector seats. In such cases high-speed photography of a model, which in suitable cases could be a human, passing through the operating cycle permits a computer-controlled digitiser to plot the movement of vital components, such as human and mechanism hinge points, and so derive deflection, acceleration and stress curves for both body and mechanism.

14.9 References

14.1 Wood, J. 'The application of ergonomics in the design of a computer console for emergency services; a case study.' Paper presented at the Sixth Congress of the International Ergonomics Association. University of Maryland, Washington DC, July 1976.
14.2 Case, K. and Porter, M. 'SAMMIE: a computer-aided ergonomics design system.' *Engineering*, January 1980.

15 Standardisation and quality in relation to industrial design

W. H. Mayall

15.1 Philosophy for standardisation and quality

Quality as defined by the European Organisation for Quality Control is 'the totality of features and characteristics of a product or service that bears upon its ability to satisfy a given need'. Any techniques that help to produce better products in this context while also minimising costs are important. Standardisation is such a technique since it contributes to cost reduction; it permits the use of production and inspection procedures which result in better quality levels; and it helps the user to maintain quality of performance in service through improved consistency in the operational and maintenance practices and spare parts availability. Aesthetic and ergonomic techniques are also relevant to operational and visual quality, and along with standardisation they should be applied to assist in optimising quality at required cost levels.

The word 'quality' has a number of interpretations. One is the 'degree of excellence' of a product, which implies a quality scale for its performance which might vary from 'good', for a product that works perfectly throughout its prescribed life, to 'bad' for a product that works inefficiently and requires frequent repair. Another interpretation, ignoring performance, concerns itself with surface finish, straightness, alignment, aesthetic merit, colour, graphics, layout and so on. This favours aesthetic and ergonomic aspects of design, but is again far too restrictive. An inspector of products in a factory may regard quality as the degree to which a product conforms with its manufacturing information, which itself involves two classes of variation, namely that specified by the designer, sometimes called 'quality of design', and that contained in the product, sometimes called 'quality of conformity'. All these narrow interpretations are dangerous and overlook the basic fact that in design, to quote G. B. R. Fielden, 'everything affects everything else'.

As an example, 'quality of design' clearly affects 'quality of conformity' and a product that is difficult to make or inspect will entail wider errors in conformity. If a machine overheats it can cause the surface finish to deteriorate, and if good appearance has resulted in inaccessibility for maintenance, performance may suffer.

Perhaps the important point to note in the EOQC definition of quality is its emphasis on the *totality* of features and characteristics that are relevant to meeting a given need. If scales of overall quality are adopted it becomes difficult to avoid suggesting that all products should be 'Rolls-Royce' in character, when clearly less expensive quality levels in design and manufacture may be more appropriate to operational requirements or market needs. However, guidance on quality levels is needed in product development specification (see Chapter 16), and this guidance will mean more to designers if presented in terms of relevant attributes, such as reliability or appearance. The quality level of the product will then be dependent on the integrated effect of these factors.

15.2 Standardisation

15.2.1 Design and standardisation

Standardisation programmes are dependent on markets and marketing policies in conjunction with the resources upon which a manufacturer may call, and these should be outlined in the product specification.

One requirement is essential, in that standardisation means interchangeability of parts within a product or product system, and this involves careful attention to detail design and tolerances.

Standards have a finite life and changes become necessary in them, affecting performance, dimensions, safety etc. The point to note here is that when change is necessary anyhow, this offers an opportunity to improve other aspects of product design, including aesthetic and ergonomic factors. An example is the Central Electricity Generating Board's substations, for which new component standards to improve performance and interchangeability were introduced in the 1950s. This opportunity was then used to create a new overall substation concept that incorporated an effective concern for environmental values (Figures 15.1a, b).

Figures 15.1a and b Early electricity substations (above) not only spoiled the environment but used non-standardised equipment. Redesign took account of aesthetic factors and modern substations (opposite) are much less obtrusive. (*Central Electricity Generating Board*)

15.2.2 Dimensional and functional modular systems

The common brick is perhaps the best-known dimensional module ever created, and demonstrates by the wide variety of ways in which it is used that modular systems can offer variety rather than the regimentation sometimes ascribed to 'standardisation'. But modules need not be identical, and indeed modular design usually means the creation of 'packages' which may be of varied dimensions or function, but which have features that enable them to be coupled together.

Modular design can have a profound effect upon the visual and ergonomic characteristics of a product. It was Roger Fry who said in relation to what we see that 'without order our senses will be perplexed but that without variety they will not be stimulated', and modular design depends essentially upon the establishment of an order in the relationship between whatever kinds of module may be envisaged. As example, the Bruce Peebles cubicle switchgear (Figure 15.2) is based upon modules having a dimensional order similar to that employed for drawers in a filing cabinet, and this type of dimensional order can certainly help to improve product appearance. Excessively rigid adherence to modules with standard dimensions can, however, produce an overbearing impression of order, but very little variety needs to be introduced into an excessively regimented system in order to encourage visual appreciation (see also Chapter 5).

Another approach is to look at modular design in terms of producing units that perform discrete functions, the units being connected together to provide a variety of services. An example is the Baker Perkins Ltd Cremeline machine for making fondants and fudges. Built originally as a single machine embodying cooking, cooling, mixing and separating functions (Figure 15.3a), this unit was redesigned as a collection of modules (Figure 15.3b) each undertaking one function. Each functional unit is produced in a range of sizes, and these can be marketed singly or built into coupled systems which can be more varied than was possible when all the functions were built into one monolithic unit. Standardised, simple and low-cost frames have replaced the large and complicated casing required for the integrated design.

Modular design can also contribute to technical development, as it permits replacement of individual modules by improved versions which retain physical interchangeability, again refuting the comment that standardisation inhibits progress. An example affecting visual and ergonomic requirements is a Pye Radio Ltd television set, for which the controls are built into three types of interchangeable module, one of which was actually designed for *envisaged* rather than immediate market needs.

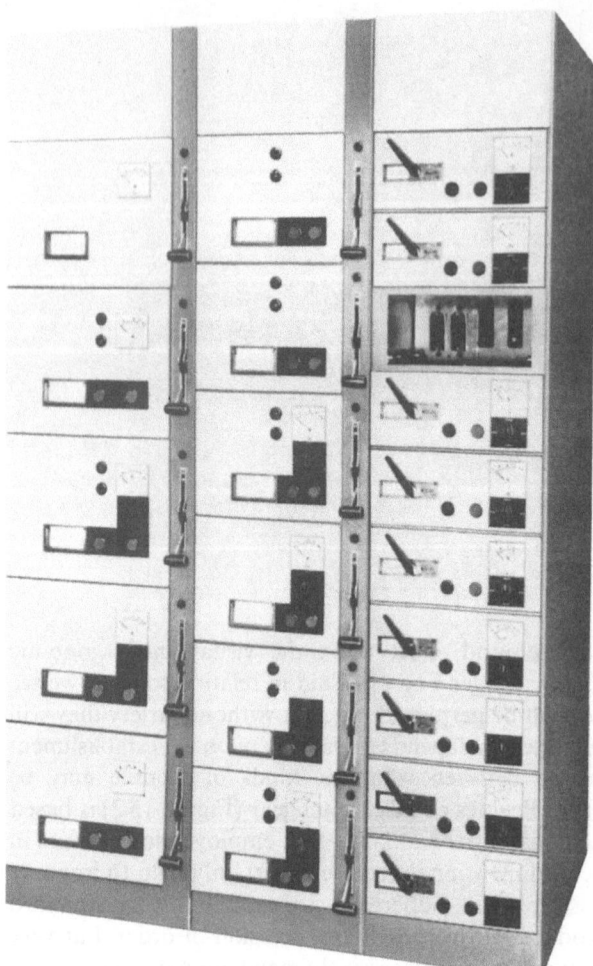

Figure 15.2 Modular switchgear, employing a filing-cabinet type of construction to assist maintenance and improve appearance. (*Bruce Peebles Ltd*)

15.2.3 Core and dressing modular systems

Another aspect of modular design is the creation of a basic core unit to which different elements can be fitted, so enabling a variety of versions to be produced. The important requirement for this approach is that the core should have sufficient capacity to cope with all expected variations in performance and usage.

An example is the Perkins six-cylinder diesel engine (Figure 15.4) in which redesign for a modular approach reduced core variations from 300 to 25, while a similar approach was taken to standardisation of the engine dressings. Despite an apparent reduction of variety, this engine still satisfied a very wide range of requirements, and combined with improvements in performance and reliability achieved by the redesign, this helped to overcome any feelings by customers that their choice had been restricted.

A 'spin-off' advantage that can be derived from modular design of advanced products such as turbines or jet engines is that this encourages the creation of specialist design teams, each responsible for a component module, which can assist in optimising the performance and cost of products. A problem can also arise, however, when the assembly has to satisfy aesthetic criteria, because, almost by definition, it is the overall and not the individual appearance of an assembly of

15.2 Standardisation

Figures 15.3a and b The Cremeline confectionery machine originally incorporated several functions in a single unit (right), but was redesigned following modular principles (below) with a separate module for each function. (*Baker Perkins Holdings*)

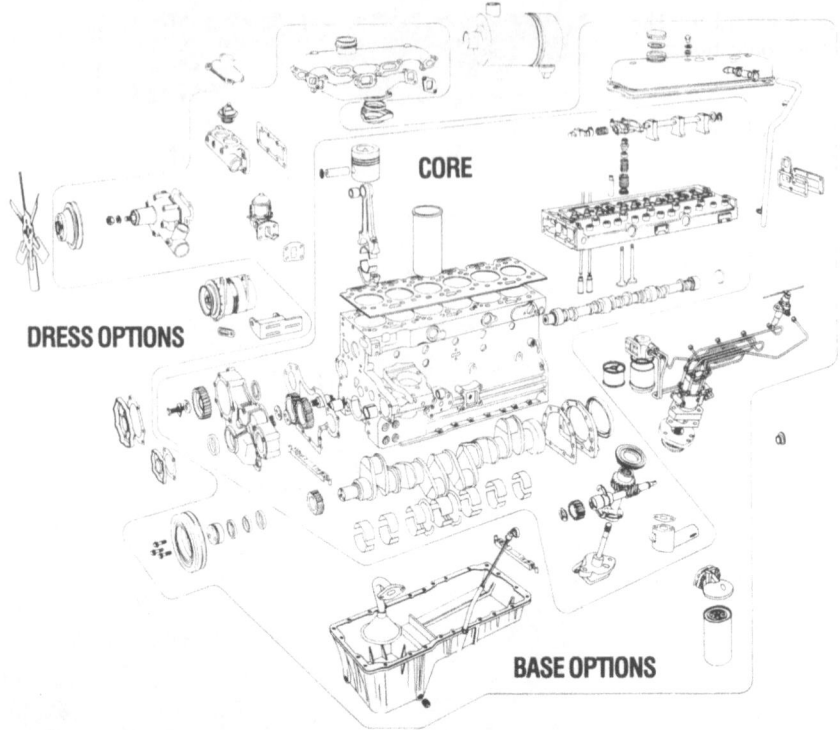

Figure 15.4 A breakdown of a Perkins six-cylinder diesel engine showing the core and dressing principle. (*Perkins Engines*)

Figure 15.5 The range of fork-lift trucks made by Lancer Boss employs standard constructional techniques for chassis, mast and superstructure, with common control consoles and drivers' seats.

units which matters. In the case of aircraft engines the aesthetic criteria may be of little consequence, although perfection of detail can achieve impressive aesthetic results as seen in Chapter 1 (Figure 1.18), but when such products are in contact with people, then overall aesthetic assessments and design decisions must be made at a very early stage in their development, and this requires a policy of co-ordination (see Chapter 18.21).

Examples of core and dressing techniques include the ICL computer (Chapter 18.11), the Vickers microscope (Figure 12.3), and the Mather and Platt motor (Figure 14.4) which departed from conventional construction by using folded steel plate and was based on a motor core and top dressings for a variety of cooling systems. In each case control over the range of options was maintained by an industrial designer, in the interest of achieving aesthetic objectives in relation to cost, ease of use and maintenance.

15.2.4 Standardisation of construction

Another approach is standardisation of construction, which is valid in its own right and is applicable even to products manufactured in small quantities, or in ranges having similar characteristics, where it can help both by improving quality and reducing cost. The Lancer Boss fork-lift truck (Figure 15.5) exemplifies this approach, which can be considered under the following headings.

(a) Standardisation of material and material sizes, possibly covering sheet, plate, bar, extruded and rolled materials, coated steel and punched sheet steel.
(b) Standardisation of hole sizes, thread sizes and joining systems.
(c) Standardisation of basic components, including bolts, screws, seals, bearings, pipe joints, cleats and electrical connectors.
(d) Standardisation of structures and assemblies, which in order to achieve aesthetic objectives might include requirements for concealed fixings, hinges and locks; standard shapes, sizes, arrangements of drillings for casings, doors, mountings and plinths; standard corner and edge forms; and standard details such as louvres and wheels.
(e) Standardisation of service systems and their installation, including wiring and pipe runs, harnesses and safety systems, including layout rules to ensure tidy appearance and accessibility.
(f) Standardisation of controls, including panel layouts, knobs, levers, handwheels, visual indicators and warning devices, all determined with close concern for ergonomic requirements and visual co-ordination.
(g) Standardisation of treatment and finishes, including colour selections for operational purposes as well as any required for house styling.
(h) Standardisation of lettering, symbols, name plates, rating plates and logotypes.

These standards should be embodied in company standard manuals, and this is a procedure best carried out by specialist staff if the scale permits. An important aspect of standardisation, design and recording is keeping it up to date in relation to product ranges and new developments, and the following parameters must be constantly borne in mind.

(a) The product's purpose.
(b) All the circumstances which the product will experience, from storage through manufacture to transport, installation, use, maintenance and replacement.
(c) The cost conditions to be satisfied.
(d) Available or specified manufacturing facilities covering parts manufactured, treatment, assembly and inspection.
(e) Specified quality levels related to performance, reliability, safety, maintenance, ergonomic and appearance requirements.
(f) Relevant international, national and client standards and codes of practice.

(g) Legal requirements.
(h) Technological and manufacturing developments, which are particularly important.

15.2.5 Standardisation of design procedures

In many cases, design procedures relating to performance, efficiency, structural strength and so on will be similar to those used in previous projects, and when this family likeness exists the preparation of standard design procedures will reduce design effort, save time and reduce the risk of error.

Ergonomic and aesthetic aspects of design systems can also benefit from standards in some special cases. For example, the logic to be used in control panel design can be standardised for particular types of system, leading to consistent quality and reducing the risk of ergonomic errors that may lead to operational mistakes. Standards for basic ergonomic requirements such as work heights and control forces will assist in all cases. Aesthetic factors cannot be covered by calculation or prescribed techniques, but nevertheless, when they are relevant, standards for form and colour for particular products will improve co-ordination. The maintenance of visual records of so-called 'appearance design', covering not only the types of product with which the designer is concerned, but also those with which his products are likely to be seen, including where possible constructional details, will also be of value.

15.3 Quality

15.3.1 Design and quality

As discussed earlier, quality is all embracing; it includes aesthetic and ergonomic factors and the interaction between them and other aspects of quality of the product. There are a variety of features in most products that directly affect quality of appearance and ease of use, and the designer must identify these and also the other features most likely to interact with them in service. Examples of features that have a direct effect on aesthetic and ergonomic performance for a 'cased' product might include the casing construction, control panel layout, fixings, mountings and so on *and the ways in which all these features are connected.* Features having an indirect effect might be the internal working elements, their interconnection system, and more especially the ways in which they are coupled to the external, 'direct' features. The sooner features directly affecting quality as it relates to any characteristic, whether it be performance, ease of operation or appearance, can be identified in the development process, the better. Only then is it really possible to study the following aspects of product quality.

(a) Capacity to undertake defined functions in envisaged environments.
(b) Capacity to sustain these functions having the interaction of all related features in mind, not only when they are performing correctly but also if and when they deteriorate to lower levels.
(c) Capacity to withstand overloads, incorrect usage and maltreatment.
(d) Capacity to withstand deterioration not only in use, but also in storage,

distribution, erection on site and of course during maintenance.
(e) Capacity for being manufactured without undue variation in dimensions, shape, finish and colour, or in such a way that expected variations will not diminish aesthetic requirements. For example, clear gaps between panels will reduce the visual effect of dimensional errors; specifying distinct colour differences may be preferable to trying to match colour finishes on different materials.
(f) Capacity for enabling quality of conformity to be achieved economically.

It is thus vital to consider the interactive effects of all related features. In very many cases, whether in terms of appearance or other characteristics, it is inadequate interface design of features such as couplings and mountings, and a failure to judge the indirect effects of parts of a product upon one another, that lead to quality problems. Having noted the main contexts in which a product's features, *and the ways in which they combine,* can be judged, it has to be admitted that the judging process is by no means easy. One possible approach is through the use of a 'design audit'.

15.3.2 Design audits and checklists

Design is a process of hypothesis and evaluation; of conceiving solutions to apparent or simulated needs and then analysing the solutions to make sure they will satisfy those needs. This process of iteration can be extensive, even to produce the design specification required before further development can proceed. For example, determining the quality levels required by a specification may involve analysing trading conditions, the character of competing products, the ways in which the product will be used, where and by whom, and adjustment may then occur as levels are viewed against cost, resource requirements, and time-scales for development, manufacture and installation.

Although detailed evaluation of product features cannot be undertaken until they are known, it is well worth while initiating a design programme together with a quality appraisal. This initial approach should then be followed by a second examination, when all the features are known, and end with a thorough examination before the product is manufactured. Marconi Avionics Ltd, who are concerned with products for which high quality is very important and also a contractual obligation, recommend this approach, with continuous monitoring between review stages, in a comprehensive 'quality manual'. There is probably no better way of identifying the nature of the techniques required for reviewing such products than by reference to this manual, with the kind permission of the company. The design audit, or review, reproduced here refers also to checklists, which can provide a valuable detail reminder of all aspects of a design that should be checked—for example stress, life, corrosion resistance, fire resistance, isolation of incompatible services, control logic etc.

Design review (adapted from Marconi Avionics Ltd)

1 Purpose

Procedures shall ensure that:

(a) all design and development staff are made fully aware of their responsibility for quality and cost targets;

(b) the results of all aspects of design appraisal are considered;
(c) all factors determining quality and cost are emphasised from the outset of the work;
(d) the development programme objectives, and progress towards their achievement, are subjected to frequent and thorough review to ensure timely identification and anticipation of problems;
(e) final design and supporting design data fully meet the requirements of the contract; and
(f) the costs and time-scales are within the contractual limits.

2 Checklists

The checklist is an important tool for Design Review Board activity. The necessity for careful evaluation and responsible checklist entries cannot be over-emphasised. The objects of a checklist are:

(a) to list all the possible factors pertinent to each major design area;
(b) to prevent inadvertent omission of any critical factor;
(c) to prevent time-consuming reviews of minor or non-critical areas by the Design Review Board;
(d) to highlight critical areas for review and action;
(e) to make evident a design problem or solution for use in other areas of activity; and
(f) to provide a record of review activity and design achievement.

Product parameter review

Areas of design examined by the Design Review Board at all stages of review shall include, but not be limited to, the following:

Cost and schedule	Quality assurance	Power supplies
Performance	Reliability	Power consumption
Configuration	Maintainability	Test methods
Mechanical design	Mechanical and electrical	Standardisation
Electrical design	interchangeability	Production engineering
Thermal design	Electromagnetic	Value engineering
Ergonomic design	compatibility	Product safety
Appearance design	Environment	Patentability
	Weight and size	

NOTE: Several of the above will have an impact on any software associated with the equipment, and on the process of component selection.

Membership and responsibilities of Design Review Board

Member	Responsibility
Divisional Manager (Acting Quality Director)	Implements the Company Design Review Board policy.
Chief Engineer (Chairman)	Convenes and conducts meeting of the Board.
Production Manager	Advises on suitability of the design for production.

Divisional Quality Assurance Manager (Co-ordinator and Secretary)	Assists Board regarding agendas, checklists, procedures, codes of practice, minutes and records.
Project Manager	Presents the design of the product of the Board and implements recommendations.
Other Members, Divisional	As required, to assist the Project Manager and the Board.
Other Members, Company	As required, at the discretion of the Chief Engineer, to assist the Board; the company Design Consultant and Chief Engineer Aviation Service and Repair Division, to be advised of all meetings of the Board.

Where ergonomic and aesthetic requirements are significant, these characteristics will have to be related in particular with configuration, quality assurance, reliability, maintainability, environment, weight and size, standardisation, value engineering and product safety. They will be demonstrated in most products by the following.

> *Construction*—with consistency of manufacturing joining or fastening techniques very much in mind.
> *External shape and treatment*—with reference to proportions, the effects of ledges and pockets on corrosion and ease of plating or painting in mind. The consistency of recommended finishes, colours and textures should also be considered.
> *Access panels*—with their location, size and ease of handling in mind with respect to maintenance.
> *Control panel layouts*—which must be examined to ensure conformity with ergonomic principles and safety.
> *Colour and graphics treatments*—which must be considered both in terms of aesthetic and ergonomic needs and conformity with the company's house style or client's specification.
> *Supply connections, mounting systems and relationships with other products.*

An example of successful design carried out through this procedure is the Marconi Avionics automatic pilot, discussed in Chapter 18.8.

15.3.3 Feedback to design

The design process does not end when manufacturing drawings and instructions have been produced. Information must be fed back to designers from manufacturing, development and prototype testing, and in particular from experience in use. Four requirements for ensuring quality levels then are:

(a) That data is obtained from all the activities which take place after a design has been completed;
(b) That the data collected is comprehensive, especially when it applies to defects or failures;
(c) That data reaches designers without bias and in an intelligible form; and
(d) That this data, the design response and the reasons for it are systematically recorded.

Feedback from user experience may well have to cover a wider field than defects and similar reactive types of report in order to provide the kind of information that will assist in improving aesthetic and ergonomic qualities. Objective studies may then be needed to see just how people treat the product and the impact of environmental conditions, vandalism, pollution, misuse and so on. Feedback is also relevant to installation, operation and servicing manuals. These should be clear, brief, simple and logical in the procedures called for, but their quality can often be improved if their editors work closely not only with product designers, but with those who use and maintain the equipment in the field.

16 Specification for the development of engineering products

Charles H. Flurscheim

16.1 The purpose of specifications for development

Most specifications are written to define the specific requirements of purchasers in relation to more or less established products, but here we are concerned with the broader issue of the development of new products. The specification for a development project must of course embrace the needs of purchasers, but should also provide guidance on financial, time and technical targets, so far as these can be determined in advance.

It is important to clarify who should be responsible for the preparation of a development specification. The selection of a product for development is influenced by what is needed, by what can be made, and by finance. A first priority is therefore to assess the needs of the market, based on combined commercial and technical viewpoints. Market needs can sometimes stimulate solutions for previously unsolved design problems, as for example in the case of radar with the invention of the klystron. The technical assessment of what can be made may sometimes suggest new developments and opportunities, but these must be checked for their market realism before they can be pursued with confidence. Financial viability depends on cost and cash flow generated by the product, estimated from the commercial assessment of likely price and sales levels, and on the technical assessment of probable development and production costs. It follows that the preparation of specifications for development must be a joint responsibility of the commercial and technical functions, acting closely together.

It should not be overlooked that industrial designers can often contribute to the definition of what is needed, and their position within the engineering function should be such that their views are considered at the earliest stages, when the specification is still flexible.

The question then arises of how flexible such a specification should be. When practicable, a complete and final project specification should be prepared before beginning development proper, but in practice flexibility in the specification as development proceeds is sometimes desirable or even necessary. This is particularly so when there is a major degree of innovation or extrapolation, when significant development and design work may be essential before the specification itself can be finalised with confidence. This then implies a two-stage procedure, in which a tentative development specification is first authorised to cover initial development, leading to a detailed specification, followed by the approval process for the complete project. 'Final' specifications should also be considered as flexible to the extent of accepting changes resulting from opportunities or difficulties that may be exposed at later stages in the development, provided that these changes are agreed through the appropriate procedures and that their effects are noted.

The preparation of a development specification is therefore a vital step towards achieving success, and should be carried out in sufficient breadth and depth for the document to fulfil three important functions.

First, it should be so compiled that, in combination with a commercial brief, it will clothe the project proposal with the marketing, financial, technical and time-scale information on which management approval to proceed can be based.

Second, it should define the marketing requirements of what is to be developed, including performance requirements, limiting criteria, quality, cost, output levels and time-scales in sufficient depth to provide design guidance and boundaries within which development is to be carried out.

Third, it should provide a methodical statement of agreed development aims against which progress in achieving performance, quality, cost and time targets can be monitored as the work proceeds. This will then make possible the early identification of unexpected opportunities, difficulties or deficiencies that arise in relation to the target objectives. Such identification will then help in ensuring prompt action by management, by reinforcing the design effort, or by modifying the specification in terms of performance, quality, time-scale or cost targets, or, if the picture is too unfavourable, by cancelling the project.

Design is a highly integrated operation, and it follows that it is not practicable to isolate the industrial design requirements, or the requirements of any other technology, in a separate specification, because of their interdependence and the imbalance that would result. All significant requirements that will influence the design should be scheduled in a single specification, although their relevance to the different aspects of design may vary in importance. If some of the design work, such as the industrial design aspect, is to be carried out externally by consultants, the specification should also include a brief of in-house design practice and notes on any special production facilities that might influence their approach.

16.2 Specification content

The principal headings relevant to the preparation of an engineering development specification are discussed below.

16.2.1 Objective

The objective of a development product should be stated so that, in combination with a commercial brief, the logic for proceeding is clear and, as the alternative is not proceeding, the consequences of this course of action should also be discussed. The principal objective of development based on existing equipment may be to improve performance, reliability, life, ease of maintenance, safety, ergonomics or appearance; to present a new commercial image; to reduce costs; or to extend the range of size or function available. If the aim is to develop entirely new equipment, this may be to penetrate existing markets or to create new markets through innovation. The general marketing philosophy is important in the assessment of viability and risk, and will also help the engineering team by providing the background against which the work will be carried out.

16.2.2 Definition of product

A general description should be given of the function of the product, how it will be applied in service, and the environments in which it will be used. This will need to be supported by schedules of types, sizes, performance and ratings to be developed in the initial programme. Where relevant the nature and quality of inputs to the system should be scheduled.

16.2.3 Variants

If the equipment will have a variety of functions or performance requirements, achieved through adding or substituting components, these should be scheduled with their purpose. Since making provision for such components will usually require additional engineering and fabrication, it should be made clear which versions are standards and which are specials for which ad hoc design and manufacturing changes are acceptable.

16.2.4 Future requirements

If additional ratings or variants, or 'stretched' size or performance versions, will be required in the future, these should be outlined, so that they can be taken into account in the initial development, with the objective of reducing subsequent design changes to a minimum.

16.2.5 Component standards

The extent of standardisation of a product has important effects on cost and reliability. Standardised components or modules should be identified in the specification, and if these could serve other products, or if they are existing modules or standard components, this should be clarified.

When company standards on components and materials exist, these should be identified, more especially when some of the work is to be carried out externally, for example by industrial designers who may not be familiar with in-house procedures.

16.2.6 Technical standards

International, national, client, in-house and suppliers' standards relevant to development should be scheduled. These may cover dimensions, performance, permitted tolerances, testing, application engineering, materials, finishes, safety, appearance, environmental aspects, transportation, quality levels, quality assurance and inspection, design auditing, insurance, and design methods or systems to be employed.

16.2.7 Special legal requirements

There are sometimes special legal requirements over and above those normally implied, and special liability risks, and these should be scheduled.

16.2.8 Design basis

Design should normally be executed in the metric and SI unit systems, but this should be clarified when there is uncertainty.

16.2.9 Quality requirements

Although many quality characteristics of engineering products are difficult to measure in precise terms and criteria for them are often taken for granted, without some degree of specification these aspects may receive inadequate attention, or on the other hand, by aiming at perfection, design may cease to be cost effective. Some of the aspects of product requirements which influence quality in service and which impose constraints on design are given below, and these should be defined in the specification so far as is practicable.

Performance. Some aspects of performance can be measured, such as efficiency, accuracy, output, vibration and noise, and these should be stated. Other aspects of performance are more subjective in character, such as stability, smoothness and response to control, and for these the specification should indicate their significance, with the required levels quantified so far as possible or relatively in comparison with other existing and similar equipment.

Reliability. Reliability can sometimes be measured in terms of failure rates, expected availability or life expectancy, or related to maintenance levels and costs. Absolute or relative guidance, in comparison with other equipment, should therefore be given on reliability levels required, since these may influence basic design, the type of reliability testing required, the extent of ergonomic assessment for control systems, the use of fail-safe techniques, and materials and finishes used.

Ergonomics—user population. Where there is an identifiable user population this should be specified in terms of, for example, special skills or lack of skill; physiological, psychological and cultural characteristics; age, sex, nationality, language occupation, captive or non-captive status, disability etc. This data can affect anthropometrics, sensing, operational, appearance and safety design requirements.

Ergonomics—control. Ergonomic control requirements, their importance and relevant standards should be identified. These may take into account the user population, the control logic to be employed, preferred mechanisms for transmitting information, the need for feedback, and the type of display to be used. With some products, more especially domestic appliances, the instability of market requirements may demand minor changes in the control function, or in its presentation, during development or at relatively frequent intervals in production. Such possibilities should be clarified in the initial specification so that the class of design selected for the components can accept minor changes with the least possible dislocation and re-tooling costs. The extent to which ergonomic testing should be carried out should also be clarified.

Ergonomics—environmental. Several environmental aspects require clarification in a specification. These include: local working conditions created by the system itself or imposed by its surroundings; broader environmental problems such as high or low temperature, altitude, humidity, wind, salt spray, blown sand, flooding, invasion by insects, earthquakes, dangerous gas atmospheres etc; and finally man-initiated conditions such as rough handling, vandalism, vibration, radiation, and the special environmental requirements of aircraft cabins, hospitals, process plants, coal mines and steel mills.

Maintenance. If there are special requirements for maintenance these should be specified. These may include, for example, design to achieve exceptionally low maintenance levels, or to extend the period between preventive maintenance procedures; ergonomic study to make maintenance easier, simpler and suitable for non-skilled labour; design to reduce the need for cleaning; the use of factory-built replacement modules for worn components; and the need for built-in fault identification systems. Precautions for safety during maintenance and associated instruction manuals may also need consideration.

Appearance. Appearance is an important aspect of quality and guidance should be given as to generalised market requirements and trends affecting form, colour, style and finish; presentation of the control system; permissible rate of deterioration of appearance in service; any special requirements affecting appearance associated with the user group or the manner in which the equipment will be used and maintained; compatibility with products with which the equipment will be associated; environmental requirements; and in-house standards affecting style and colour schemes.

Graphics. Graphics contribute both to appearance and to ergonomic efficiency, and directly affect the clarity of presentation of information. Aspects that should be covered include relevant graphic standards and type; requirements for instructions and name plates; requirements in relation to language and symbols, especially for export, and possibilities for standardisation. In-house facilities should also be identified.

Safety. Safety embraces all the modes of failure that could lead to accidents. It is always an overriding consideration and relevant safety standards, including any legal and special requirements, should be identified. In terms of design specification this may include the need for coding, interlocks or guard systems; acceptable radiation exposure levels; the extremes of population size which the design must accommodate and the action to be taken to make any restriction acceptable; and the requirements for a statistical assessment of risk, and permissible levels of failure. Design for safety will also be aided by reference to the nature and cause of known accidents associated with similar equipment in the past.

Transportation. Product design may be affected by dimensional or weight restrictions in transport, and by the need to restrict damage initiated by exposure, vibration, or shock associated with mishandling in transit.

Storage. Some products have to be stored for long periods, and special design attention may then be necessary to control deterioration.

Quality assurance. If development of the product is to be subject to external or internal quality assurance procedures, these should be identified. A special aspect of QA is the procedure known as the design audit. The specification should clarify whether a design audit is required, and the degree of independence it should possess. The industrial design aspect of the product in particular may require auditing with respect to control, maintenance and safety.

16.2.10 Production levels

Output, including batch and annual production levels and also the expected number of years the equipment will remain in production, will influence the cost of tooling that can be justified. Estimates should therefore be given of production levels for the product and its variants. These should take into account the effects of any work under licence that might modify the production level of parts.

16.2.11 Work in progress

The value of work in progress necessary to achieve the specified production levels and delivery times should be estimated. Design for minimum cost of large components that require extensive processing, or have a long material procurement time, may not always be compatible with delivery time and WIP cost targets, and it may therefore be necessary to consider alternative and more expensive design.

16.2.12 Obsolescence

If the development replaces existing products, these should be identified, together with any components that will still be required in the new product, so as to provide advance warning to the production staff of future changes in demand.

16.2.13 Product cost

A target is required for the total manufacturing cost of the product, its variants and components, including labour, material and overhead costs, and in relation to proposed production levels. If identifiable parts of the products, such as those comprising the man/machine interface, are to be designed by external consultants, it may be helpful to give separate cost targets for them.

If future cost reduction for a subsequent model is envisaged, the target costs should be given to permit estimates of its engineering and tooling costs to be made. If lifetime cost is relevant, guidance should be given on the ratio between lifetime cost and development and production costs.

16.2.14 Direct development costs

Development costs are a factor in assessing the viability of a product and they also provide a measure of the technical effort required. Costs for development tend to be underestimated because sufficient margins are not included to cover unexpected technical difficulties or the need to retrace steps in order to take advantage of improved possibilities exposed as the design proceeds. To be realistic, estimates should always include an allowance for such possibilities, the size of which should be related to past experience.

In order to assist in the monitoring processes, development costs should be divided into appropriate phases.

> *The design phase* includes the provision of essential research-based information; engineering and drawing-office work; external design consultancy that may arise, for example for industrial design; quality assurance design auditing; construction of models as an aid to design; and associated ergonomic studies.
>
> *The prototype development phase* includes the manufacture of prototypes and type testing.
>
> *The design for reliability phase* consists of long-term reliability testing.
>
> *The production phase* consists of production tooling.
>
> Finally, the *post-production phase* covers license fees for use of patents or design information.

If a subsequent cost-reduced design is envisaged, estimates should be given for the cost of value analysis engineering and the re-tooling required.

If lifetime cost is relevant, estimates should be given for the additional reliability engineering, ergonomic studies and extended life testing that might be necessary.

16.2.15 Indirect development costs

In addition to direct costs, indirect development costs arise which may or may not be charged for accounting purposes to development, but which are relevant because they erode profit margins. Although often disregarded, these costs can be significant, and they should be estimated from past experience, adjusted for innovation (which tends to increase them) or improvement in quality assurance procedures (which tends to reduce them). They include: learning costs, which arise in the initial stage of production as the consequence of teething troubles, or setting new piece-work rates; costs arising from the fact that initial production levels may be below those estimated; rectification costs carried out free under guarantee or to retain goodwill; and the costs of preparing operating and maintenance instructions, which can be initially expensive, although they will in the longterm reduce subsequent maintenance and lifetime costs.

16.2.16 Time-scale of development

The time required for development is an important factor, influencing the launching of the product on the market and also the total development cost. A timetable should therefore be prepared giving estimated completion dates for the various stages in the development procedure. Some overlap between these stages is, however, inevitable. Such a schedule will facilitate the monitoring of progress. Typical stages in a project will include:

Design phase
Completion of any basic information from research or similar external sources necessary to permit design progress.
Completion of performance calculation and general concept arrangement.
Completion of any design concept work provided by external sources, including industrial design.
Completion of all detail design and manufacturing information for prototypes.

Prototype development phase
Completion of manufacture of prototypes. Completion of type test programme.

Design for reliability phase
Completion of any prolonged reliability testing programmes and associated design modifications.

Production phase
Completion of manufacturing information for production.
Completion of manufacture of special tooling.
Completion of manufacture of the first batch of products from production tooling.
Completion of any check-testing programme for items made from production tooling.

16.3 Discussion

The extent to which any particular development should be specified in detail will of course depend on its importance, complexity, risk and novelty. Experience suggests, however, that there is often insufficient analysis of what is really needed before the full development programme commences, that specifications therefore tend to be inadequate in the guidance they provide, and that in consequence project efficiency is impaired, so that costs, time and performance drift away from the targets that are required and are possible.

A special problem arises with industrial design because interface requirements are sometimes omitted altogether from the initial specification, which then results in this aspect of design being relegated to a low priority and to a later stage in the development, when the difficulty in making changes to components already in an advanced state of design, and the cost and delay this involves, inhibit adequate consideration and implementation of interface requirements. The specification of these controlling factors will ensure that they are considered in depth when the design is still flexible, and will encourage their detail development in parallel, rather than in series, with the rest of the engineering work.

The use of standards, when relevant; the definition of service conditions; clarification of acceptable life expectancy, failure risks and levels; the approach to safety and design audit procedure; and the effectiveness of the specification in interpreting these and the important aspects of quality discussed above, including those of industrial design, will all contribute to the improvement of the product, the elimination of deficiencies and errors in design, and consequential troubles.

As a spin-off, an adequate specification procedure will also assist, by reducing risks and by providing evidence of the care taken in the design process, in the area of product liability, which with the hardening of international legislation on this subject is becoming a significant aspect of total product costs.

The not inconsiderable work involved in the preparation of adequate specifications should therefore be accepted as an essential component of the development procedure. The extent of this effort is likely to be very small in comparison with the benefits that result from a comprehensive definition and follow-up procedure.

17 The integration of industrial design and engineering

Antony Gibbs and Charles H. Flurscheim

17.1 Introduction

Engineering design, like other major functions, requires organisation, and within the terms of reference that are relevant here this involves three important issues. First, it requires a detailed design brief against which the development team can work, as discussed in Chapter 16. Second, it requires the organisation of the design team itself.

Ever since the Industrial Revolution, continuing sophistication in engineering and manufacturing techniques has led to a technology of such complexity that no single individual can hope to grasp it all. For this reason the team may not only have to include individuals conversant with the different disciplines within engineering, but also specialists from other disciplines such as industrial and graphic design and marketing. Obviously the team will consist largely of in-house staff competent in the techniques needed. Any technique that cannot be covered in-house must be covered by consultant specialists. It is important to note, in passing, that the value of such support should be acknowledged, since the reputation of specialists is largely dependent on such acknowledgement.

Finally, organisation of engineering design requires that the operating procedures of the design organisation will ensure proper integration of all the technologies involved.

The introduction of industrial design and other techniques and their marriage with engineering, so essential to achieve balanced design, is just one special aspect of the overall responsibility for the management of design, and the issue here is how this expertise should be provided and integrated with engineering.

There is popular belief that design by committee leads to mediocre results. However, if each member of the team confines his contributions to his specialisation, the ideas will not get diluted. If suggestions are not in conflict but are linked, a better relationship between the experts themselves will develop, allowing for cross-fertilisation of ideas. For example, an expert may not be able to solve a problem because he is so close to his subject; but a suggestion from another member of the team may catalyse his thoughts and lead to a solution.

If the expertise of a consultant is required, then he should become a full member of the team. Furthermore, each member of the team should be of sufficient rank so that, first, subject to the co-ordinating responsibility of the team manager, he can make decisions without recourse to a higher authority, and second, each member of the team must be of equal status, thus ensuring that any decision made is in the interest of the project as a whole, avoiding problems arising from rank-pulling and empire-building.

The contribution made by industrial design may be substantial, comparable with that made by the engineering-based technologies used in designing the principal features of a project; or it may be secondary, comparable with the many engineering technologies that may be required only for specialised and local design problems. According to the nature of the product, thermodynamics, electromagnetics, electronics, hydraulics, aerodynamics, acoustics, optics and industrial design may be required in major or in support roles, but in all cases they can have important repercussions on performance and quality.

Because of the width of their field, engineering teams, unless very large, are likely to be equipped in depth only with the major engineering-based technologies appropriate to their product, and to a more superficial extent with those in a supporting role. Because of the small amount of technical training engineers receive in the field of industrial design, their knowledge of this subject tends to be very restricted. The extent of the theoretical sciences needed to cover the principal engineering-based technologies inevitably restricts the extent to which peripheral subjects can be studied within the limited time available in degree courses. In consequence the training of engineers is rarely, if ever, sufficient for them to carry out industrial design in its major role and, unless it is supported by postgraduate courses, is usually inadequate even in the support role. Experience in the field is, of course, some substitute for academic training, and engineers who have acquired experience in specific industrial design problems can extend their effectiveness. Certainly there are engineering teams who can achieve excellent industrial design in their own right, especially in the field of ergonomics, as is witnessed by some of the examples discussed in Chapter 18. It must always be remembered, however, that it can be very expensive in terms of time and uncertainty to employ a person who does not thoroughly understand his discipline. It is always better to get advice from an expert rather than to muddle through.

On the other hand the extent of engineering knowledge that is required in order to contribute usefully to the industrial design of a project increases as the man/machine interface moves from external appearance aspects towards the control of complicated machines and systems. The training and experience of industrial designers in engineering varies widely, and those who work in the field of product design will be more effective if they can communicate on the same wavelength as the engineers with whom they have to co-operate. One of the objectives of this book as stated in the introduction is therefore to help engineers to achieve high standards of industrial design. In the general case, however, and especially when industrial design is a major consideration, high standards are more likely to be achieved with less experimentation and error if appropriate specialised assistance is integrated with the engineering-based teams.

Another objective of this book is therefore to help engineers to decide when specialist assistance should be sought by providing a better understanding of the man-machine interface problems. In addition to the immediate help provided by the introduction of industrial design specialists, there is an additional educational benefit for the design team in its approach to future man/machine problems.

17.2 Methods of integration

In practice the man/machine interface problems amenable to solution by industrial design procedures vary widely. In heavy engineering, for a steam turbine for example, several hundred engineers and draughtsmen will be involved for several years in designing a new machine. The industrial design contribution can influence the design of major components such as the low-pressure turbine cylinders, their steam flow paths, the alternator stator shell, removable enclosures for turbines or exciters, and the arrangement of many associated components such as hydrogen-cooling controls, lubricating pipes, instrumentation and cabling. Industrial design capability should therefore be available locally in the design office, whether provided in-house or by external sources, throughout the design period, so that the problems requiring attention can be identified and dealt with as they arise and before design is too far advanced. The engineering and industrial design problems are in fact so interdependent that it would be quite unsatisfactory to permit any segregation of them. At the other extreme, simpler forms of engineering-based consumer equipment may require only a small design team, for a period of perhaps months rather than years, designing apparatus in which it is sometimes possible to separate out the components that form the interface with the user from those that are engineering-based. It may therefore be possible to design interface components in a separate and remote office, provided that close collaboration is maintained to ensure optimisation and compatibility. This situation also applies to products based on microprocessors when these are closely related with the visual presentation of control.

There is, therefore, no single option for implementation of industrial design which will meet such different criteria effectively, and possible alternatives are discussed below. The overriding principle in selecting a method must be that, the greater the difficulties in separation of man/machine interface design from the engineering-based components, the greater the need for close integration between engineering and industrial design. The closest integration is achieved when the responsibility for industrial design is carried by a member of the design team, and this becomes more necessary with increasing complexity. He can then identify ergonomic and aesthetic problems as they arise, and before design has progressed so far that adjustment becomes difficult or impossible. The industrial design responsibility may then be carried either by a member of the design team who is experienced in industrial design techniques, more especially relative to his own fields of products, or by a professional industrial designer co-opted onto the design team for the period of the project. If industrial design is to be carried out by the former method, the responsible individual should be clearly identified, as would be the case with other specialist techniques. He should be of a status that will enable him to influence the industrial design aspects of the design specifications, design concept and detail design as it proceeds. It should be remembered that, while industrial design training courses will assist him, they are not a substitute for the expertise of a professional industrial designer.

Such an approach possesses the inherent advantage of complete integration, but has the disadvantage of relatively restricted industrial design training and breadth of experience. Co-opting a professional industrial designer onto the design team will also ensure maximum integration, but conversely, as this approach is most appropriate with complicated machinery, the industrial designer should in turn possess an adequate engineering background.

An alternative is to employ a visiting industrial designer on a part-time consulting basis, so that he can provide advice and contribute to design proposals, to be executed in detail by the company's engineering design team. When such consultation is concentrated on a particular design, is sufficiently frequent, and starts at the specification stage, very good results can be achieved without the risks that arise when the interface components and engineering of the machine are carried out in different offices. A modification of this scheme in large companies is to use an external consultant in an educational manner by spreading his advice more thinly over the whole engineering activity. This will have the effect of raising the overall industrial design appreciation of engineering design teams by improving their general understanding. This approach may, however, slow down the design programme, as training designers on a project can lead to mistakes, and if the work of trainee industrial designers is not thoroughly vetted by a qualified man, the quality of the final product may also be reduced. Intermittent consultancy on any particular product is therefore not likely to achieve outstanding results, but a worthwhile improvement in general appreciation of industrial design can be expected over a period of years.

Another and widely used method is to employ external consultants with basic work being carried out in their own design offices, thus permitting the maximum concentration of their experience and facilities on the design problems involved. This type of organisation can be very successful when the work is primarily the ergonomic arrangement of components for the man/machine interface, the detail design of which is largely established. This arises in a wide variety of systems, such as design for control and safety of armoured vehicles, machine tools, agricultural implements or control panels. When the detail design of the interface and control components becomes an important part of the project, this system will work best when the nature of the product permits these to be designed with a considerable degree of isolation from the functional mechanism they support. This is possible in projects such as domestic machines, already referred to above, when the mechanical couplings between interface enclosure and mechanism are simple, and the control coupling is mostly electric wiring; or in instruments such as microscopes in which the mechanical design of the structure has relatively little effect on the design of the optical components. The difficulties with this method are likely to increase with the degree of mechanical interdependence of machine and interface, because of the communication problems this then generates between the separate design teams working on the same components.

Finally, a system which retains close integration and which has been successful in some large companies that have enough design offices to warrant the expense, is to organise an in-house industrial design department to serve all the design offices. Such a department can maintain close contact with the engineering design teams, and can build up a company approach to industrial design practice. This method of operation can provide for the temporary transfer of industrial designers from the central department to work within engineering teams in the way suggested above, or in suitable cases it can design the complete man/machine interface. It can also provide support services such as an index of standard components and materials that are used in the man/machine interface, and of bought-out or in-house detail components such as hinges, handles, covers, controls, switches and rating plates.

17.3 Discussion

The quality achieved by industrial design in engineering products is responsive not only to the competence of design teams but to the technical environment in which they operate. Success is more likely when it is recognised that the disciplines of engineering, industrial design and commerce are inextricably interwoven throughout the design process. A product's success will depend on its appearance and the way it interfaces with man; on how well it functions; and whether it appeals to the right section of the community and is available at the right price. If it fails in any one of these aspects it will drag the others down with it.

Management has therefore a major contribution to make to the environment in which design operates by insisting on the value of a high standard of man/machine interface design, and by encouraging the use of specialist assistance when this is appropriate.

18 Engineering product design, some solutions in practice

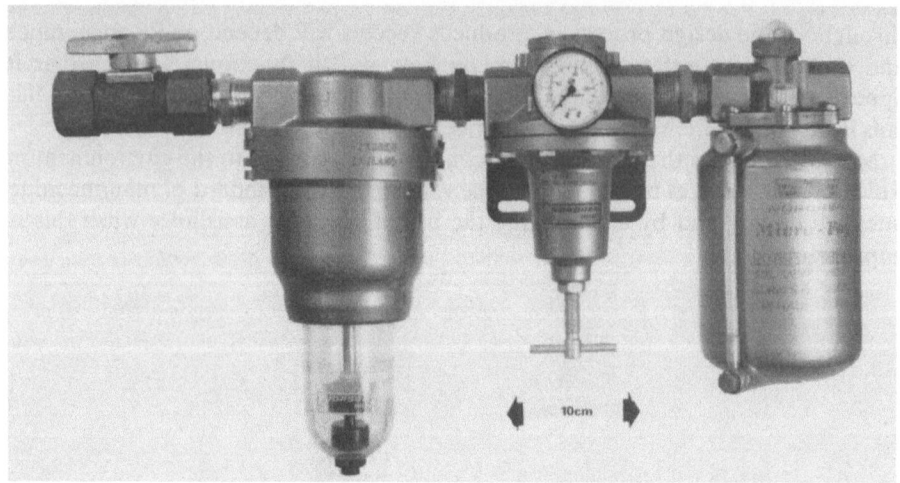

Figure 18.1.1 Norgren compressed air processing equipment assembly before redesign.

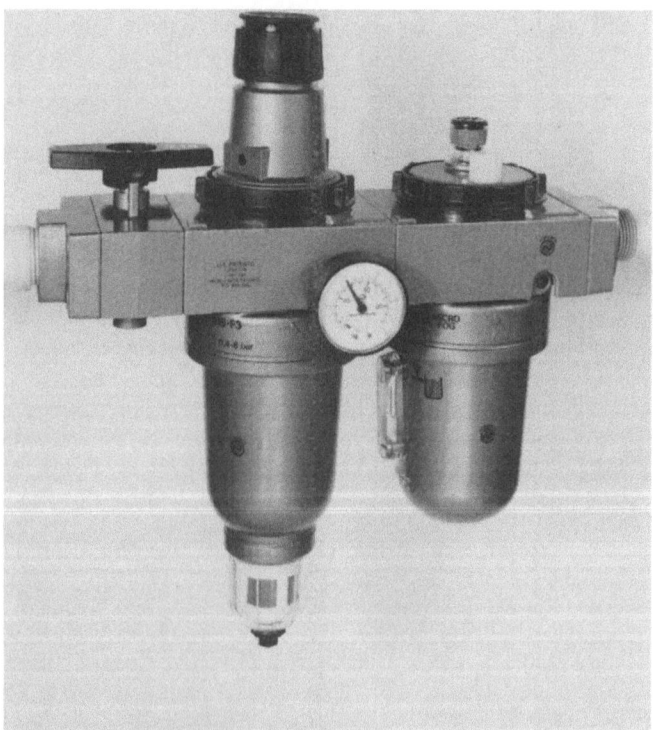

Figure 18.1.2 General view of Olympian compressed air processing equipment after redesign. This assembly performs the same functions as that shown in Figure 18.1.1.

18 Engineering product design—some solutions in practice

Charles H. Flurscheim

In this chapter a wide variety of products are considered which have combined good design and service experience with commercial success. Since engineering and industrial design are interlocked, both are discussed here in order to clarify the background to the interface problems and the solutions adopted.

Some of these products have won awards for merit. The Design Council Awards, including the Duke of Edinburgh's Designer's Prize, as it now is, are judged on the totality of the design combined with performance in service, and industrial design makes an important contribution to these criteria.

The Author wishes to acknowledge the help he has received from the designers and organisations responsible for these products, but the views expressed here are his own.

18.1 Olympian plug-in system of compressed air processing equipment

    Manufactured by: C. A. Norgren Ltd, Shipston on Stour, Warwickshire
    Designed by: Norgren Design Team
Design Council Award 1974

Objectives. Compressed air processing equipment for factories used to be assembled by screwed couplings as shown in Figure 18.1.1, in which there are eight screwed connections which have to be airtight. The system looked untidy, and required a wide range of couplings and adaptors, with many different threads, to be stocked by suppliers and users. Inspection meant dismantling and reassembly, often a tricky or unsafe operation in inaccessible positions.

Objectives of the development programme included improving maintenance facilities and safety, appearance, ease of cleaning, and reduction of stock levels and system costs.

Design. A typical assembly of the developed product performing the same functions as 18.1.1 is shown in Figures 18.1.2 and 18.1.3. The modular design includes a lockable stop valve which vents downstream pressure to atmosphere in the shut position, and rectangular die-cast air bus-bar units capable of accepting one, two or three components, which can be assembled as required. All air seals are by O rings, the only screw connections being for the incoming and outgoing air lines. The adaptors for these will accept copper or plastic pipes of all normal sizes, are interchangeable, and are either inserted and pinned or bolted on to the frame yoke immediately before despatch. Individual components are plugged in and sealed by O rings, and this reduces the material content of the component bodies. The elimination of screwed interconnections decreases the potential leakage points, and avoids alignment problems between units and variations in mounting dimensions after maintenance, which were inevitable with the older system.

Assembly or dismantling of components, as shown in Figure 18.1.4, is by manually screwing up or unscrewing a clamp ring which pulls the body into

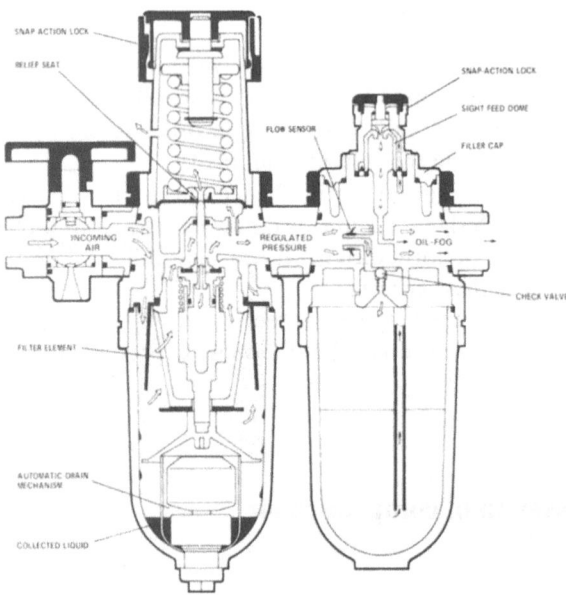

Figure 18.1.3 Sectional view of the Olympian assembly shown in Figure 18.1.2.

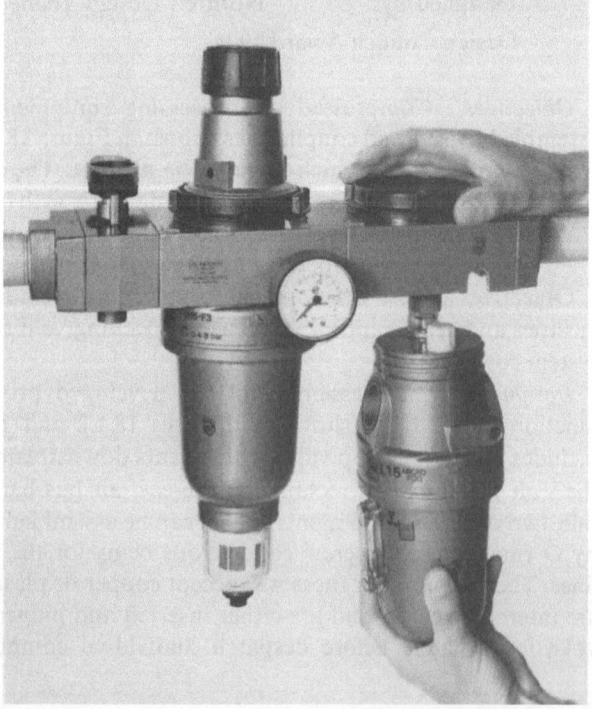

Figure 18.1.4 Simple dismantling and assembly are features of the improved design.

position against tapered faces carrying O rings at incoming and outgoing air positions, and no tools are required. Any unit can therefore be withdrawn for inspection without disturbing any other unit or seal. The modular style permits assembly in virtually any arrangement, pre-drilling for wall mounting since the dimensions can be pre-determined, and the use of modular brackets providing rear entry for the compressed air supply when necessary. It also permits the provision of a spare unit with a blanking plug, which passes air directly from the inlet to the outlet port.

The new design makes maintenance easier and safer, reduces stocking to one third of previous levels, presents a good functional appearance based on a compatible combination of curved and rectangular forms, and has reduced costs.

18.2 Mark II yarn top detector for textile machines

Manufactured by: Trip-Lite Ltd
Martin House, Gloucester Crescent
Wigston, Leicester

Designed by: K. A. Jordan, Managing Director,
B. Tasker, Development Engineer, and
Professor Frank Height, Design Research Unit
Consultant Designer

Design Council Award 1976

Objectives. A Mark I detector, shown in Figure 18.2.1, had been in successful service worldwide for about 20 years, and the objectives in developing the Mark II detector, shown in Figure 18.2.2, were to improve further the ergonomic performance and appearance and to reduce costs. It was felt this could be best achieved by detailed study of the operational cycle while retaining the proven principles of the Mark I.

Design. The yarn, travelling at high speed, tends to snake and whip itself around any projection, and to set up harmonic vibrations which adversely affect performance. Study of these problems and overall ergonomics led to the following improvements, shown in Figure 18.2.3 a and b.

Efficiency of operation was improved by the introduction of a simple oscillation-damping buffer (c), elegantly achieved by modifying the existing pivoted stirrup to make it out-of-balance, so that its weight tends to damp high-frequency oscillations of the dropper wire and the yarn itself.

The introduction of a self-locating yarn lock (a) on the sweep shoe further improved operation, by locking the yarn in its running position above the shoe.

The introduction of a wedge-base lamp (e) eased lamp replacement and increased illumination by 30 per cent, which was amplified further by a faceted housing which ensured more effective distribution of the light, reduced the risk of damage, and improved signal visibility, all at reduced cost.

After a thread break, rethreading and resetting the detector mounted high above the machine can be awkward, and the wider opening and clean face

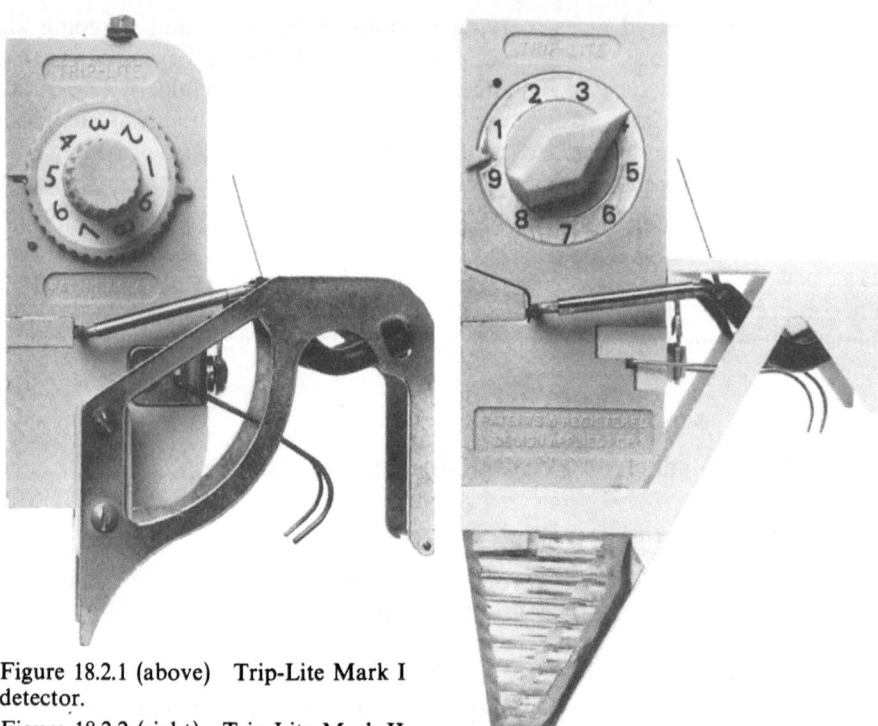

Figure 18.2.1 (above) Trip-Lite Mark I detector.

Figure 18.2.2 (right) Trip-Lite Mark II detector, after redesign.

Figures 18.2.3a and b (below) Schematic arrangement of the Mark II in up and down positions.

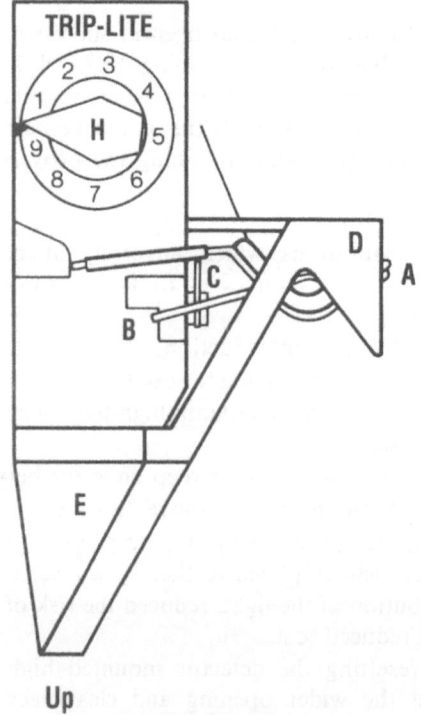
Up

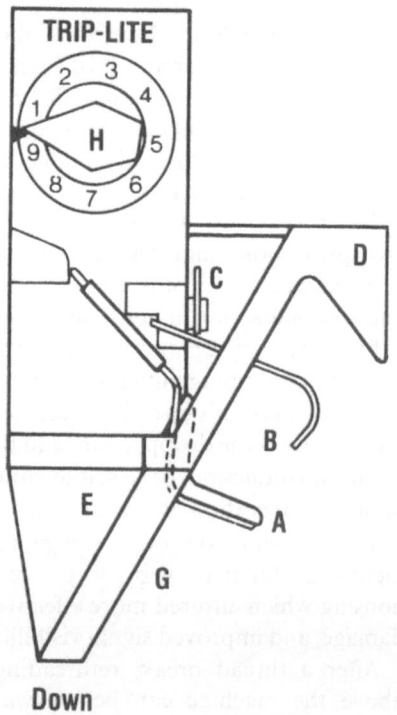
Down

provided by the new lamp housing and sweep guard ensured a much easier path for this operation.

Another ergonomic improvement was to redesign the analogue and numerical calibrated tension control knob (h) to provide clearer graphics and easier adjustment. A completely new commercial image was provided by generally improved functional form, by the use of moulded nylon components and by the careful introduction of colour, with the body in light beige, a faceted amber plastic lamp housing and black scales on a white background.

Value analysis led to the use of injection moulded, snap-fitted components and self-locating metal parts, eliminating all screws and rivets and halving the assembly time, and the use of an integrally moulded nylon terminal loop and ratchet wheel, replacing six separate components in the earlier model. The total number of parts was reduced by 40 per cent.

The close integration of engineering and industrial design in this project improved performance, ergonomics and appearance and also reduced the cost of the product significantly.

18.3 Cuplok scaffolding system

Manufactured by: Scaffolding (Great Britain) Ltd
23 Willow Lane, Mitcham, Surrey
Designed by: SGB Design Team
W. Blank, Group Development Engineer
Helmond, Holland
J. de Leeuw, Helmond, Holland
A. Bierhoff, Helmond, Holland
B. Smits, South Africa
S. N. Thomas, Mitcham

Objectives. The objective here was to develop a scaffolding system for the building industry which would be simpler to assemble than the conventional arrangement of clamps, nuts, bolts and wedges, and which would therefore save in erection and dismantling times and costs.

Design. The resulting Cuplok system permits structures of virtually any configuration to be assembled from two basic types of components, vertical 'standards' each fitted with pre-located Cuplok fasteners at fixed intervals, and horizontal 'ledgers' and 'transoms', identical except for length, which have forged steel ends designed to fit the internal configuration of the Cuplok fasteners.

Each locking device has a fixed lower cup welded onto the vertical standard and a movable inverted upper cup which can be raised a few centimetres and held in this position by turning it in either direction, when it engages with the upper face of a stop welded to the standard. When lowered, it can be forced downwards by clockwise rotation and cam action between its inclined upper face and the lower face of the welded stop.

The simple assembly process (Figure 18.3.1 a and b) involves setting up a vertical standard, raising the upper cup of a fastener, locating the forged ends of transoms and ledgers in the lower cup, lowering the upper cup and so enclosing the upper part of the forged ends and securing these components in position. The upper cup

Figures 18.3.1a and b   Stages in assembly of the Cuplok system showing (top) joint assembly and (above) joint locking. (*Owen Lawrence*)

18.3 Cuplok scaffolding system

Figure 18.3.2 (left) A detailed view of a Cuplok assembly.

Figure 18.3.3 (below) Cuplok scaffolding erected on site. (*Handford*)

is then rotated further by a hammer blow on the lugs provided, forcing it downwards and ensuring a positive and rigid connection. Dismantling is by reversing the rotation.

The complete structure can be adjusted in height by means of threaded jacks at the base of the supporting standards, and individual platforms can be adjusted by local jacks forming part of the fork heads that locate them.

All components are galvanised to ensure long life and to avoid rusting, which makes scaffolding unpleasant to handle and can stain stonework.

There are no small loose parts to be lost or dangerously thrown about during erection or dismantling.

Such structures may be in use and visible for several years, and the galvanised finish and tidy detail help to improve their appearance (Figure 18.3.2).

This simple design is the result of ergonomic study of the work movements and skill required with scaffolding, both of which have been significantly reduced.

18.4 Cash dispenser

Manufactured by: Chubb & Sons, Lock and Safe Company Ltd

Designed by: Jack Howe, Industrial Design Consultant, and Chubb Engineering Designers

Electronic
components by: Smiths Industries Ltd Access Systems Unit

Design Council Duke of Edinburgh's Prize for Elegant Design 1969

Objectives. To design a machine to enable bank customers to obtain money at any time using a coded identification system and to debit the transaction.

Because the machine pays out money, a first priority was security, and since customers have to identify themselves in sufficient detail to enable the machine to carry out the necessary checks, simplicity of control and clarity of instructions were essential features. The appearance and general character of the machine, its reliability and robustness, had to inspire users' confidence and survive extended exposure to the weather when projecting through an opening in a wall, with the operating facia sealing the aperture.

Design. Since the machine (Figure 18.4.1) represented a new concept in vending equipment, the design was approached with an open mind uninfluenced by previous practice. A coded card identification system was adopted, and the machine will not operate unless the card is the correct size, substance and thickness, is coded correctly and is inserted correctly. If this first check is satisfactorily completed, the machine then instructs the customer to key in his personal and identification number by pressing digital buttons in correct sequence. The dispenser then verifies this number and delivers an envelope containing cash, and the customer's account is debited. In Mark I form the card was retained as proof of a transaction. This involved returning it by post, with consequential delay and cost, so in the Mark IV the card is returned directly to the customer after completion of the cycle.

Many people operating a machine of this type for the first time are afraid of doing something wrong, with the result that they often do so, for example

18.4 Cash dispenser

Figure 18.4.1 (left) A general view of the Chubb cash dispenser in use.

Figure 18.4.2 (below) Schematic assembly of the cash dispenser mechanism.

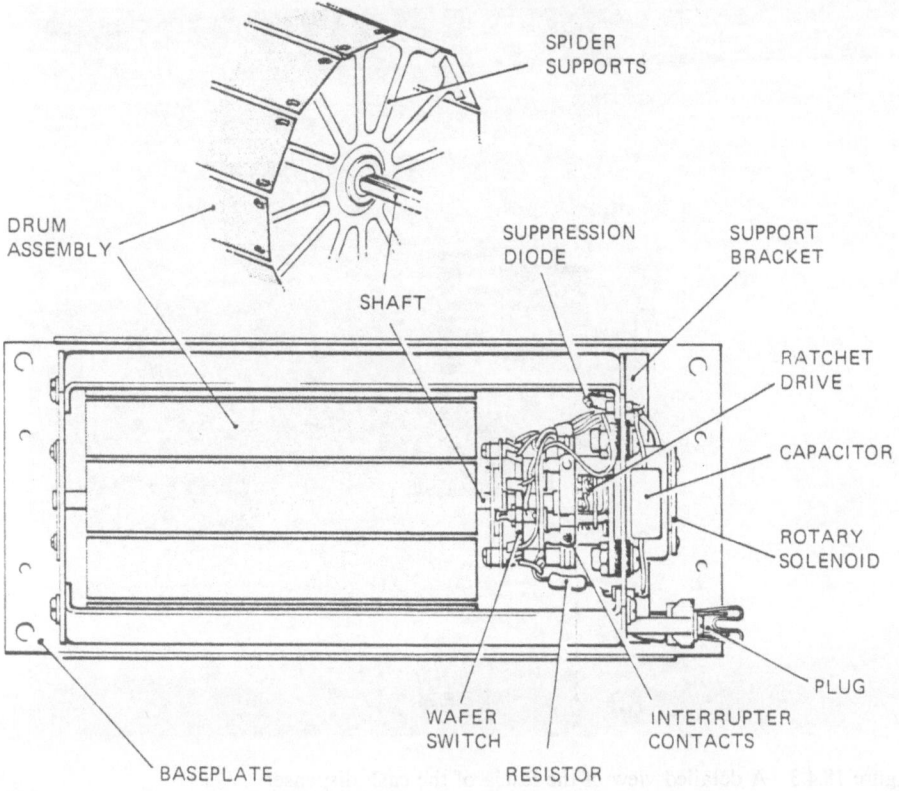

forgetting their identity number, and the machine was therefore designed to allow two mistakes in keying in the correct identity number, after each of which it instructs the customer to try again, but it invalidates and retains the card after a third mistake in the interests of security.

A problem arose in early prototypes because issuing ten £1 notes in a plastic container was expensive in container costs and the capacity of the dispenser safe was unduly restricted. Again, modifications were made to use continuous paper envelopes, each capable of containing up to ten bank notes, which increased storage capacity up to the insurance security limit of the safe and reduced packaging costs.

Another environmental problem arose in early models because the operating instructions were given by selective illumination of fixed legends from the back of the machine, and in bright sunlight these sometimes became almost invisible. A modified design presented all information at a single window by means of instruction legends such as 'Insert card', 'Wait' etc, screen printed on 0.08 mm thick Melinex strips carried by lightweight spiders forming a twelve-sided drum. This was rotated by a stepping motor consisting of a rotary solenoid, the armature converting axial to rotary movement by means of steel balls engaging with spiral grooves situated between the armature and solenoid housing (Figure 18.4.2). The combined axial and rotary movements are then used to engage and operate a ratchet drive which rotates the drum to the next position and resets the switching ready to accept the next rotary step. The low inertia of the drum permits it to be rotated through the twelve positions in a revolution within approximately one second, and by changing the Melinex strips, many different languages can be used.

Figure 18.4.3 A detailed view of the fascia of the cash dispenser.

Turning to general construction and appearance, Jack Howe designed the facia in stainless steel to be weather and vandal resistant, and with very simple lines, free from ornamentation, which harmonise with both modern and traditional architecture, and which do not date (Figure 18.4.3).

Perhaps because of their simple and unostentatious appearance, these machines have not been vandalised in the UK, and such restricted damage as has occurred elsewhere, such as glue thrown over the controls, would be difficult to eliminate by design. In environments involving risk of blown sand, sea-water spray or invasion by insects, some additional protection might be necessary, such as slight pressurisation of the safe with a blower motor.

18.5 Domestic press

Manufactured by: Elna, Tavaro S-A, Geneva, Switzerland

Designed by: Elna Design Team

Objectives. This hand-operated electrically heated domestic press was designed to afford a high standard of ironing with both large and small items, in all the usual household fabrics, to be easy and safe to use, convenient to carry and to have a good appearance.

Design. The press is shown in Figures 18.5.1 and 18.5.2 and comprises a chassis which supports a padded ironing board, and a pivoted arm with two handles which carries a flexibly mounted heated shoe. The handle on the left is fixed to the pivot arm and is used to close and open the press, which prevents the operator making inadvertent contact with the heated surfaces with his left hand. The handle on the right locks the press closed and applies pressure. It is hinged on the pivot arm (Figure 18.5.3), rotating on needle roller bearings, and carries a pair of needle rollers biased by a compression spring against co-operating cam tracks fixed to the chassis. During closing the geometry is such that the spring is compressed, resulting in a biasing torque towards the open position, the operator providing the force to overcome this with his left hand. The pressure lever is locked relative to the swinging arm by the cam linkage throughout the closing movement and is released only when the heated shoe meets the ironing board carrying the work load, preventing incorrect sequential movements by the operator. While the press is closed the pressure lever can be swung downwards; initial movement locks the press shut, and further movement builds up the required pressure to the order of 45 kilograms. The cam linkage releases the spring slightly during the application of pressure, and in consequence virtually no effort is required from the operator for this function. To open, the pressure lever is raised against a slight resistance and when this movement is completed the press is unlocked and opens under spring bias Figure 18.5.4 a and b.

The principal energy input is therefore made by the left hand during the closing cycle, but because of the high efficiency of the linkage this effort is relatively small, and because virtually no effort is needed to apply pressure, the effect on the operator is to minimise his assessment of the total effort required which helps to reduce fatigue. This impression of easy operation is reinforced by the convenient ergonomic design of the movements and controls, such as the handles, which have

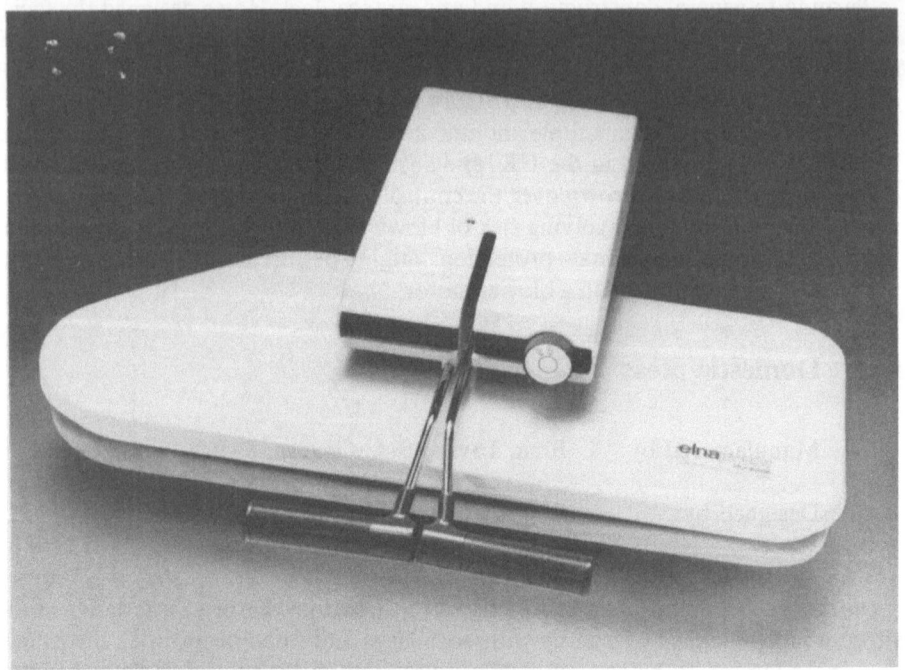

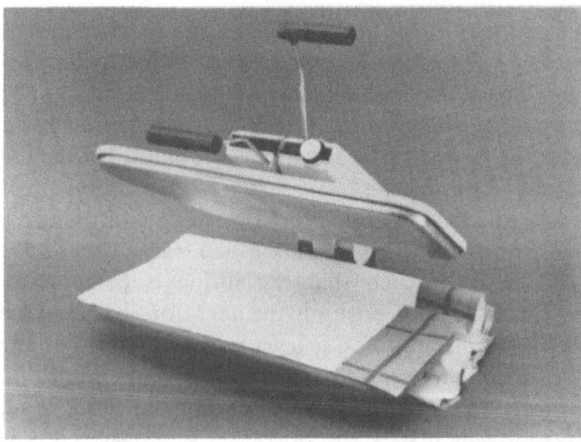

Figures 18.5.1 and 2
General views of the Elna domestic press in closed and open positions.

small annular grooves to prevent the hands sliding sideways, and by the clear graphics on temperature controls.

The design of the pallets allows large items to be folded in the space behind them and worked forward step by step, and a minimum of experience is required to avoid contact with the heated surfaces, the risk being reduced further by a thermal insulation cover protecting the shoe edges. An additional safeguard can be provided which sounds a buzzer and cuts off power should the press be left closed beyond the safe time.

To minimise weight and reduce risk of corrosion the main components are in die-cast aluminium alloy, finished in a semi-matt mushroom stove enamel. The ergonomics and clean appearance make it a pleasure to use, and when locked closed by the key mechanism provided, it can be carried safely with one hand.

18.5 Domestic Press

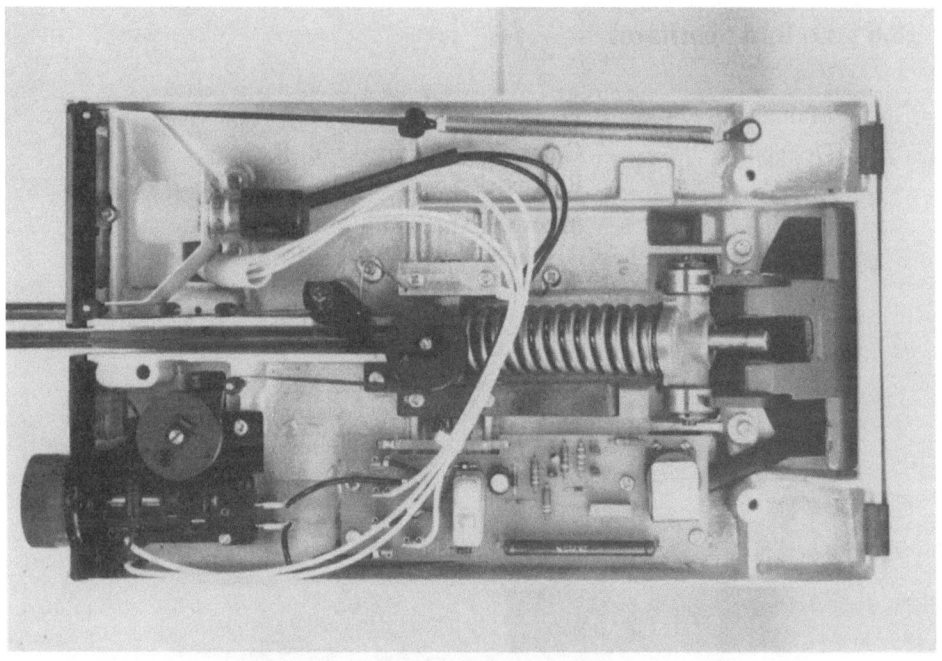

Figure 18.5.3 Press base, showing mechanism details.

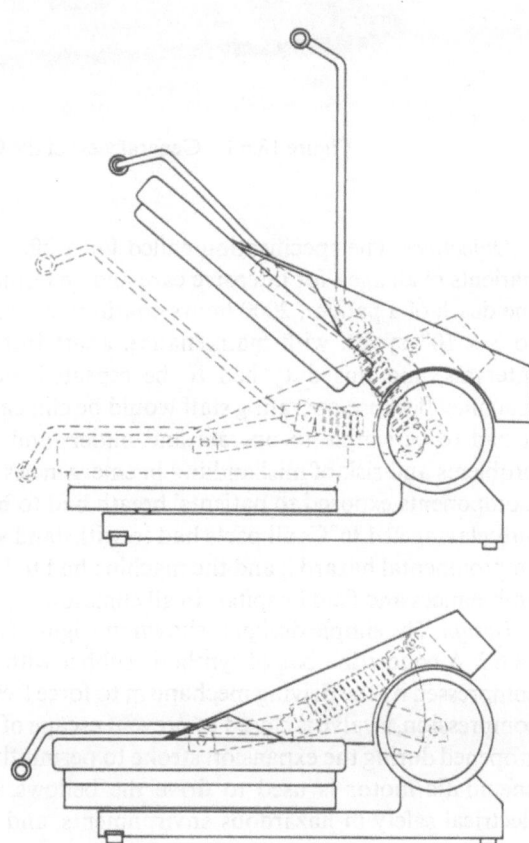

Figure 18.5.4 (right)
diagrammatic arrangement
in open and shut positions

18.6 Oxford ventilator

Manufactured by: Penlon Ltd, Radley Road, Abingdon
Designers: Company's designers—B. R. Sugg,
R. Russell-Smith, B. G. Chapman, G. I. Wond

in co-operation with consultants
A. Odell, R. Morgan, R. Gray, A. P. Smallhorn,
all at the time with Woudhuysen

Design Council Award for Medical Equipment 1976

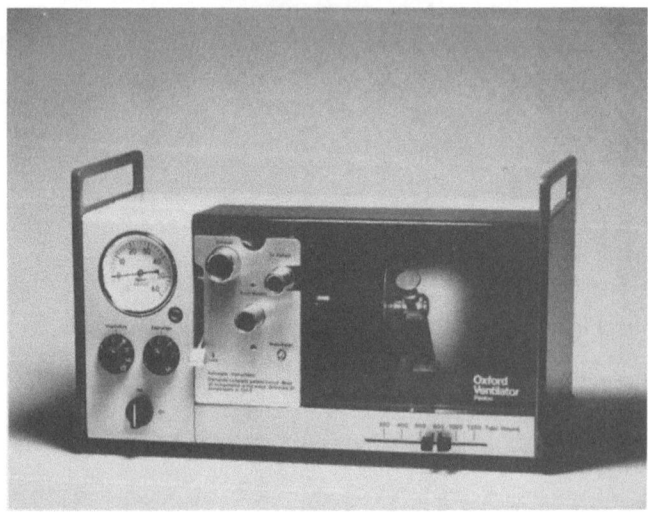

Figure 18.6.1 General view of the Oxford ventilator.

Objectives. The specification called for a life-supporting breathing machine for patients of all ages, for intensive care and anaesthesia. Because failure could lead to the death of a patient, 2000 hours continuous running was required representing 2 to 3×10^6 cycles, with maintenance, apart from daily checks, at three-monthly intervals, and this duty had to be repeated many times during the life of the machine. Because operating staff would be clinically rather than technically skilled, it had to be simple to operate and understand, and because of accommodation problems and risk of mishandling in emergencies it had to be compact and robust. Components exposed to patients' breath had to be suitable for sterilising by steam autoclaving at 136°C; all parts had to withstand solvents, saline, blood and similar environmental hazards; and the machine had to be suitable for service in hospitals, ambulances and field hospitals in all climates.

Design. The simple design is shown in Figure 18.6.1 and schematically in Figure 18.6.2. A concertina bag of synthetic rubber with one-way valves is expanded and compressed by the driving mechanism to force fresh air into the patient's lungs. On compression, a valve is closed to prevent escape of air from the lungs, and this valve is opened during the expansion stroke to permit the patient to breathe out. A linear pneumatic motor is used to drive the bellows, eliminating problems related to electrical safety in hazardous environments, and this can be driven by any of the

18.6 Oxford ventilator

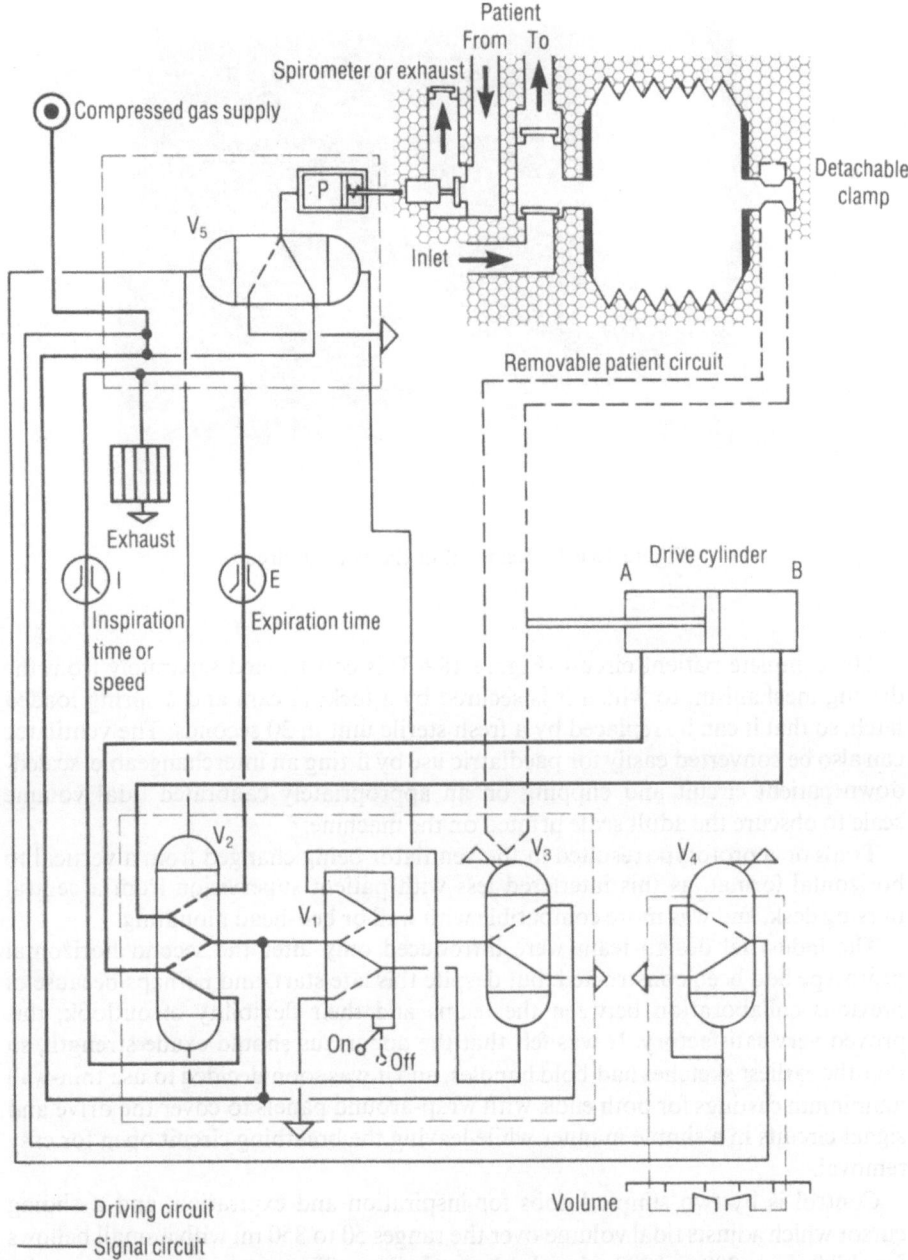

Figure 18.6.2 Schematic arrangement of driving and patient circuits.

dry gases available, usually by oxygen. All bearings are PTFE; the cylinder works unlubricated and the Kay pneumatic valves use light alloy spools running in polythene seals with very small quantities of 'Fomblin' PTFE-based lubricant (by Montecatini) and have a life of 10^8 cycles, with the complete system affording no fire risk in an oxygen atmosphere. For ease of servicing, most of the pneumatic connections are provided by drilled metal blocks—one replacing 13 tubes—and all demountable connections are coded by numbers. All non-return valves are identical, are mechanically coded to prevent incorrect assembly for direction of flow, and are sealed in position by O rings.

Figure 18.6.3 Removal of the patient circuit.

The complete patient circuit (Figure 18.6.3) is constructed separately from the driving mechanism, to which it is secured by a locking cam and a spring loaded latch, so that it can be replaced by a fresh sterile unit in 30 seconds. The ventilator can also be converted easily for paediatric use by fitting an interchangeable, scaled-down patient circuit and clipping on an appropriately calibrated tidal volume scale to obscure the adult scale printed on the machine.

Trials on a prototype resulted in the ventilator being changed from a vertical to horizontal format, as this interfered less with patient supervision from a central nursing desk, and was more compatible with wall or bed-head mounting.

The industrial design team were introduced only after the second horizontal prototype had been constructed, but despite this late start, and perhaps because of previous collaboration between the teams and their flexibility of outlook, this proved very satisfactory. It was felt that the apparatus should exude strength, so that the earliest sketches had bold handles, and it was soon decided to use thin-wall aluminium castings for both ends, with wrap-around panels to cover the drive and signal circuits in a simple manner while leaving the breathing circuit open for easy removal.

Control is by two simple knobs for inspiration and expiration, and a sliding cursor which adjusts tidal volume over the ranges 50 to 350 ml with a small bellows for children, or 200 to 1200 ml with a larger bellows. The cursor design presented a difficult problem, solved by using the cursor itself to slide a pneumatic trip valve to the required control position, the nylon cursor being designed with two projecting arms (Figure 18.6.4) with slots engaging with the metal chassis edge, so providing a friction lock in the relaxed position, released by pressing the two arms together.

On ergonomic grounds there are no gas connections at the back. The drive gas can be connected at either end, the patient gas connections being made on the front panel. The length of the unit had to be restricted to about 50 cm in case rack mounting should be required and during development an ergonomic mistake resulting from this became evident. Because of the restricted internal space, when a technician tried to remove the breathing circuit without switching off, his thumb

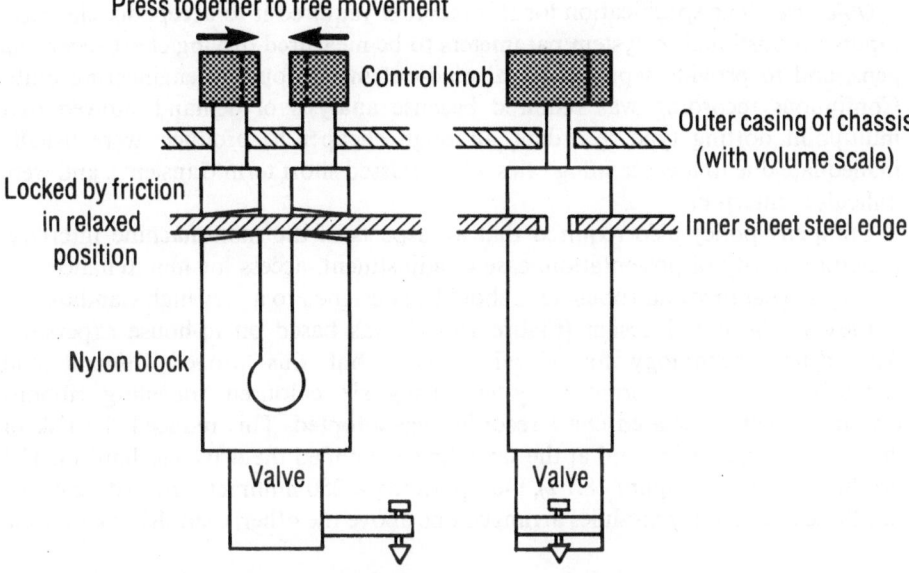

Figure 18.6.4 Diagrammatic arrangement of the cursor design.

became trapped against the end casting and remained trapped until the pneumatic driving pressure of the pump was released. Safety interlocks were not practicable to cover this risk, so the inside of the end casting was relieved to leave room for a thumb. One final ergonomic nicety was adjustment of the noise level to provide a quiet and reassuring rhythmic sound so that the medical staff could both see and hear the operation of this functional and tidy-looking instrument. Reliability is now well proven, and the only warning device found necessary has been an alarm to indicate accidental disconnection of the tube between ventilator and patient.

18.7 Series 320 multi-pen potentiometric chart recorder

Manufactured by:	Chessell Ltd, Worthing, Sussex
Product concept:	B. Chessell
	Dr M. J. Somerville
Mechanical design:	C. A. E. Newell
	C. K. Cleal
Electronic design:	Dr M. J. Somerville
	C. R. Williams
Industrial design:	D. E. R. Tustin

Design Council Award 1977

Hanover Fair Product Design Award 1978

Objectives. The specification for this recorder required it to accept six electrical inputs proportional to system parameters to be measured driving chart recording pens, and to provide separate visual read-out in appropriate engineering units. Continuous recording was required because analysis of demand showed that multipoint dotting type recorders, although cheaper to produce, were usually inadequate due to low scanning rates which missed short term transients, and were difficult to interpret.

Company policy also required that all aspects of the man/machine interface, including clarity of presentation, ease of adjustment, access for maintenance and testing, appearance and robustness, should be designed to a very high standard.

Design. The initial design (Figure 18.7.1), was based on in-house experience with drum technology for visual displays but this proved difficult and expensive, and an alternative system using six coloured travelling ribbons matching their associated chart recorder was adopted. This reduced the risk of incorrect interpretation and at the same time simplified the drive mechanism. The production design (Figure 18.7.2), incorporating a 250 millimetre recording chart, has these tape display modules arranged one above the other, their driving motors,

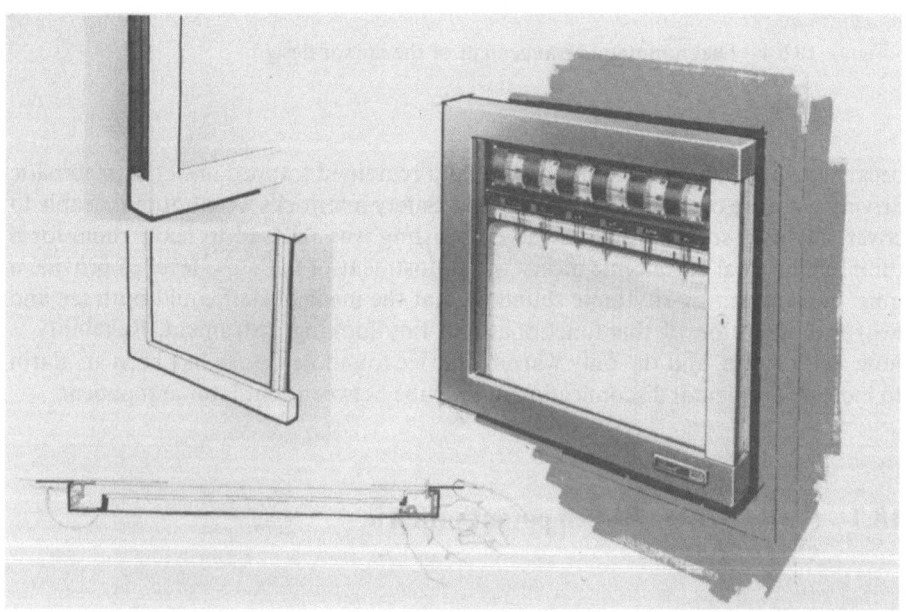

Figure 18.7.1   Initial sketch designs for the Chessell recorder, featuring drum display technology.

projecting downwards, being arranged in different lateral locations so as to avoid mechanical interference when stacked. Each module has its own scale of units, standardised for world-wide use by reducing variety and by employing symbols instead of names, clarity being achieved by the use of Helvetica typeface. The individual scales can be changed by loosening two screws, and they are held in position by rigid plastic frames free to expand, avoiding the risk of warping.

Up to six chart pens employing bright sharp colours operate in different planes, each can use the full width of the chart, and individual pens can be lifted from the paper by push-buttons on the front face. This operation requires spring loading on

18.7 Series 320 multi-pen potentiometric chart recorder

Figure 18.7.2 Front view of Series 320 design, using ribbon displays.

two planes, achieved by a twisted cantilever spring. Should any of the pens jam mechanically, their driving clutch is designed to slip and generate a noise to draw attention to this failure. Ink-priming buttons are mounted adjacent to the lift button for each pen, incorporating a small hole which is closed off by the operator's finger so that, when the plunger is depressed, pressure builds up in the priming valve and forces the ink into the capillary system, but with the finger removed the system is vented to atmosphere and cannot flood.

Because of the risk of mechanical damage when in the open position associated with front access doors hinged as in the early prototype (Figure 18.7.1) a design was adopted based on Victorian 'up and over' bookcase doors, which could be swung upwards and then slid back inside the instrument case (Figure 18.7.3). An ergonomic detail is that this door is spring loaded forward when unlatched,

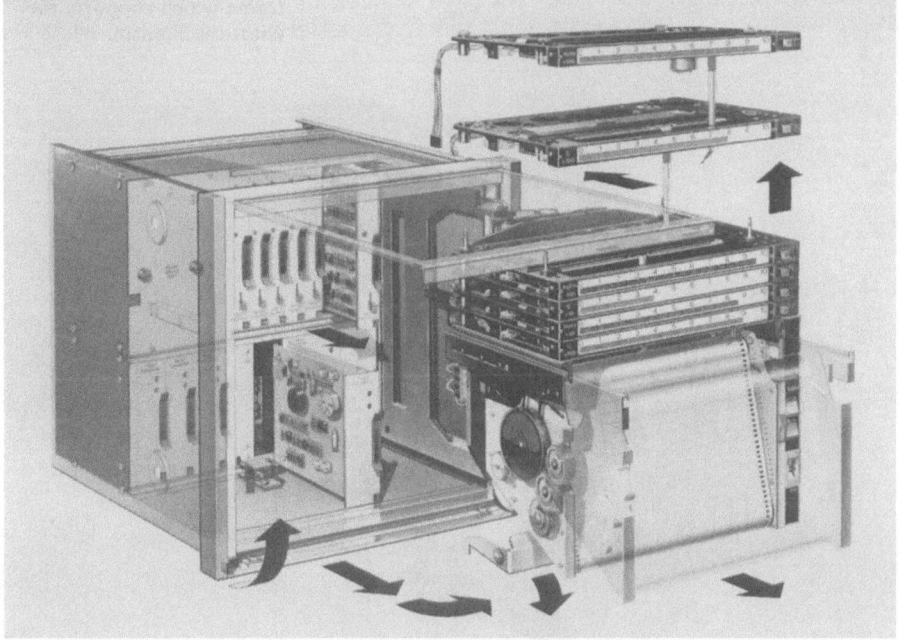

Figure 18.7.3 Exploded assembly view of the recorder.

suggesting the subsequent movements of rotation and pushing back. Similarly, when unlatched the chart cassette swings forward, inviting the operator to lift, rotate and pull out this component, giving primary access for changing ink and chart paper (Figure 18.7.4).

The chart mechanism can be switched off independently from the display; it can have a ten-speed electronic gearbox and be synchronised with the speed of the process it monitors; and it can have a fan-fold cassette combined with a unique collapsible bottom trough, enabling the chart to run out in front of the instrument so that the operator can view many metres of traced history. Secondary access for complete maintenance is obtained by pulling forward a 'total access' latch situated underneath the chart system, when a torsion spring reacting on a lever thrusts the entire mechanism forward, enabling the operator to pull it out of the housing enclosure, and pivot it on a triple-axis jointed frame (Figure 18.7.5), which also exposes all the electronic modules accommodated at the rear of the instrument case.

Figure 18.7.4 (left) Access to the chart cassette is quick and simple.

Figure 18.7.5 (below) Detail of the triple-axis frame which supports the chart mechanism.

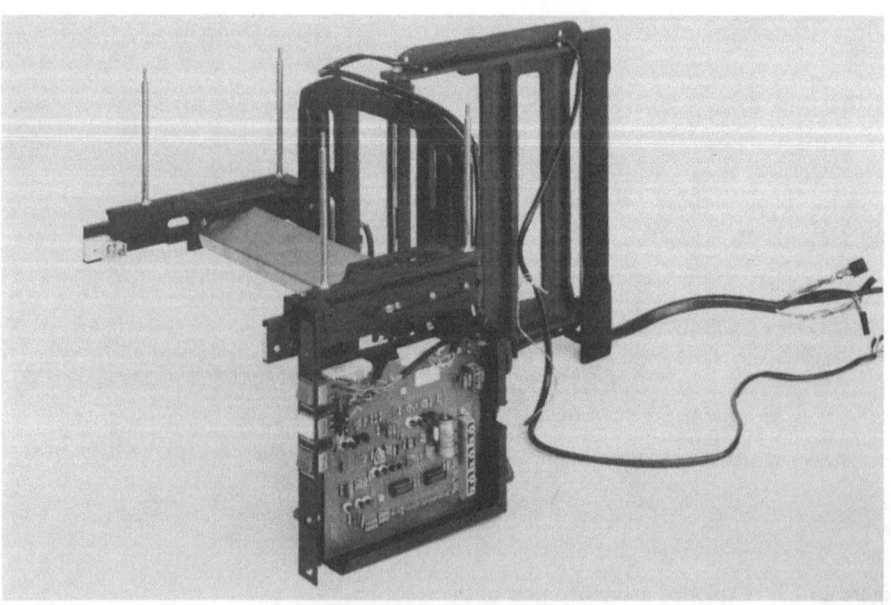

A unique feature is the calibration socket near the chart drive mechanism, which enables the operator to remove any of the input modules from the rear, plug it into the socket, select the appropriate test mode at the rear of the instrument and apply simulated inputs to the test socket, enabling calibration to be carried out much more simply than usual.

The robust construction is based on aluminium sections and sheet metal, requiring only simple bending operations, which dictated a hard edge approach to design. The mechanism is designed to operate without lubrication, which could lead to troubles with paper dust originating from the charts.

The effectiveness of the design policy and its execution using aesthetic, ergonomic and graphic techniques has contributed to an expanding world-wide demand for this instrument.

18.8 Automatic flight control system for Concorde supersonic airliner

Designed and manufactured by: Marconi Avionics Ltd, Rochester, Kent, England and SFENA, Villacoublay, France

Objectives. The ever-increasing sophistication of the control functions in aircraft, and the consequential proliferation of dials, warning lights and controls, presents the pilot with scanning and communication problems which can only be overcome through advances in electronics, but safety and efficiency in flight are also largely dependent on the quality of the ergonomic design. The automatic flight control system (AFCS) discussed here was developed to meet the advanced control problems of the Concorde supersonic aircraft.

Design. The control system developed comprises an array of system computers and their electromechanical actuators, shown in Figure 18.8.1.

An integrated automatic pilot and flight director provides automatic control from initial climb, through cruise to automatic landing. It embodies monitoring techniques that ensure 'fail soft' operation—that is, the system will itself diagnose any internal failure, announce this to the pilot, and automatically disconnect itself from control for all modes. In addition, where safety is involved during final approach and automatic landing, it provides automatic failure survival by means of a stand-by channel. The flight director enables the pilot to monitor the automatic flight path control against basic instrumentation, and permits flight director control in all modes of cruise and approach flying. The system also has computation facilities to enable the pilot to assess the serviceability and fitness of the whole system to proceed through automatic control to touch down. An automatic 'go-around' facility is also incorporated for use in the event of an automatic landing being aborted.

A three-axis autostabilisation unit improves the natural handling characteristics of the aircraft and hence passenger comfort. An autothrottle system controls the engine thrust according to air speed or Mach number. Electric trim provides pilot-operated pitch trim in manual flight and reduces transient control movements when the autopilot is disengaged. A safety flight control provides warnings and corrective action in the event of inadvertent high angle of incidence flight.

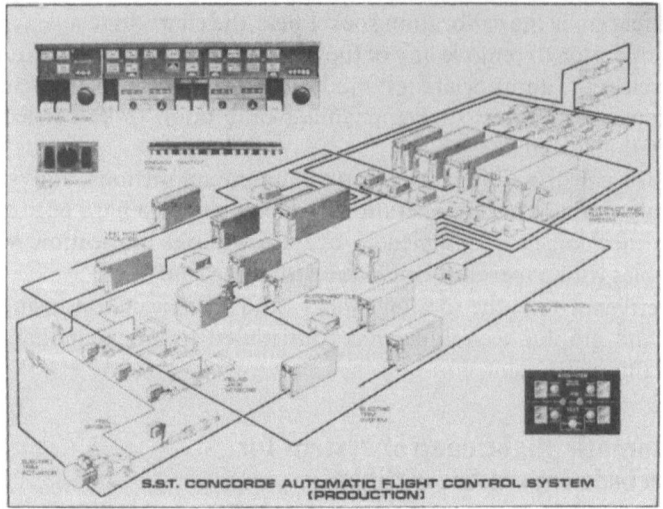

Figure 18.8.1 The computer and actuator array for the Concorde automatic flight control system.

Turning to control, an engage switch panel is mounted in the cockpit roof panel. This comprises switches for all the systems which have to be set up prior to take-off and which, except in case of failure, do not need any further pilot action until after landing. A datum adjust unit comprises those facilities which are required by the pilot to manoeuvre the aircraft through the autopilot and autothrottle systems. It is located in the centre pedestal, between the two pilots and remote from the AFCS. This provides a good ergonomic arrangement should the pilot need to control the aircraft through adjustment of the 'pitch-hold' and 'heading-hold' modes over long periods, and also facilitates accurate setting under turbulent conditions of flight.

The main control centre of the AFCS is the pilot's control unit (Figure 18.8.2) which is mounted immediately below the glare shield in front of and midway between the two pilots (Figure 18.8.3). The principal controls provided include: system engage switches for the autothrottle, flight director and autopilot; Mode selector push-buttons; heading-track VOR course selector, providing VHF radio guidance using transmissions from ground-based transmitters; altitude selector for 'altitude acquire' and 'altitude alert'; and indicated air-speed selector for autothrottle.

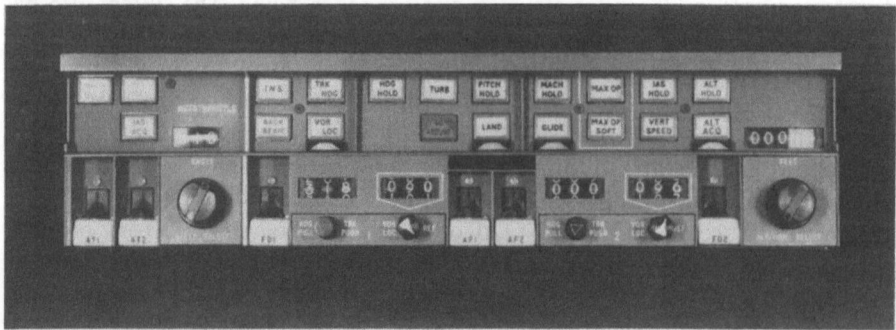

Figure 18.8.2 AFCS pilot's control unit. (*Marconi Avionics Ltd*)

18.8 Automatic flight control system for Concorde supersonic airliner

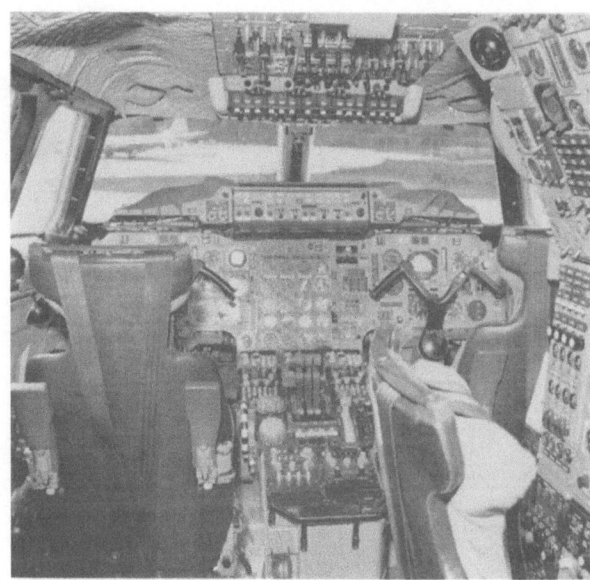

Figure 18.8.3 AFCS cockpit installation.

Through these the pilot's controller indicates 'prime' functions for preset modes and provides a wide range of 33 modes of flight operation which can be selected, including functions such as the control of aircraft acceleration from 5000 ft to supersonic cruise altitude, working within the limiting flight envelope defined by Mach number, speed and temperature. The automatic landing mode is selected by a single push-button which initiates all the capture and approach and hold functions associated with instrument landing, and causes in-flight testing of the critical systems to be carried out automatically prior to settling into the final approach. The system then progressively introduces interlocks and tightens up monitoring as altitude is reduced. The complete automatic system is in fact designed to reduce the risk of error, as compared with the accident history of manually controlled landings, by a factor of ten.

A warning and landing display indicator indicates the available capability of the automatic landing system, and is mounted in front of the pilots underneath the glare shield.

An integrated test and maintenance system (ITEM) permits continuous in-flight monitoring of the complete AFCS. It is also used for ground testing, to identify faults in any 'line replaceable unit', which can then be quickly replaced so as to render the aircraft serviceable in the minimum time, and information stored in-flight by the test system can be read out to make this process more reliable. Every electronic line replaceable unit in the AFCS has built-in test circuitry and the ITEM co-ordinates these test facilities and, using two simple computers, one for each segregated half of the AFCS, controls the testing, which is operated and displayed from a small push-button control panel on the flight deck.

All components on the pilot's control panel are modular, and controls are grouped on the panel face by their function—speed on the left, direction in the centre and height on the right. All systems are duplicated, with the directional controls having duplication of numerical presentation of course selected. The others have electrical duplication, as each switch operates twin microswitches in parallel, controlling duplicated systems through duplicate wiring harnesses.

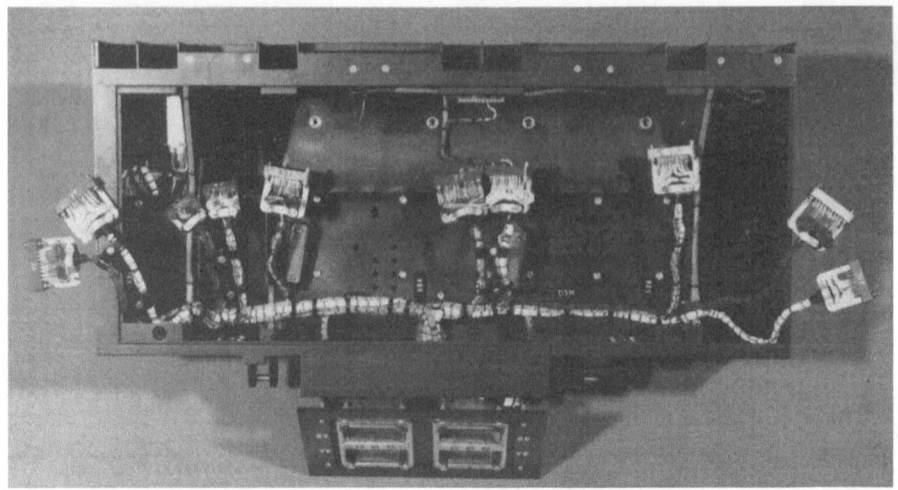

Figure 18.8.4 (above) panel assembly of AFCS pilot's control unit. (*Marconi Avionics Ltd*)

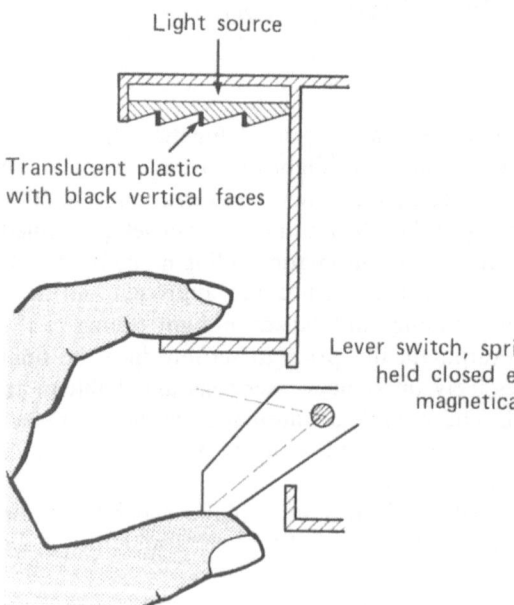

Figure 18.8.5 (left) Diagrammatic cross-section of switch panel, showing light diffusion and steadying ledge.

During development, rotary mode selectors were changed to push-buttons to ease pilot operation and scanning of the operational status of the system. In order to prevent the possibility of a mechanical jam in a push-button rendering the whole system inoperative, mode selections are made by momentary contact of the push-button switches. Successive selections of different switches update an electronic store in the relevant computer. System engagement is by means of electromagnetically held lever operating switches, which are retained in the 'on' position all the time the internal computer monitoring does not detect a fault. If a fault is detected, the electromagnet releases the switch which is automatically returned to the 'off' position by means of a return spring.

The panel is fabricated in light alloy from flat sheets, shaped with interlocking tabs and corresponding slots. After assembly, the tabs are twisted to lock the components, the whole is dip brazed and the excrescences are then cut off and ground smooth (Figure 18.8.4), a system which produces accurate, rigid and light units in relatively small numbers without special tooling, other than the preparation of punch tool programmes for each component. The panel finish and graphics are selected by the user, in this case matt black and grey.

Visibility for control selection at night has been provided by local lighting from an overhanging shelf forming part of the top of the control panel, which encloses a number of small bulbs which throw light downwards and across the panel front through a translucent plastic diffuser. To prevent light being thrown forward into the line of vision of the pilot, this diffuser has a saw-tooth cross-section, with the vertical edges facing the pilot coated in black (Figure 18.8.5). Another feature of the panel design is the provision of a ledge above the row of lever-type switches, so arranged that in turbulent flight conditions the pilot can steady his hand by placing his fingers on the ledge and operating the switch below with his thumb. The knobs controlling height and speed settings have small handles to facilitate rapid and accurate adjustment, and these can be folded back and spring locked when not in use.

This system has contributed to the flying qualities of the Concorde not only through its engineering quality, but through good ergonomics and the ordered appearance of the cockpit it helps to create.

References

1 Moulton, M. 'Avionic displays—a man-machine interface.' *Electronic Equipment News*, September 1975.
2 Howard, R. W. 'Automatic flight controls in fix wing aircraft. The first hundred years.' *Aeronautical Journal*, November 1973.
3 George, R. A. 'Automatic flight control.' *Flight International*, 8 May 1975.

18.9 National grid control centre for the CEGB super grid system

Designed by: CEGB Transmission, Design and Systems Operation Department in collaboration with E. G. M. Wilkes and J. W. Ward (now with Wilkes and Ward) Industrial Design Consultants, Horsham, Sussex
Desks manufactured by: Hyclamet Ltd Horsham, Sussex

Objectives. This control system was required for surveillance and control of the CEGB 400/275 kV super grid and associated power stations which, when the design was initiated in 1968, included approximately 174 power stations, with 56,000 MW of generation capacity, 200 EHV substations, 100 EHV circuit-breakers and 2000 isolators, requiring about 9000 automatic indications of positions, load and alarm conditions. The control system was specified to display on demand the state of each component of the network in terms of relevant real-time parameters, such as power generated, power flow, voltage and position, and a number of computer-calculated conditions, such as network security, based on

Figure 18.9.1 General view of CEGB control room, showing grid mimic display and control desks. (*Central Electricity Generating Board*)

actual conditions or on a variety of assumed failures. It also had to provide communication facilities for the operators to issue advance or corrective operating instructions.

Design. The developed system comprises a control room (Figure 18.9.1) containing a mimic diagram, three control desks, which can receive information via Post Office telephone lines or from a separate computer room containing on-line computers for storing and processing incoming information, and a communication room with input and output terminals and teleprinters.

The control room mimic diagram shows the complete 400/275 kV network, but because of their size, the substations can be presented only in block form, showing the line terminal circuit-breakers. The diagram is nevertheless unavoidably very large, and could have formed the end wall of the control room, but it was felt it would be less oppressive if it were constructed as a free-standing panel, with gaps of about 40 centimetres between it and the floor and the ceiling.

The three control desks (Figure 18.9.2) are similar and are occupied by the national control engineer, who exercises overall supervision and control of the entire system, and by two engineers responsible respectively for system load supervision and system switching duties.

The desks were designed in rectangular sheet-metal form in a low profile to minimise visual interference with the mimic diagram. Each desk has four cathode ray displays behind a horizontal working surface, originally in oiled teak, but later changed for practical reasons to plastics laminate with a teak finish. The CRTs are angled to aid vision, are mounted on trolleys withdrawing from the rear on guide rails, are recessed into the desk so that the operators look slightly downwards at them, and are shielded from downward light by cowls. Chart recorders, which have to be inspected daily without disturbing the operators, are mounted on a linkage system so that they can be swung free from the back of the cabinet and then

18.9 National grid control centre for the CEGB super grid system

Figure 18.9.2 Diagrammatic layout of control desk, showing installation of CRT unit and (detail) horizontal detailing of panel join.

rotated. The horizontal recessed feature (shown in detail in Figure 18.9.2) was introduced between the top and bottom sections of the panels to break up the otherwise rather large visual mass, and also to provide a natural joint line for the cabinet work.

The desks have duplicate keyboard controls for selecting the type of presentation of the grid system information for the CRT displays. Each keyboard comprises an alarm and interrogation panel and has several groups of keys. The first row links the keyboard with any one of the four CRT displays on the control desk, and when depressed an internal lamp is lit as soon as the computer can accept the new work load. The next bank of keys selects a function to be displayed, and the third row selects between alternative details. As the cathode ray tubes can display only a minute section of the grid system at a time, they are designed to sweep to other areas under control of the selector keys, or by means of a tracker ball system which rolls the display to any sector required. The result is that the display gives the impression of being a very large grid map, which can be moved in any direction at will, viewed through an aperture the size of the CRT screen. The scale of the display can also be adjusted to show greater detail, down to the complete circuit-breaker and isolator positions in a substation. Having four display tubes, the operating engineer can call up four system characteristics simultaneously should this be necessary. Visual alarms are given by six red indicator lamps, duplicated in each desk, which flash when system alarms are received, and acknowledgement by the operator results in the cause of the alarm being displayed on the screen. The desks also carry instrumentation and switching panels in their wings together with facilities for direct telephone communication to every regional grid control centre. The telephone base is a special design in GRP to accommodate two standard Post Office handsets.

The basic colours adopted were light grey for the large areas with dark grey surrounds for displays and green walls, which affords a restful impression. The metalwork was painted to a very high specification to allow for heavy use.

Noise in control rooms can be disturbing, and teleprinters and similar equipment were located so far as possible in separate rooms, but a noise problem exists with some types of CRT displays, which produce a 16,000 cycle note. This is not heard by older staff but can be an irritant to younger operators with a wider range of hearing.

The use of CRT displays in control rooms in which the staff have to read diagrams and write up logs presents lighting problems which may not arise when, as for example in air traffic control, their duties are principally in the visual supervision of screen information, when semi-darkness can be accepted. In order to provide a flicker-free display for the CRTs with a picture change rate of $16\frac{2}{3}$ frames per second, an L.4 phosphor was used on the screen having a 100 ms decay time. This gives a bright yellow trace, but with some reduction in tube life. The general illumination of the working area of the desks was provided by a bank of fluorescent tubes embedded in the ceiling and fitted with egg-box louvres to direct light downwards, and this produced a satisfactory compromise for viewing the tubes and instrumentation and for writing. The viewing distance of the tubes was governed partly by the character size available, and partly by the need for working space for the operator. It is usual to interpolate viewing distance from the guiding rules that a 2.5 cm character can be read satisfactorily at a distance of about 9 m, but in this case the proportional distance was extended slightly to allow for long sightedness in older men, although this introduced some conflicting problems for users of bifocal lenses.

This unique system, controlling the largest integrated power network in the western world, has proved very successful in ensuring network security and availability of power supply, and in generating and transmitting electric power at minimum cost.

Reference

1 Dillow, J. W. *Data Reduction, Transmission and Presentation Particularly CRT Displays, at the CEGB National Control Centre.* Institute of Measurements and Control.

18.10 True Data interactive display terminals

Manufactured by:	Grundy & Partners Ltd
	Bond Mill, Stonehouse, Gloucestershire
Designed by:	Grundy & Partners Ltd and
	Electrical Research Association Ltd
Industrial Design Consultants:	Tony Gibbs
	Antony Gibbs, Industrial Design

Objectives. This specification required a desk-mounted machine for use by entirely untrained staff which, in combination with a computer, would accept and store information and on request search for and display answers to questions posed to it, relative to a memory store. This concept called for a large matrix of light-emitting diodes on which could be superimposed questionnaire forms relating to the information stored, the forms having a perforation adjacent to each question through which an LED could be seen. In the same way that a tick or cross is placed in a box next to the question on a written questionnaire, so an affirmative or negative could then be indicated by changing the state of the LED from 'off' to 'on', using a light reactive pen.

In operation the information thus fed into the device was to be stored on a cassette memory tape in the computer. The device could then analyse the input information compared with information already on the tape, and select matching, or nearly matching, information from the stored file. For example, if an estate agent owned this device, he might design his questionnaire in such a way that his potential client could state on the questionnaire the price bracket, number of rooms, area of garden, district in which he wished to live and so on, up to 200 parameters. The machine could then be instructed to search the files for a match to these requirements. If no perfect match was located the machine could then be asked to select nearly matching properties. In theory the machine could select from all the properties in the country within three minutes.

Of special interest here, the specification identified ergonomics and styling as the operating areas to be covered by the industrial design consultant, and enlarged on these in detail as including the following aspects.

Simple line.

Covers to be removable to give access without the need to disconnect cables, unless by plug and socket.

All operations to be carried out from a single seated position.

Ergonomic sequence and requirements to be reflected in the design.

Colour scheme to be neutral, so as not to conflict with other equipment that might be used nearby.

Industrial design consultants to interface with the requirements of the engineering team and of the marketing department.

Design. The production design, with its array of light-emitting diodes, is shown in Figure 18.10.1, and with a typical questionnaire overlay in position in Figure 18.10.2. The combined team was set up immediately after laboratory R and D had made it possible to draw up the specification, and this enabled the industrial design consultants to influence the initial basic design in terms of logic requirements. In particular it was decided that the operator should know not only the sequence in which he should be working, but also the effect of his last input to the computer, which would help to direct his next action or choice of action, a feature very necessary with inexperienced operators. A group decision was therefore made to enable such feedback information to be generated by the device, and a Boroughs 32-digit alpha-numeric plasma display was selected. Another area that introduced problems for the operator was that of software, in which the designers were using Basic, but it was evident that this permitted ambiguities because of code abbreviations required in fitting phrases within the 32 digits available. By close co-operation with the software designers it was possible to suggest more readily understandable phrases that would fit within the digital constraints.

Another problem was the need for a series of controls that would instruct the machine to select weighted answers in the event of perfect matches not being found. It was therefore suggested that a biasing device should be introduced, so that weighted matches could be selected by the computer if 75, 50, 25 per cent, or none of the input parameters were matched. This raised the problem of devising a

332   18   Engineering product design, some solutions in practice

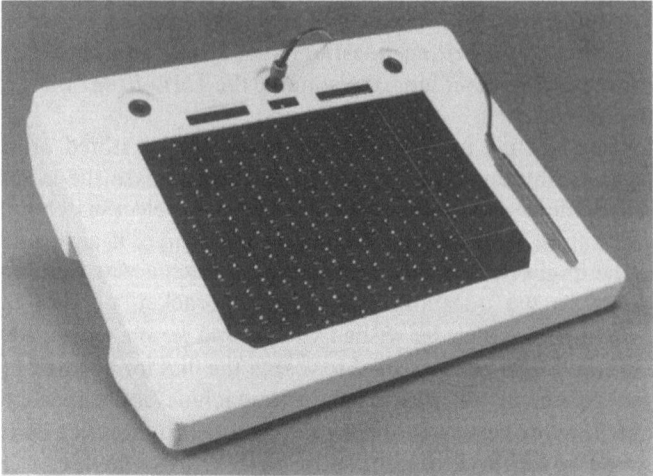

Figure 18.10.1   Production version of True Data display terminal.

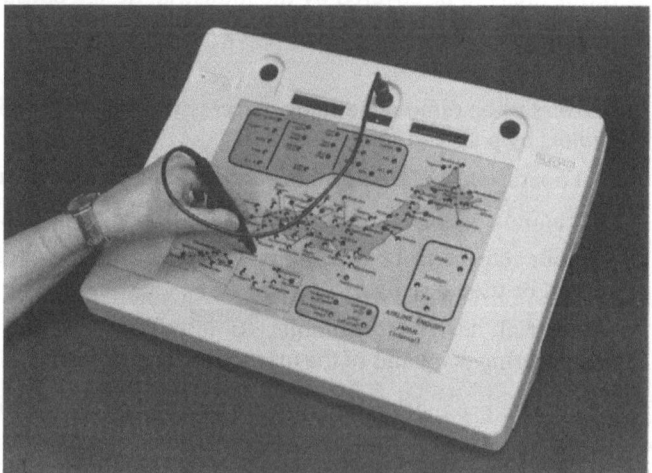

Figure 18.10.2   Display terminal with questionnaire in position.

Figure 18.10.3   The opened display terminal, showing casing, circuit boards and reverse of matrix.

heading which would suggest the meaning of these controls to an inexperienced operator, ultimately solved by using 'match', 'probable', 'improbable' and 'mismatch'.

In order that users should not be put off by too much information, it was decided that a 'secret until lit' display system should be used. The display area was therefore constructed in matt but transparent plastic, coated with black on its underneath surface giving the effect of opacity. However, when illuminated from the rear, lettering reversed out of the black background is revealed. By putting a transparent colour lacquer over the surface it is then possible to achieve the effect of colour lettering when illuminated. It was hoped that this technique could be used to cover and protect the LED matrix, but it was discovered that it would scatter the light from the LEDs to such an extent that their operation would become unsatisfactory, and as a result the LEDs had to be arranged to protrude through the matrix.

Considerable experimentation was carried out on the different types of LEDs available. A reversible LED giving red and green emission was considered, but the green, like the other yellow and green LEDs tested, was too weak in normal daylight conditions. The next problem was to select the correct LED to give a response over a wide enough variation of angles, so that irrespective of the angle of the light pen a reaction could take place. Too wide an angle would mean that the pen might trigger more than one LED, if held too far away from the matrix. Eventually a moulded matrix of lenses was designed which focused the emission of the LED to a suitable angle, mounted the LEDs securely, and provided a flat first surface on to which the overlay could be placed without damaging the LEDs. The pen initially had a simple transistor amplifier in its head which fed information back to a suitable amplifier in the device itself. Problems in reliability of the light pen device were subsequently resolved by employing fibre-optics techniques and moving the amplifier from the top of the pen to the root of the fibre-optic circuit.

The angle of the matrix panel was carefully considered, bearing in mind the way the operator would hold the pen, and the result turned out to be more like an old fashioned clerk's desk than a typewriter. Various low-volume manufacturing techniques were considered for the panel case, finally developed as a foamed polyurethane moulding, which achieved a desirable external form combined with moulded internal lugs, locating blocks etc, the only major post-moulding operation being to paint the component. The top of the case surrounding and retaining the matrix frame had to be robust, and vacuum-formed aluminium alloy painted to match the base and produced by Superform Metals Ltd provided freedom of design with high durability at comparatively low capital cost (Figure 18.10.3.)

The pen lead plugs in at the centre of the panel, and the pen itself is parked in a hole, on either the left or right of the panel, depending on left or right-handed operation. Because of its stiffness, the fibre-optic cable stays in a loop and does not become entangled with itself. The base moulding is designed in such a way that its wedge form can be used as a desk, or it can be reversed, which makes wall hanging feasible. The design has also taken into account the option of panel mounting to suit the requirements of equipment installers using fitted consoles.

As a result of the success of this project a new design was commissioned aimed at a much larger market, and learning from experience, this was to be designed to be much flatter.

18.11 ICL 2903 Computer system

    Manufactured by: International Computers Ltd
                                Putney, London
    Designed by: ICL Design Team in association with
                                London & Upjohn
                                Industrial Design Consultants
Design Council Award 1977

Objectives. The specification for this machine, of which over 3000 systems had been sold by 1978, was for a compact, easy to use, general purpose digital computer, primarily intended for first-time users, using an established programme language and capable of extension to meet a very wide range of requirements varying in size by a factor of the order of ten. The system was to be free standing, without the need for under-floor cable connections, and to operate without air conditioning.

Design. The ability to provide a wide range of systems from the same basic machine has been made possible by two technical developments: the production of a powerful and compact processor, and the use of micro-code software which is built into the machine.

Micro-code controls the operation of the machine in accordance with predetermined instructions, compiled by the manufacturer, to perform the functions required by the customer. Thus, under the control of the micro-code, the machine can be made to perform a variety of functions from payroll calculations to controlling a steel mill.

An important feature of the machine which results from the use of micro-code is what is known as direct data entry. This allows data to be entered directly into the machine's main storage from up to eight separate terminals, each of which can be located many miles from the machine. Validity of the data is automatically checked as it is keyed in and displayed on the video screen at the terminal to permit visual inspection by the user. Micro-code also permits the data in the main store to be examined and updated at any time, even while the machine is performing another function. This is achieved by means of a feature called 'roll-in, roll-out' in which the on going job is rolled out while the information fetching procedure is rolled in. When this operation is completed the original job is rolled back and continues as though nothing had happened. The overall operation of the machine is aided by the use of a simple, easily learnt programming language, RPG 11.

Advanced engineering ensured that the total kilowatt loading was reduced to a level where the need for air conditioning was eliminated.

From the point of view of the industrial designers, the project developed as an interesting departure from the norm—a series of tall free-standing units—as possible methods of meeting the fundamental marketing requirements for solid floor installation were investigated. Free-standing cabinets have always assumed a false floor for interconnection and a number of means were considered for combining the basic elements above floor level, from which the final solution evolved as shown in the outline drawing (Figure 18.11.1), and in the photograph of a recent production variation (Figure 18.11.2), comprising a 2903 processor, memory storage, video console and line printer. The floor plan, although 'wrapped round'

18.11 ICL 2903 Computer system

Figure 18.11.1   Outline drawing of ICL 2903 computer system.

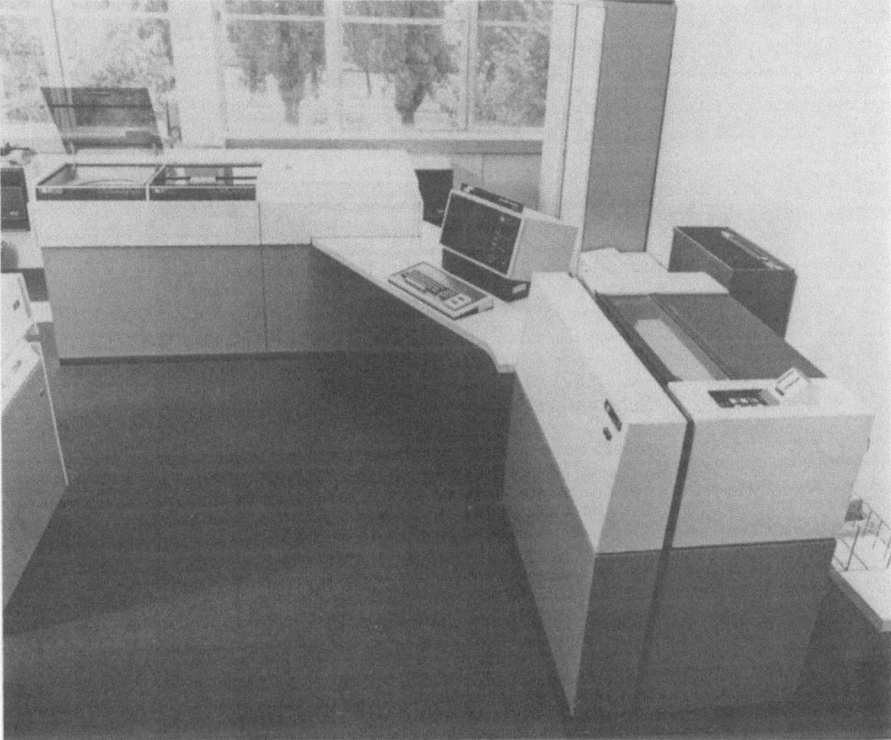

Figure 18.11.2   The production version of the system installed in an office

the operator, was in fact based on a right angle to allow siting near the walls of a normal room plan. The elements were finally organised so that the processor was placed in the junction of the main cabinet-desk wing, with a link duct for the necessary cables running under the rear of the desk wing. Individual units can be uncoupled electrically using plug-in connectors.

The working surface in Formica was arranged at a height of 700 mm. Local storage for documents and manuals has not been provided on the basis that staff are likely to keep the area round the machine tidy and that local storage would tend to detract from the very high level of appearance and order achieved. The control panel layouts were developed to a modular grid in order to reduce variety. Experiments showed that currently available LEDs, which had deeply recessed light sources, were visible over a wide viewing angle when placed behind a tinted window with the captions silk screened first surface.

The colours employed were matt black, Argentine grey and tango, the main surfaces being spatter finished, with smoked acrylic transparent covers. Lettering is in very clear Helvetica script, except for the '2903' logo on the end unit, which was designed to support the aggressive marketing aims of the company.

In the past, computer systems have tended to dominate man. This design shows what can be done in creating a more friendly atmosphere in computer offices, combined with improved ergonomic operational efficiency.

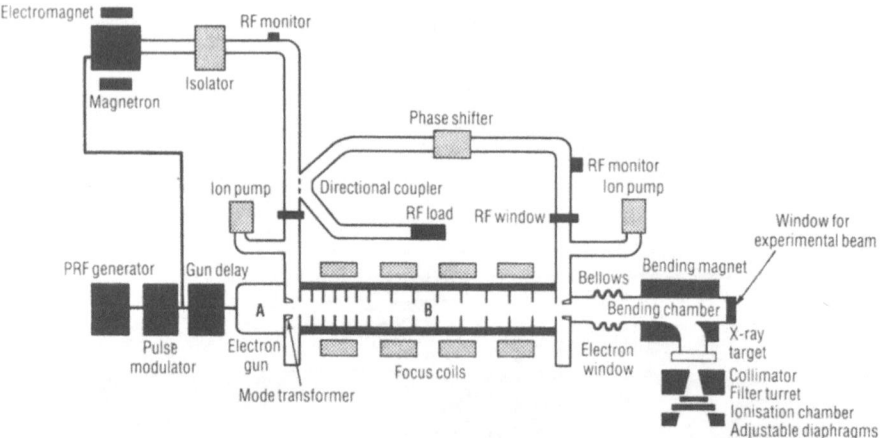

Figure 18.12.1 Schematic arrangement of the SL75-20 linear accelerator.

Figure 18.12.2 The accelerator assembly under construction.

18.12 SL75-20 linear accelerator

    Manufactured by: MEL. A division of Philips Electronic
                                      & Associated Industries Ltd,
                                      Manor Royal, Crawley, Sussex
    Designed by: T. R. Jarvis, Technical Manager
                                      C. I. Henderson, Project Leader
                                      K. R. Ayling, Senior Development Engineer
                                      B. S. Driver, Senior Mechanical Engineer
                                      C. S. Ashman, Senior Development Engineer
                                      P. Trussler, Philips Industrial Design Centre,
                                      Consultant of Styling
    Design Council Award 1978

Objectives. This linear accelerator for the treatment of cancer had to provide in one machine the choice between x-ray treatment, particularly suitable for deep-seated tumours, and electron beam radiation, for tumours of intermediate or superficial depth, with extremely accurate control of the radiation beam in terms of dose, shape and positioning, with either static radiation or with the beam rotating around the patient. Special requirements were that the controls should be simple to understand and operate and be fail safe, and that the equipment should inspire confidence in the patient, rather than the fear that can be induced by oppressive machinery in the difficult environment created by cancer radiotherapy.

Design. The system developed (Figures 18.12.1 and 18.12.2) comprises a high-energy electron generator mounted on a rotatable annular drum which also carries a gantry arm supporting a beam-forming unit at its outer extremity. The gantry arm projects through a panel into the treatment room, containing the treatment support system for the patient. The remainder of the machine is in a machine compartment, separated from the treatment room by a partition (Figure 18.12.3). Switchgear and basic controls are monitored by the hospital technician from a control centre outside the theatre area, and treatment is carried out from a separate control room protected from radiation.

Figure 18.12.3 Treatment room installation of the accelerator.

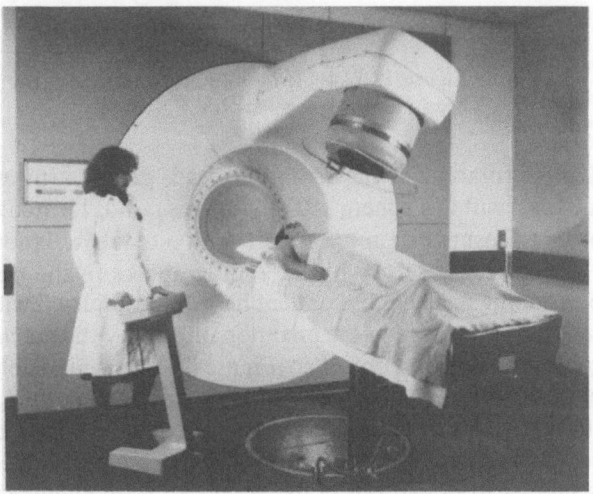

The beam generator system comprises an electron gun generating electrons and a simultaneously pulsed English Electric (GEC) magnetron providing the electron accelerating power. The resultant flow of electrons and radio frequency waves are fed into a wave guide accelerating structure. The wave energy pulse can be selected at 2.5 or 5 megawatt power, for a duration of 2.5 microseconds, and this accelerates the electrons in the accelerator to an energy level of the order of 20 MeV. These electrons, travelling at a speed very close to that of light, are then deflected through 90° in a bending magnet, and controlled in their beam shape by adjustable motor-driven heavy metal diaphragms. When treatment is by electrons, further control is introduced by fitting electron applicators between the beam outlet and the patient. For x-ray treatment, a heavy metal target is inserted in the radiation head on which the electron beam impinges, causing the generation of x-rays.

The mechanical system provides the rotational movements shown schematically in Figure 18.12.4 in addition to vertical and horizontal linear adjustment of the table-top position. The gantry can be rotated 360°, the radiation head can be rotated on its axis to twist the rectangular beam to any specified angle, the patient's support can be adjusted on all three axes and rotated about the vertical axis. The

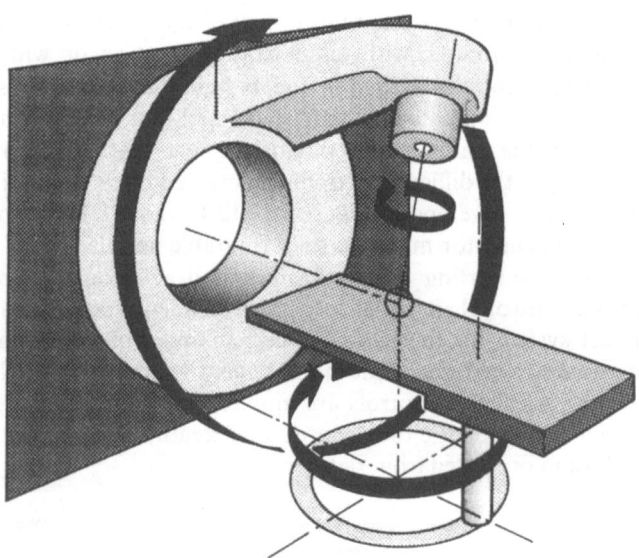

Figure 18.12.4 Rotational movements of the gantry, radiation head and patient support.

axes of rotation of the radiation head, gantry arm and patient support always intersect within a sphere of 2 millimetres radius located on the identified tumour, in order to ensure maximum concentration of the radiation beam.

Treatment is set up by the radiographers in the treatment room. Mechanical positioning of beam and patient have to be adjusted with considerable accuracy to take full advantage of the precision offered by the machine control, using a mobile gantry movement controller (Figure 18.12.5 and visible in 18.12.3) and a control position on the treatment support table. It might appear ergonomically more efficient to combine these positions, but in hospitals treating perhaps 100 patients a day, two radiographers are required working simultaneously in setting up and this is simpler from two positions. To aid in beam alignment, a pair of back and front

18.12 SL75-20 Linear accelerator

Figure 18.12.5 The mobile gantry movement controller.

mechanical pointers can be set up which define the position of the isocentre and, for safety, these pointers are held in position magnetically. Light-beam pointers are also provided for the same purpose. To avoid the risk of mechanical interference between gantry head and patient support, the gantry carries a flexible touch guard which stops movement if it is deflected, extinguishing a 'systems ready' light on the movement controller. Further movement of the gantry is then possible only by the operator using a touch guard override switch.

The gantry movement controller supervises gantry rotations and positioning of the diaphragms by means of thumb wheel controllers, which are speed responsive to displacement from their centre position. The movement controller also carries a digital display of the position of the components it supervises, and a 'system ready' light, illuminated only if the power supply is on and there is no mechanical interference. The controller also switches an illuminated projection onto the patient which displays the shape and position of the radiation beam settings, and a scale of distance from the beam source to the focus skin, providing a visual check

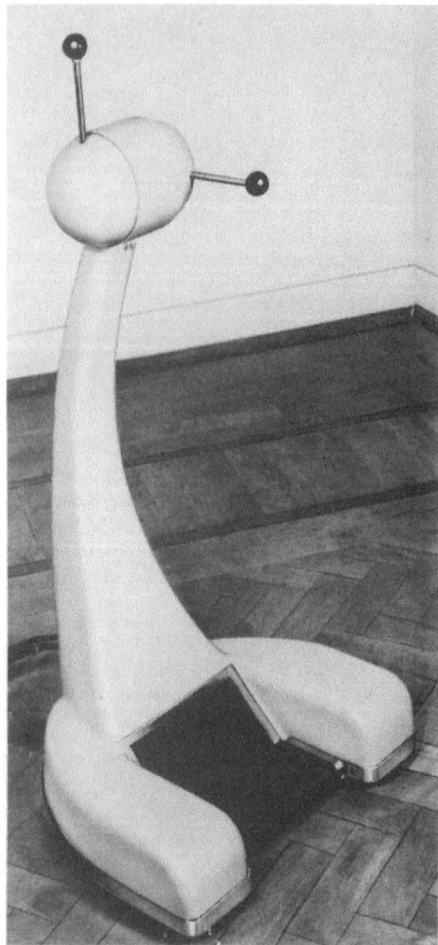
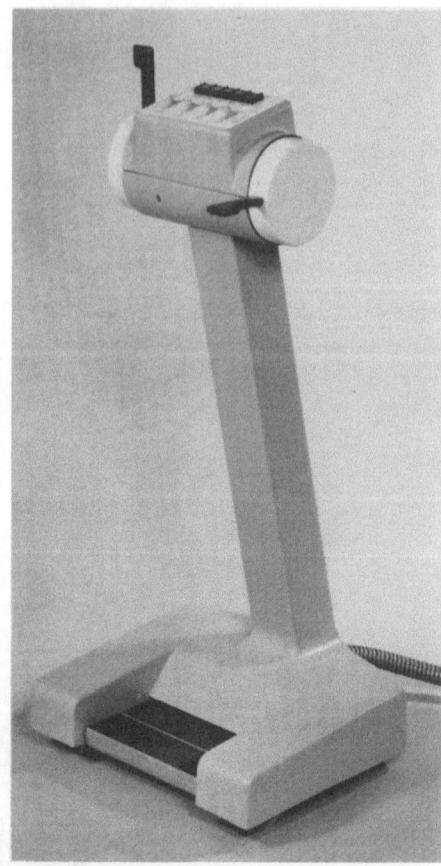
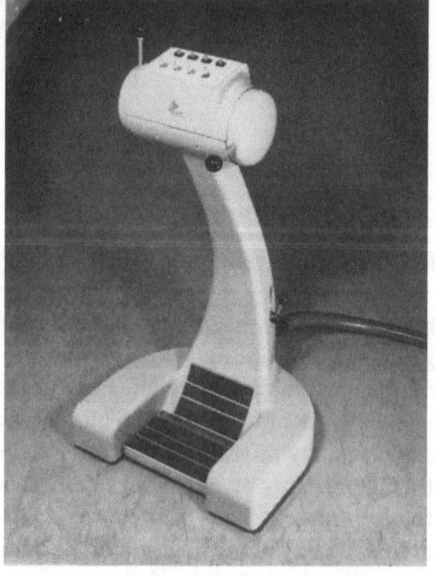

Figures 18.12.6a, b and c Three earlier versions of the gantry movement controller shown in Figure 18.12.5.

of the planned treatment. To improve definition, a treatment room light dimmer switch is also provided. The duties of this gantry controller have increased over the years, and style has also responded to changes in fashion, as shown in Figures 18.12.6 a, b and c.

The character of the treatment (x-ray or electron beam), energy level and stationary or rotating gantry treatment, are selected by push-buttons on the treatment selector control panel, mounted on the facia surrounding the machine. These push-buttons remain depressed when operated, but are illuminated only when associated machine interlocks, such as target position and wedge filters, have been completed. If x-ray treatment is selected, heavy metal wedges can be used to modify the character of the beam and these, being rather vulnerable, are stored in the treatment selector. The specified wedge is inserted into a slot in the radiation head, and when correctly installed energises a coding identification circuit. If the electron mode is selected, the specified electron applicator is fitted to the radiation head and connected to the code identification circuitry. Fitting these applicators requires adjustment of the field defining diaphragms, which is carried out automatically through the code control, and the diaphragm display then indicates the applicator field size on the movement controller. At this stage the set up procedure is completed by pressing a time delay push-button linked with the treatment room door. The radiographers then leave the treatment room and operate the safety interlocks on the doors, which de-energise a 'room door' warning light on the treatment control console in the control room. Visual supervision of patients is then provided by radiation-proof windows or closed circuit television.

The treatment set up is then confirmed and carried out from the treatment control console. The dose and time are set by a series of switches and, in the case of the rotational arc treatment, selection of a safety stop angle. Rotation, wedge, filter and energy push-buttons are then depressed, and if confirmation of setting is correct these will light up along with the start push-button. To initiate treatment it is then necessary to continue to press down the energy confirming push-button, as well as the start push-button, when initiation of treatment is confirmed by illumination of both red stop and amber interrupt push-buttons, and by analogue indication in the beam energy meter. If necessary the treatment can be aborted by means of the stop push-button, or interrupted with a view to continuing later by means of the interrupt push-button.

In addition to the safety confirmation procedures, there are back-up systems that terminate the treatment in the event of a failure in the control system, voltage level, divergence of field, or in gantry movement. In the event of power failure the treatment ceases immediately, but the dose display meter remains illuminated for a short period to enable the dose which has already been delivered to be recorded by the radiographer. It is of interest to note that the next generation of treatment control consoles now being developed (1978) will be operated by a typewriter keyboard associated with a digital display replacing the present analogue system.

Turning to general arrangement and appearance, machines of this class have been mounted in the past completely within the treatment room, and their appearance, complication, size and noise can create a sense of fear in the patient. As mentioned earlier, the SL75-20 has been designed so that only the gantry and radiation head are in the treatment room. Since the gantry rotates, it is provided with a circular GRP end cover which can rotate with it within the fixed end panels. The gantry itself has well shaped GRP covers, the standard colours being off-white with pale grey panels, and this reduces to a minimum the visual impact of the machine on the patient. At the same time the electronic and mechanical components have been made very accessible for maintenance, since the covers that would normally be necessary if these components were in the treatment room can be omitted.

This elaborate and expensive apparatus is made in small numbers, so that design (for example the use of GRP covers as opposed to the alternative of expanded polyurethane) has been influenced by the need to keep tooling costs low and to retain the flexibility necessary for design changes. Extensive ergonomic study has led to a control system which is relatively simple to operate, is clear in presentation and embodies a high level of safety protection and fail-safe duplication. The design is also an example of how attention to both concept and detail can contribute to improvement of the working environment.

18.13 Charnley-Howorth sterile operating theatre

Manufactured by: Howorth Air Engineering Ltd
Lorne Street, Farnworth, Bolton, Lancs
Designed by: F. H. Howorth in association with
Sir John Charnley, CBE, Professor of
Orthopaedic Surgery and Director of Research
at Wrightington Hospital, Wigan
Design Council Award 1978

Objectives. In surgical operations the settling out of airborne bacteria-laden dust particles imported into the operating theatre by staff, or generated within the theatre itself during the course of the operation, can lead to serious consequences for the patient and as much as a twenty-fold increase in the total cost of treatment. Sir John Charnley recognised that, if extreme cleanliness could be achieved by environmental control above the high levels normal in operating theatres, these risks should be significantly reduced. To achieve this he invited Howorth Air Engineering to co-operate with him in designing a 'clean air' operating enclosure to be erected within a modified surgical theatre, specifically designed in the first place for the now widely used hip replacement procedure which he himself had developed.

Research had demonstrated that bacteria do not exist in the air on their own, but that they ride on particles, or are in clusters of the order of three microns in diameter or larger, and that particles smaller than one or two microns in diameter remain suspended by brownian movement, so that filtration of incoming air to remove all particles above this size could be expected to achieve a radical improvement. To reduce bacteria generated locally within the clean air enclosure, access to it was to be restricted to the surgeon, essential assistants, and that part of the patient's body which had to be exposed for the operation, the anaesthetic team and all other staff associated with the operation being situated outside. Measures would then be necessary to deal with bacteria generated by the residual sources

18.13 Charnley-Howorth sterile operating theatre

Figure 18.13.1 General view of the Charnley-Howorth sterile operating theatre in its plate glass enclosure form.

which could not be eliminated. Since surgery involves great manual skill and precision, and is exhausting, any system would have to avoid interference with the surgeon/patient operational interface and efficiency.

Design. The armour plate glass enclosure developed (Figure 18.13.1) extends from the ceiling to the floor of the operating theatre. Fans provide a downward flow of air at controlled temperature and humidity through filters mounted in the roof of the enclosure, supplying about five complete air changes per minute, corresponding to an airflow rate of the order of 0.4 metres per second. This rate was chosen as the optimum balance between the minimum needed to maintain clean air conditions and the maximum that would avoid excessive turbulence, which could cause even the smaller particles passed by the filter system to be driven onto the wound, and which could cause excessive cooling of the wound due to evaporation.

The filtered air is distributed evenly through a series of cloth bags which cover the whole area of the roof of the sterile theatre, producing, so far as is practicable, laminar flow to avoid turbulence. Because of discontinuities in the flow system, caused for example by the patient's body, some turbulence is inevitable, but other discontinuities in the airflow are avoided.

The pressurised air input to the enclosure escapes through narrow gaps between the bottom of the glass walls and the floor of the theatre and a small outflow is arranged between the vertical joints in the enclosure walls at the corners, to eliminate still air pockets at these points. The entrance to the enclosure for part of the patient's body and operating table is sealed off by sterile curtains, which also permit a small outward flow. The service hatch through which sterile packages of operating tools are passed in sequence to the surgeon is left fully open, permitting a high outward airflow to prevent any reverse flow from the external environment.

There remains the problem of contamination within the clean zone by the patient and the members of the surgical team. An exhaust point is positioned on the operating table in the perineal area to establish a negative pressure under the patient drapes so the wound cannot be infected by airborne contaminants from that source. The members of the surgical team wear face masks and special gowns extending from the top of the head down to within about 30 cm from the floor. The gowns are made from specially woven and proofed lightweight material (170 g/m^2) of low permeability to air and impermeable to water, blood, etc. Air is evacuated from under the gown through a small tube connected to the lower part of the face mask and passing down the back of the wearer to the floor where, after a length to provide freedom of movement, it connects to a group manifold and centrifugal exhaust fan. This body airflow improves cooling and oxygenation of the wearer and provides a negative pressure under the gown and face mask so that any air leaks are inwards and, therefore, the wearer does not emit contaminants. Temperature and humidity in the operating room are controlled by the general air-conditioning system. In certain operations where two-pack cements are used, the ideal curing temperature is 20°C. Many operating rooms cannot be kept as cool as this in warm weather, in which case a packaged cooler unit may be added to the general air conditioning system, which also accelerates the downflow of air from the sterile operating unit.

System safety is maintained by duplication of fans and extractors, so that in the event of failure 50 per cent flow conditions can be maintained, and power supplies are also duplicated when economics permit.

A further development to meet wider applications was made possible by the amendment of the UK regulations forbidding recirculation of air in surgical theatres. This permitted the economic use of higher airflow rates, provided that, in order to counter some increase in contamination risk, the filter system was improved to remove particles down to about 0.3 microns diameter. Using this recirculation system, the enclosure was redesigned to use hinged transparent polycarbonate sides which could be removed easily in a few minutes when the enclosure was not required for major surgery, and this also enabled the equipment to be constructed in modular form for easy packaging.

The latest Mark 4 'Exflow' system (Figure 18.13.2) has been developed by Howorth Air Engineering with the objective of providing a sterile air zone for operating with or without the enclosure panels, according to the requirements of individual surgeons. The problem of peripheral entrainment of contaminated room air with the panels omitted could not be overcome by the airflow pattern of the existing units, and an air pattern like the open end of a trumpet was created. This flows downward into a clean air zone approximately 2.8 metres square and then outwards round the perimeter. In the interest of safety, the new unit has four fans working in parallel, built into the modular roof assembly.

This sterile operating theatre equipment is an excellent example of a concept and its development directed in this case at a specialised form of environmental control. Interference in operating efficiency arising from ergonomic or body working conditions has been avoided; the use of transparent enclosures has permitted staff to follow the procedures for training purposes without adding to the risk of contamination; as regards the primary objective, records at Wrightington Hospital, Wigan, show that with the same surgical team patient infection has been reduced from 8.9 per cent, itself a good record, to 0.3 per cent by using the enclosed clean air system; and working without restrictive enclosures has more recently been made possible.

18.13 Charnley-Howorth sterile operating theatre

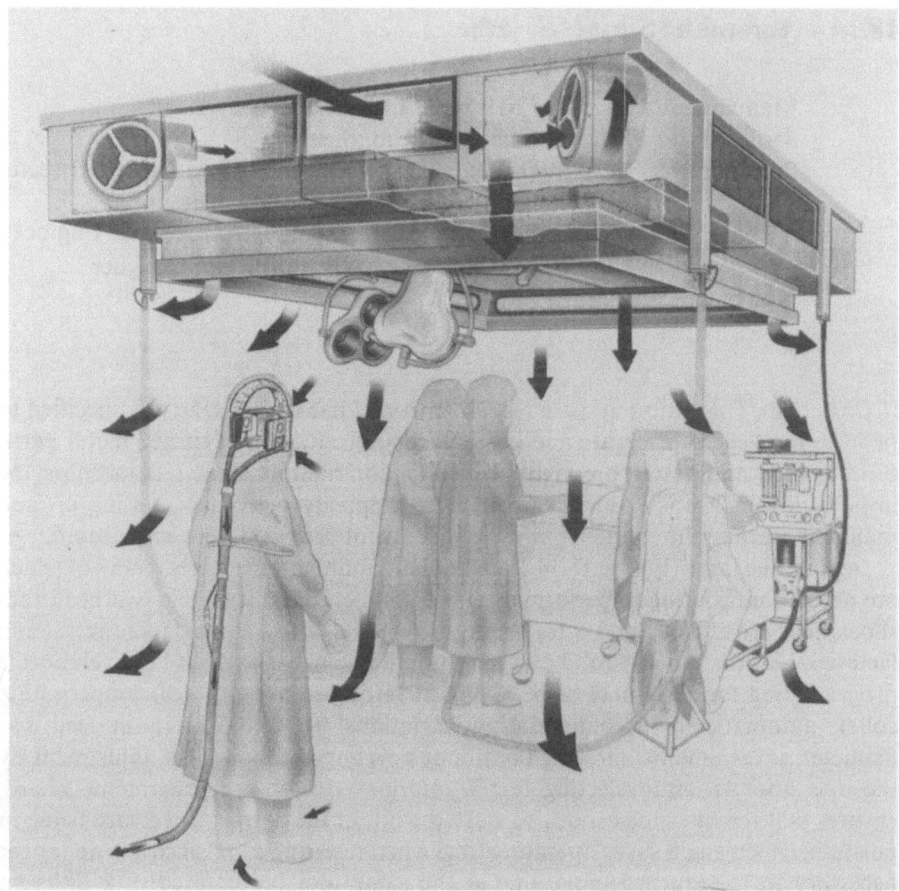

Figure 18.13.2 Schematic arrangement of the Mark 4 'Exflow' system, which dispenses with enclosure panels.

References

1. Charnley, J. 'Sterile air in operating room.' *British Operating Theatres*, London, July 1972.
2. Howorth, F. H. 'A prevention of airborne infection during surgery.' *Proceedings of International Symposium for Contamination Control*, Swiss Federal Institute of Technology, Zurich, October 1972.
3. Charnley, J. 'Clean air in the operating room.' *Cleveland Clinic Quarterly*, vol 40, no 3, 1973.
4. Howorth, F. H. 'The laminar flow in operating rooms.' *International Symposium, Lyon, France*, June 1973.

18.14 Automatic Sprint 'S' lathe

    Manufactured and   EMI-MEC Ltd, Charlwoods Road
    designed by: East Grinstead, Sussex
    Designed by: D. Love, Chief Designer and Managing Director
               R. Pritchard, Chief Electrical Engineer
               P. Hatchman, Assistant Chief Electrical Engineer
               A. Hutchinson, Assistant Chief Designer
    Design Council Award 1977

Objectives. This automatic sequence-controlled bar turret lathe was specified to provide high-speed, accurate and cost-effective production of turned metal parts. Since quality and output are highly dependent on man/machine relationships, the importance of flexibility and performance, simplicity of control, setting up and maintenance, and the character of the environment provided were emphasised.

Design. The lathe (Figure 18.14.1) embodies a number of features, many of which are novel, contributing to performance and good ergonomics. Twin overhead tool slides, rear cross slides and a traversing front slide, all with automatic retraction facilities, enable them to be used simultaneously without the elaborate programming formerly necessary, saving in setting-up time. A self-compensating collet, automatically accommodating variations up to 0.75 mm in feed bar diameter, saves time by allowing continuous cycling without collet adjustment by the operator. An auto-selecting turret, interlocked with the programme board, ensures that the machine cannot be started without the correct turret face being in position, which again saves operating time when resetting jobs, since it is no longer necessary to re-sequence tools, should the same tool be required in a different position. The turret will also index from station to station in any order. A traversing front slide enables progressive cuts to be made to reduce the diameter of long lengths of bar while the turret performs other machining operations. Because one of the two axes of slide movement required is built into the basic machine, an inexpensive hydraulic auto-copying unit can be fitted to the front slide. Automatic or manual selection of six forward and reverse speeds can be made once the desired speed range has been set up manually from the five ranges available.

The control console (Figures 18.14.2 and 18.14.3) is neatly built into the front of the machine, protected by a lift-up transparent plastic cover, and includes an EMI-MEC relay sequence board, a control board, and a needle-valve panel for speed control of hydraulic slide movement, with switches and valves ergonomically positioned in relation to function.

The programmer uses a diode pin plug board through which manual set-up of the panel is converted to a sequential programme, and study of the machine response and operator response required, together with clarity of presentation of controls and graphics, has made it simple to use. By placing a dial pin in the required socket, the selected function will take place, and by placing several pins in the same vertical column many functions can take place simultaneously, such as the cross slides, vertical slides and turret operating together. On completion of each main function the instruction column is cancelled and the next column energised, permitting uninterrupted overlapping of various machine operations, contributing to flexibility and speed. Plugboard indicator lights identify the live

18.14 Automatic sprint 'S' lathe

Figure 18.14.1　General view of the EMI-MEC Automatic Sprint 'S' lathe. (*EMI*)

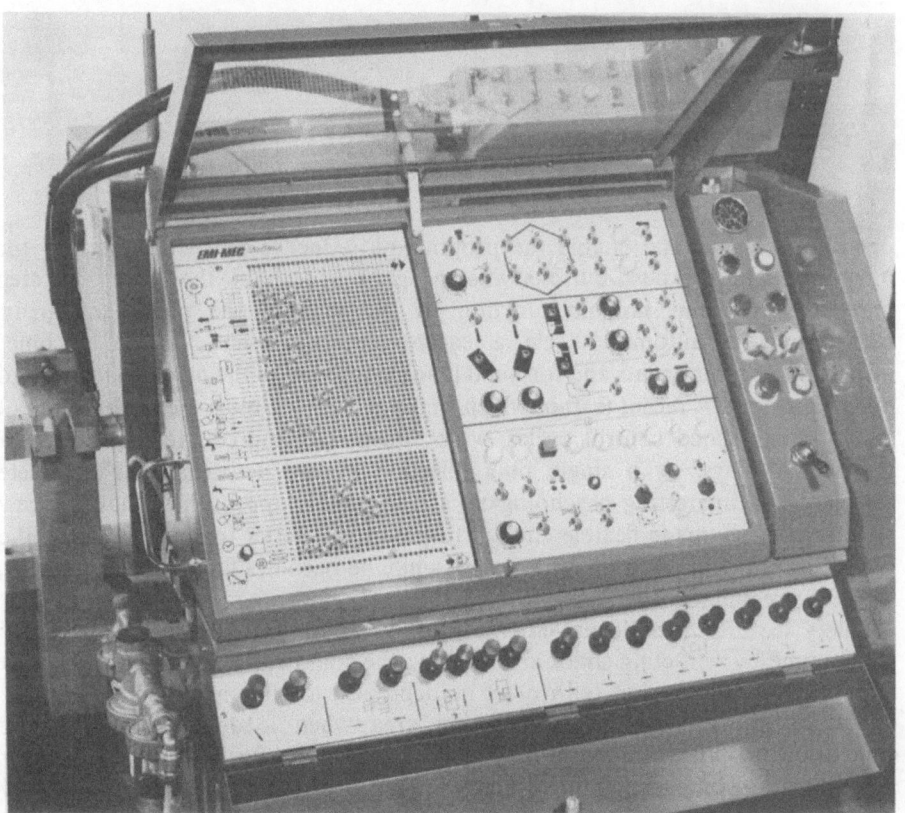

Figure 18.14.2　Detailed view of the lathe control console. (*A'Court Photographs Ltd*)

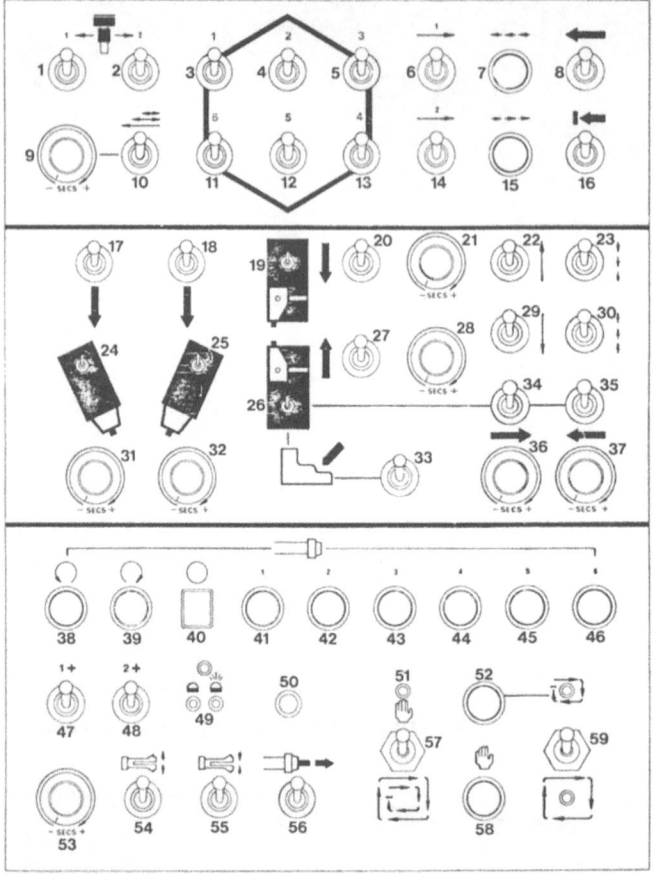

Figure 18.14.3 Diagrammatic arrangement of control panel.

operating column and enable the programme sequence to be followed by the operator. Hand step push-buttons, when programmed, enable the operator to step manually from one column function to the next in the plugboard array, interrupting the automatic cycle, a feature especially useful for 'on machine' inspection of components and for hand loading of parts to the machine for second operation work. A recent feature which simplified the set-up is the replacement of several complicated control cycles, which formerly required the insertion of a number of plugs, by a single plug selection system activating the complete sequence. An interesting psychological aspect of this approach is that the new system has been made flexible enough to permit the setter to continue using the older conventional programming system while he becomes familiar, step by step, with the streamlined facilities.

Safety features include an emergency stop push-button, which homes the programme by retracting slides and braking the spindle; interconnection of start and stop controls so that both must be depressed to start in order to keep the operator's hands clear of the working area when loading; linking of the chuck control push-button to the dual brake circuit for maximum operator protection; and halogen lighting, which avoids the stroboscopic effect produced by fluorescent lighting which can be dangerous with rotating machinery.

Silencers and filters on the compressed air system and damping on the bar feed tube contribute to the operational environment by reducing noise and preventing oil mist venting to atmosphere. The lathe bed design and hinge slide covers also contribute to cleanliness by ensuring that lubricant-covered swarf falls away to the back of the machine so that it can be readily removed and by protecting the sliding components.

Removable covers provide good access for maintenance to the complete air-hydraulic system and to the modular electronics. Built-in LEDs and code numbering assist in identification of circuits and with fault finding, and modular design enables components to be changed.

The machine is clean, tidy, aesthetically attractive and gives a visual impression of the functional efficiency it provides.

18.15 Type 180 horizontal boring machine

Manufactured by: Kearns-Richards Ltd
a Division of Staveley Machine Tools Ltd,
Broadheath, Cheshire

Designed by: Kearns-Richards Design Team in association,
for industrial design, with
R. Satherby, Satherby Design
5 Dryden Street, London

Objectives. This very large boring machine was specified to have a table loading capacity of 20,000 kg and to be modular so as to provide several alternative traversing capabilities in the vertical, longitudinal and table traverse modes and in revolving table size; to provide a flexible combination of both manual and automatic control; to have built-in measuring accuracy of 0.002 mm and to provide infinitely variable feed rates over a wide range for all five axes of movement, and infinitely variable speeds for spindle and facing heads for machining operations.

Design. The impressive result (Figure 18.15.1) weighs about 47,000 kg and has a 11 kW motor for transverse and 37 kW for rotation. All the principal modules are fabricated in cast iron, in rectangular format. From the point of view of torsional stiffness the vertical column might have been lighter in circular form, but a rectangular shape with internal ribbing has been used, matching the natural format of the remaining components; the vertical guideways for the spindle frame are integral with this column, and the counter-balance weight, connected to the spindle frame by twin chains, is housed internally.

A number of features help to reduce wear and maintenance. The guide faces are ground to the high degree of accuracy required and those on the vertical column are also hand scraped, both to improve retention of oil on these vertical faces and to provide the traditional appearance associated with machine tools. Furbishing takes time, but certainly adds to the visual impression generated here. The traversing drives are all carried out by high-efficiency recirculating ball screws which reduce the power required and improve accuracy, more especially when pre-loaded nuts are employed associated with ground threads, eliminating

Figure 18.15.1 The Kearns-Richards Type 180 horizontal boring machine in operation.

backlash, pitch error and wear. All sliding surfaces are fitted with non-metallic low friction material with automatic lubrication metered by a system that responds both to time and to the extent of relative movement, ensuring adequate lubrication but avoiding flooding and consequential cleaning problems. Automatic temperature control of lubrication is also provided for the spindle assembly in the interests of thermal stability and accuracy. Stainless steel telescopic covers protect sliding surfaces, and these have neoprene wipers themselves protected against hot metal chips by stainless steel guards.

Thyristor control in combination with gear changes enables the DC motor to develop full power over a wide range of output speeds. The gear changes in small machines can be arranged with magnetic clutches, but the power loss with large machines becomes excessive, and a sliding gear change is employed here, involving a complicated procedure which is carried out on demand by an automatic hydraulic system.

Reliability and ease of maintenance of high capital cost equipment are vital, and conservatively rated solid state controls and components therefore have been used in this case. A significant part of maintenance time is fault diagnosis and a major improvement has been made here by the development of a continuously cycling fault monitor system. Experience indicates that the most time-consuming problems are concerned with such items as hydraulic valves, limit switches etc. The fault monitor checks the state of these items continuously and, if faulty, indicates directly the circuit or component which is suspect. A secondary fault diagnosis system checks the electrical control circuitry itself.

Safety of operation is inevitably a problem with large machines should the operator become exposed to rotating cutters or traversing movements. A safety emergency stop is provided on the panel face. Precautions such as switching-off mats laid in areas which should not be entered while the machine is running, or the use of a suspended safety stop line surrounding the perimeter of the machine, have been experimented with, but they can produce a false sense of security, and the provision of several emergency stop switches located strategically in established perimeter positions may be the most effective emergency provision.

Figure 18.15.2 Detailed view of the machine control panel.

18.16 MF 30 agricultural drill

Manufactured by: Massey-Ferguson UK
                           Maudsley Road, Coventry
Designed by: A. J. Bailey, Project Manager
                           R. S. Sargeant
                           M. Dean
                           F. Ward
                           G. Holmes
                           H. B. Rogers
                           D. Brook
Design Council Award 1976

Objectives. These machines, which cut grooves or 'drills' in farm soil and meter seed and fertiliser into them, were specified to be in modular form to meet a range of sizes from 13 to 31 drills in parallel with the choice of two spacings between them; to accommodate three alternative coulters (the cutting tools) to match different soil conditions; to provide accurate and adjustable seed metering and distribution; and to be exceptionally easy both to operate and maintain.

Design. An MF 30 machine with 20 drills is shown in Figures 18.16.1 and 18.16.2. It consists of a seed and fertiliser hopper (A) mounted on a two-

18 Engineering product design, some solutions in practice

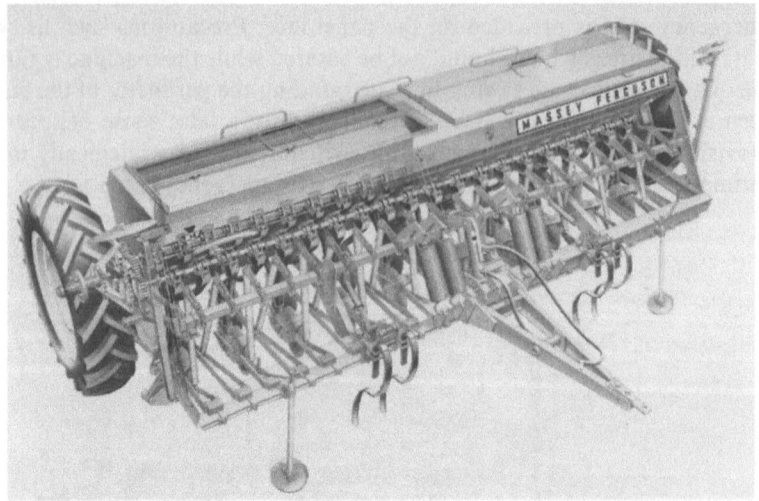

Figure 18.16.1 General arrangement drawing of the MF 30 drill.

Figure 18.16.2 The drill in use.

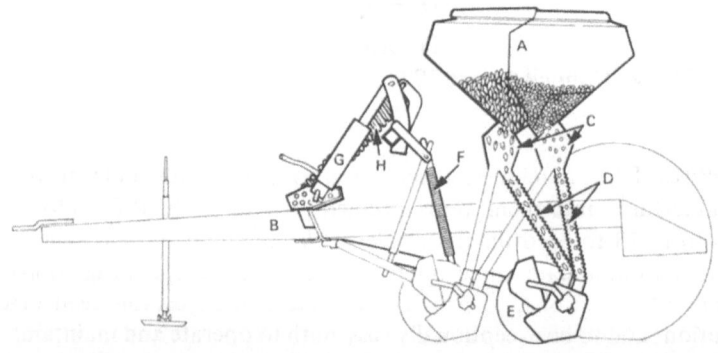

Figure 18.16.3 Schematic arrangement of the drill mechanism.

wheel chassis, fitted with a draw bar (B) for towing by a tractor, which also supplies the hydraulic power needed for control of the drill. Seed and fertiliser are metered separately for each drill through an assembly of meter units (C) driven by a chain and gear system from the land wheels, through an automatic clutch which engages only when the coulters are in the lowered position. The rate of delivery is adjusted by means of a quadrant lever with a rate scale. The seed and fertiliser are metered and force-fed by the units through flexible rubber tubes (D) to the coulters (E) which distribute them in the bottom of the drills. The coulters are forced into the cutting position by a single-acting hydraulic ram, which rotates a control beam carrying levers which are individually coupled to the coulters by compression spring thrust units, and are retracted on release of hydraulic pressure by the return springs (H). The depth is controlled by prior selection from eight positions for the pivot of the hydraulic ram, a simple operation requiring the repositioning of a single pin. In the preferred system of operation the ram extension is controlled by varying the oil pressure, monitored by the operator from his driving position on the tractor, and a marker on the ram gives him a clear indication of the coulter pressure, which is related to depth of penetration. The actual depth of cut will vary both with hydraulic pressure and with the nature of the soil, and if the coulter hits a hard patch the operator can increase pressure temporarily in response to the depth indicator in order to maintain a constant depth.

The hopper has a metal partition which can be reversed or removed by undoing a few clips to give storage ratios of seed to fertiliser of 1 : 0, 1 : 1 or 1 : 2. A platform is provided near the hopper to make loading easy and each of two half lids can provide temporary support for sacks when filling the opposite half of the hopper; the trailing edges of the lids are designed to prevent condensation dropping into the hopper when they are open, and a small flap is positioned in the front of the lid to prevent seed entrapment. Windows permit the operator to see when the hopper needs refilling from his driving position. All metering parts in contact with seed or fertiliser are nylon or are nylon coated to resist corrosion, and the metering system has a protective rain flap with a continuous nylon hinge for use in bad weather.

Safety features include non-slip platform surfaces, easily detachable covers over the gear drives, the absence of finger traps on all lids, and an instruction book which gives clear warnings and procedures for carrying out maintenance of the powerful coulter return springs.

Cleaning and maintenance are simple. The hopper is designed to run itself empty. The meter units are mounted on a single shaft (two on the largest machine), and are accessible and easy to remove with snap-in action when this is needed either to change the size of flutes or for cleaning. The flexible feed tubes are also accessible and can be cleared easily should they become blocked by damp fertiliser. There are, however, still a number of points that require greasing by hand which have been retained because many operators in the past liked their machines to be greased occasionally since this reassured them that they were receiving adequate maintenance. Although self-lubricating components cost more, attitudes are changing and maintenance-free systems may be incorporated in the future.

Agricultural equipment often gives an impression of crudity, but here appearance reflects the quality of the machine and the attention given to balanced arrangement, ergonomic and detail design.

18.17 Type 725 CM truck-mounted telescopic boom crane

Designed and manufactured by: Cosmos Crane Co Ltd
Cotes Park Estate, Somercotes
Nr Alfreton, Derbyshire

Designed by: D. Hassall, Managing Director
J. W. Johnson, General Manager
M. Chaplain, Hydraulics Engineer
J. Wilson, Chassis Engineer

Design Council Award 1977

Objectives. This crane was required to meet a wide range of performance and international legal specifications, including 32 m boom extension, 8 metric tonnes lift at up to 8 m radius, 360° rotation of crane on its truck, retaining the same lifting capacity, left or right-hand mounted truck and crane control cabins, power-operated outrigger, and a high level of man/machine interface design.

Design. The crane is shown in Figure 18.17.1. Of special interest here is the operational control. Both cabins were ergonomically studied to ensure ease of operation, comfortable seating, good ventilation, vibration and noise suppression, all-round vision, and comprehensive instrumentation including an automatic safe-load indicator. The crane cabin rotates 360° with the jib and has a sliding door for safe access. Crane control is by two joysticks (Figure 18.17.2) ergonomically placed for ease of operation, using a separate hydraulic servo-system for the control functions. The two joysticks control four modes of movement simultaneously as shown in Figure 18.17.3, offering light, precise and logical operation. A dead man's handle pushbutton is incorporated in the head of one joystick and a pressure boost control button in the other, this being used when handling loads close to the maximum line pull.

Figure 18.17.1 The Cosmos Type 725 CM telescopic crane.

18.17 Type 725 CM truck-mounted telescopic boom crane

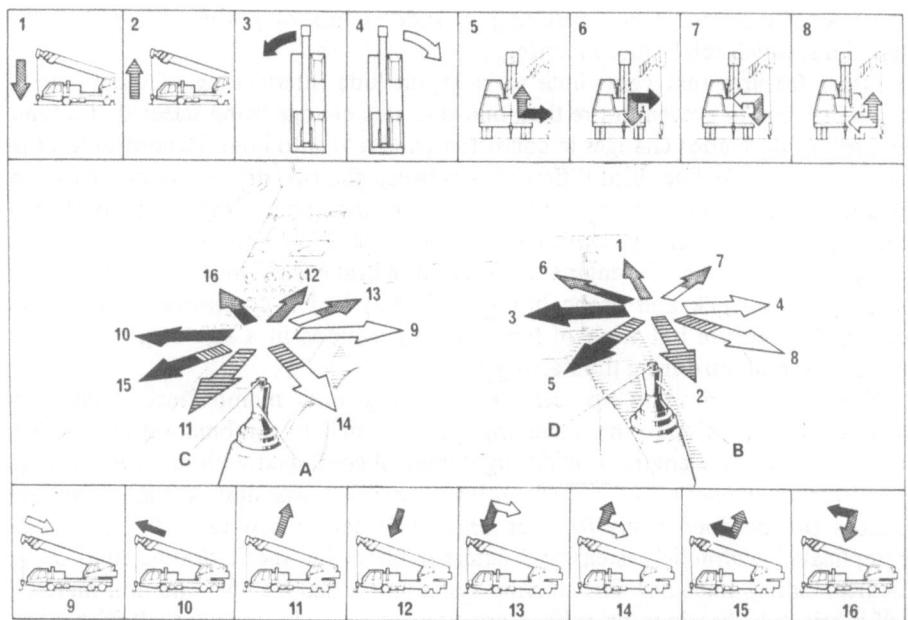

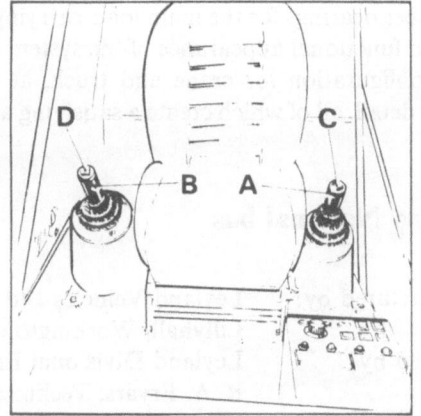

Figure 18.17.2 (above) Diagram of movement modes resulting from joystick operation.

Figure 18.17.3 (right) Driver's seat and joystick arrangement.

The crane boom comprises four high-tensile steel box sections continuously welded inside and outside at all corners, which, when compared with space welding, affords better corrosion resistance and a cleaner appearance. The boom is extended and retracted by a hydraulic jack operating through a factory-lubricated and prestressed chain system providing automatic proportional extension to each of the four sections, which simplifies control. The power-operated outriggers are interlocked to prevent operation when the crane is in use and are controlled from ground level, enabling the operator to examine ground conditions as he positions their feet.

The main hydraulic power system adjusts pressure automatically to match the load demand, but the crane driver can adjust speed of operation through a foot control pedal, which governs pump speed. If the total hydraulic power demand of several functions working in parallel exceeds that available the system reduces the power supply to the various demands by an equal proportion, which again simplifies control as compared with the usual practice in which the oil flow takes the line of least resistance, resulting in sudden loss of power to some of the functions, which can be disconcerting.

Other features that contribute to safety include interlocking of the 12-speed truck gearbox to prevent more than one gear step change being selected at a time, so preventing sudden changes in conditions, and a lock to hold the control lever in neutral; a lock for the third differential between the two driving axles for use on bad ground; interlocks to prevent operation of the crane if the engine is started from the truck cabin; a hydraulic fluid flywheel lock at 600 rpm between the engine and gearbox, enabling the engine to be used as a brake when moving downhill, and making both truck and crane brakes fail safe; and a compressed air system designed to ensure initiation of functional operations in a safe sequence as air pressure is built up during the starting process.

Features contributing to ease of working and maintenance include a servo-assisted rack-and-pinion steering system which, in combination with a very short overall crane length, provides light control combined with an exceptionally small turning circle; a jib head which lowers below horizontal, so that re-reeving pulleys can be done easily from ground level; colour coding of the inevitably complicated hydraulic system and external filters for the hydraulic fluid as in medical kidney machines. The steering system is 'sealed for life', and non-metallic self-lubricating bearings or rubber are used extensively, including British Steel Corporation laminated plastic pads for the retractable jib box sliding bearings, and Spherilastik rubber bearings for the main load-carrying bogey rocker beams.

The integrated functional appearance of the system is reinforced by repetition of similar cabin configuration for crane and truck, at different levels, and by the simple lines and detail, all of which create a satisfying aesthetic whole.

18.18 Leyland National bus

Manufactured by: Leyland Vehicles Ltd
                           Lillyhall, Workington, Cumberland
Designed by: Leyland Divisional Engineering
                           R. A. Fryars, Technical Director
Design Council Award 1974

Objectives. The specification embraced a wide range of international requirements, legislation and design rules, including left and right-hand drive, one-person operation, and a variety of accommodation arrangements, ranging up to 72 passengers, in two lengths of vehicle, and a high level of design for comfort and safety of the driver and passengers.

Design. The industrial design involvement for this bus (Figure 18.18.1) was implemented through the following procedure [18.18.1].

> *Concept study.* Definition of the population for which the vehicle was intended, and the design outline of the vehicle. Comparison with competitive vehicles, identification of countries where the vehicle would be marketed and their legal requirements. Analysis of passenger make-up relative to criteria such as age group, distribution of use, disabilities and mobility, body dimensions and characteristics.

18.18 Leyland National bus

Figure 18.18.1 The Leyland National bus.

Control study. Evaluation using basic representation of proposed driver's 'package', with a seat, steering-wheel, pedals, control location, with the application efforts and entry conditions and with representative drivers.

Mock-up. Representation of internal and external styling; study of entry conditions, and check on driver's package and control location; check on accessibility of units and systems for servicing interior airflow and aerodynamics. Check on view out, mirror field view and legal requirements, using the pivot centre of the human torso (Figure 18.18.2) as a guide. Mock-up updated as design advanced to provide a dimensional standard.

Preproduction vehicles. Dimensional check related to the mock-up standard established. Validation of position and performance of controls, instruments, the general package, ride and handling by dynamic trials.

Figure 18.18.2 Ergonomic parameters for the bus driving position.

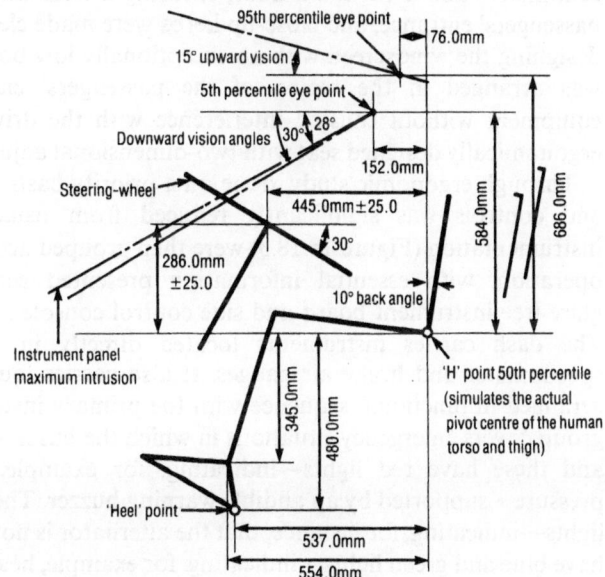

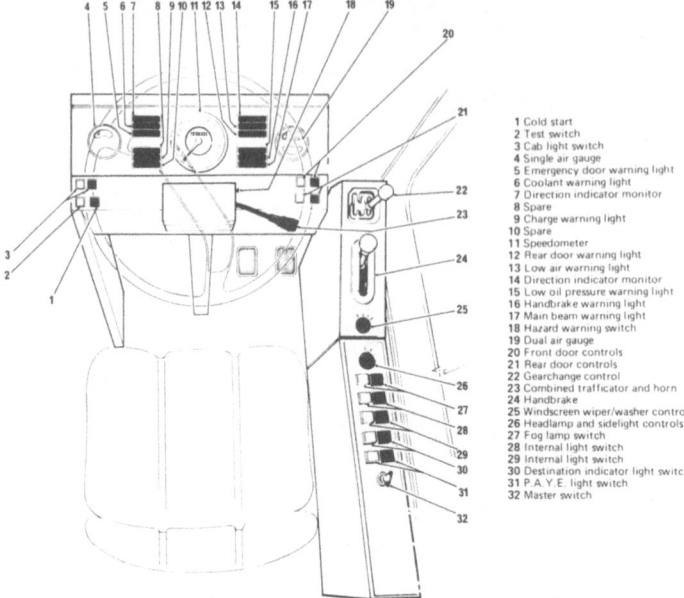

Figure 18.18.3   Controls and instrumentation arrangement.

Production vehicles. Re-check on production modifications affecting dimensions and legal requirements relative to the mock-up.

Evaluation in the field. Validation study of a number of production vehicles under service conditions with operators.

The design incorporated a number of features contributing to ease of control, safety and comfort. The floor level of the driver's compartment was raised 25 centimeters above the main floor, ensuring a clear and unobstructed view of the passengers' entrance, and close-up kerbs were made clearly visible to the driver by designing the windscreen with an exceptionally low bottom edge. Adequate space was arranged in the region of the passengers' entrance for fare collection equipment without causing interference with the driver, and the driver has an ergonomically designed seat with two-dimensional adjustment.

Through ergonomic study, done on a priority basis, the number of instruments and controls was significantly reduced from usual practice. Controls and instrumentation (Figure 18.18.3) were then grouped according to priority of use in operation, with essential information presented centrally, on a matt black, glare-free instrument board and side control console and on the steering column. The dash carries instruments located directly in front of the driver, with speedometer and brake air gauges. It also carries four blocks of warning lights arranged in functional sequence with the primary instruments and controls. One group covers emergency situations in which the bus must be stopped immediately, and these have red lights—indicating, for example, loss of brake operating pressure—supported by an audible warning buzzer. The second group have amber lights—indicating, for instance, that the alternator is not charging. The third group have blue and green lights—indicating, for example, headlamp operation. The final

18.18 Leyland National bus

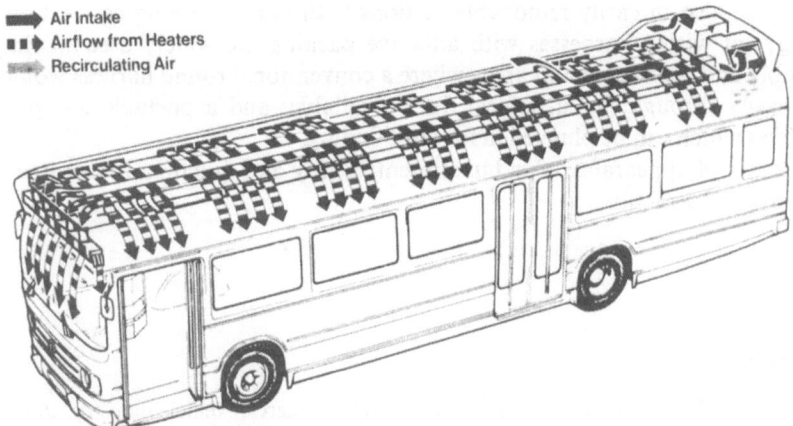

Figure 18.18.4 Heating and ventilation systems.

group comprises fare collection, as specified by the client, and is mounted in the area next to the passenger entrance.

Manually operated driving controls are mounted on the steering column, or adjacent to it on the sloped console. Lighting and master switch controls are on a panel at the side of the driver. Controls needed before moving off are mounted on the lower part of the dash.

The heating and ventilation unit is mounted and controlled from the rear roof and distributes a downward airflow across all windows, body sides and doorways (Figure 18.18.4) keeping windows free from mist. It can provide up to 40 changes of air per hour.

Design for safety had to take into account the event of a head-on collision, for which the requirements of driver and passenger are to some extent in conflict, and a compromise was adopted. In this the structure under the floor at the front of the bus is rigid, and the steel panels provide the driver with a very stiff encapsulation to prevent penetration, but the steering-wheel, steering linkage, binnacle and facia are all designed for progressive collapse. Behind the driver, the integral body chassis structure will also collapse progressively, reducing shock on the passenger compartment, and the structure will withstand roll-over without collapse. The suspension system maintains a horizontal attitude for the bus under all braking conditions, reducing risks to standing passengers. Details that contribute to safety include an illuminated entry step about 350 mm above ground level, and hand grip rails that take into account the needs of handicapped people, using oval steel tubes 32×18 mm covered in Doverite to increase the slip force that can be resisted along the rail.

Audiovisual alarms and interlocks prevent the bus being driven above 5 km/h when any doors are open; headlamp failure results in an automatic change-over to dipped beam, and vice versa; and circuit-breakers are incorporated for overload protection, which enables unfaulted circuits to be reestablished by the driver.

Design for maintenance includes elimination of corrosion for a vehicle life of the order of 15 years (which also contributes to safety) which is achieved by galvanising, phosphating and epoxy coating all the components and several thousand rivets of the integral body chassis prior to assembly, followed by undersealing with epoxy pitch and spraying with wax film after assembly; non-structural panels are designed for easy replacement in case of damage; control panels are also in easily removable sections with multi-pin plug connections for wiring; flat wiring harnesses with adhesive backing are widely used to simplify assembly and replacement in areas where a conventional round harness would not pass easily through the structural ducts available; and a portable test panel is provided which can be plugged in for fault finding.

In terms of appearance, this bus presents a tidy and functional form free from flamboyant distractions.

Reference

1 Brooks, B. M. 'The role of an ergonomist in the commercial vehicle industry.' *Journal of Automotive Engineers*, February 1974.

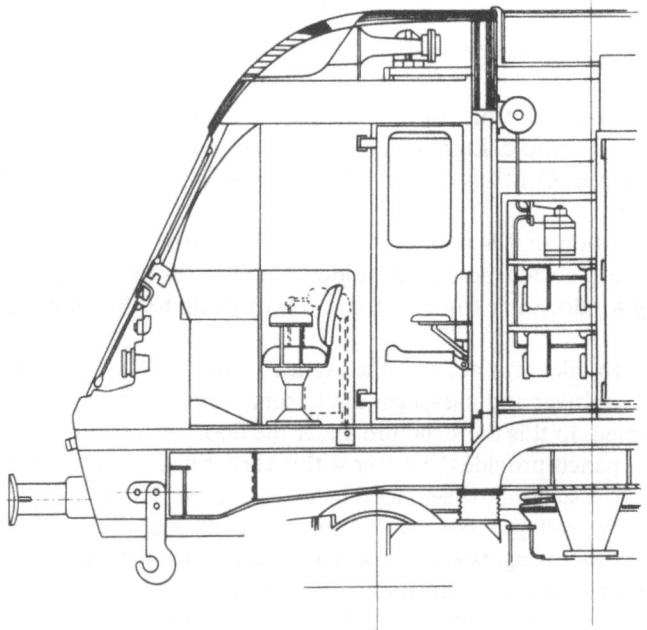

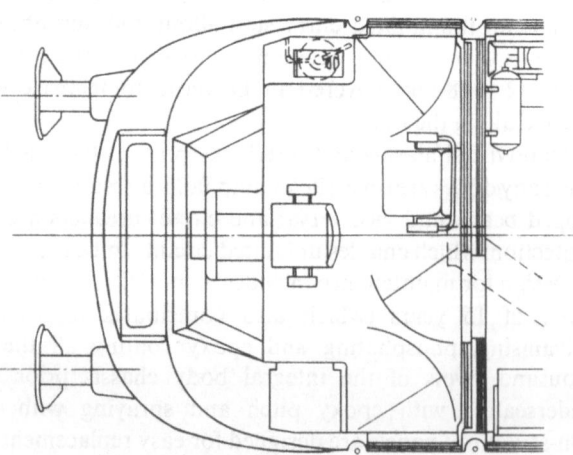

Figures 18.19.1a and b Elevation and plan views of the HST cab prototype design.

18.19 Driver's cab, High Speed Train

Designed by: British Railways Board
Consultant for external design: Kenneth Grange
Built by: British Rail Engineering Ltd
Design Council Award 1978

Objectives. The specification for this diesel-electric high speed train covered two power cars, each complete with diesel/alternator, electric motors, and driver's cab, with seven passenger cars coupled between them. The train was to be driven from either power car, the second power car being operated by remote control; to be capable of running continuously at 200 km/h, with an all-up weight of 400 tonnes; to operate on existing main line tracks, within the constraints imposed by existing signalling and allowable dynamic forces; and the very high safety standards intrinsic in all British Rail equipment were to be retained, as were the requirements for the type and character of maintenance procedures. The driver's cab was to accommodate a driver and co-driver, and to provide an environment which compared favourably with existing practice.

Design of cab. The first prototype layout for the cab is shown in Figure 18.19.1 a and b with the driver positioned centrally and the co-driver offset and behind, which permitted a wrap-round arrangement of the controls in the middle of the cab, with room for auxiliary controls and equipment at the sides. The drivers' union, however, expressed a preference for the driver and co-driver to sit side by side, and for an additional side window to permit direct sideways viewing of the platform, in addition to the access door windows which had already been included. The cab was redesigned to meet this viewpoint for the production version of the High Speed Train (Figure 18.19.2a and b).

The normal requirements for the driver's vision can be met satisfactorily with a viewing range of the order of 15° to either side. Vertically it should be possible to read signals down to 10 metres range and this can be met with an upward angle of vision of about 25°. Downward vision requires a compromise between seeing enough of the track and seeing too much, which can be disconcerting when moving at high speed. A satisfactory solution is a downward angle of vision of 15°, which permits the track to be seen to approximately 10 metres in front of the cab. One consequence of the rearrangement of the driver's cab was an increase in the width of the front window and, combined with the extra side windows, this increased the total glass area and solar gain in the cab making air conditioning essential.

The front window has to resist impact with missiles thrown up from the track by trains, or dropped from bridges by vandals, and this was met by the use of 250 mm-thick Triplex laminated glass. A consequence of using such a thick glass was that it had to be optically flat, to prevent visual distortion and double images of signal lights, and the same problem also limited the permissible angle of slope of the screen. The slope adopted was the maximum acceptable, to improve appearance and reduce aerodynamic noise levels and contamination by insects.

The external shape of the cab was therefore influenced by a number of factors. The dimensions and slope of the front window were determined as described above, as were the requirements for two side windows. Because the rearrangement of the drivers reduced available space for auxiliary equipment, some of this had to

18 Engineering product design, some solutions in practice

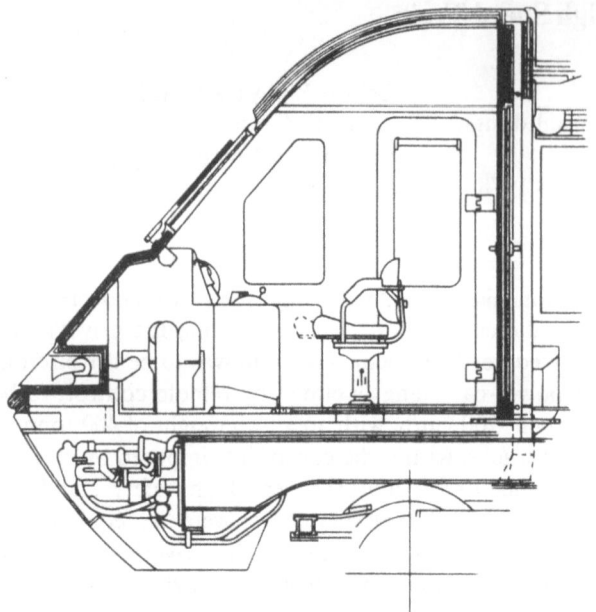

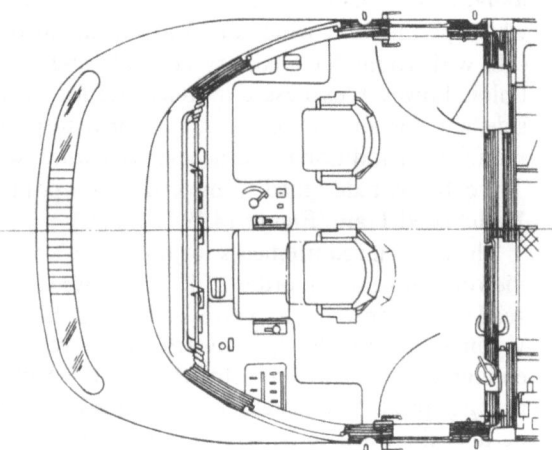

Figures 18.19.2a and b Elevation and plan views of the HST cab production design.

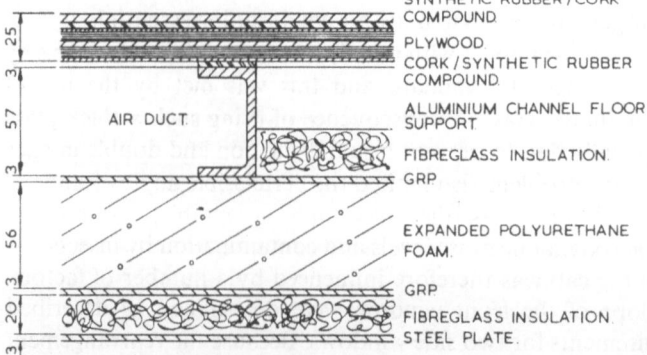

Figure 18.19.3 Diagram of cab floor construction.

be accommodated in the cab nose below floor level, so that a large available volume was an advantage in this position. The original design had the usual buffers, but the need for these in normal operation does not arise with a unit type train. It was acceptable for the bottom panel of the nose to collapse, should the train overrun stationary buffers, with the shock forces being taken by the steel underframe within the nose enclosure, so buffers were omitted. This made more space available, reduced aerodynamic noise, and contributed to improved appearance. The headlights were also submerged within the nose structure, smoothed over with transparent covers, again in the interest of noise level and appearance, but at the expense of some reduction in accessibility for maintenance. Another consequence of these changes was to increase turbulence at the sides of the front window, because of the less efficient junction form caused by the increased width, but this was compensated for by a radical improvement in flow over the top of the window, associated with the increased rake, made possible by elimination of the buffers. The cab module was formed as a single-piece GRP moulding, incorporating polyurethane foam pumped between its inner and outer shells while still in the mould. This form of construction, used in the interest of noise reduction and heat insulation, incorporates the necessary reinforcement to resist missiles.

Noise in the cabs of trains that travel at high speed can be a serious problem, but has been restricted in this case to 85 dBA at a speed of 200 km/h, a low level achieved by detail attention to all aspects of noise generation and damping. Some aspects of this have already been referred to; additional factors of importance include flexible mounting of the cab module on the underframe, with interposed rubber and noise insulation in the floor and bulkheads by expanded polyurethane and PVC slabs. The floor construction (shown in Figure 18.19.3) also includes a multi-layer noise insulation system and attention to the manufacture of the gear train teeth in the motor/axle drive system, directed at reducing pitch circle errors below the already very low level normal in locomotive gearboxes, reduced noise generated at source.

The possibility of reducing track-induced noise levels by using driving wheels in which the rim is flexibly mounted on the wheel hub was investigated, but was found not to be fail-safe in character and was not therefore adopted.

The cab temperature is maintained in the range 19 to 23°C at the driver's head level, and up to 3°C higher at floor level, by the air conditioning system.

The control console is shown in Figure 18.19.4, the top panel being formed as a single-piece GRP moulding, coloured grey in line with British Rail standards, although a two-tone scheme would have been preferred by the designers. The wrap-round arrangement of the prototype cab was retained for the controls in the production cab, so far as space permitted with the revised seating, with brake controls and instrumentation on the left, power controls on the right, and the driver's safety pedal in the centre.

The design of this safety device has been updated to take care of possible causes of loss of protection which have been experienced in the past. The driver has now to hold the pedal depressed and release it momentarily once a minute, and if he should fail to do this a bleep device is energised, followed four seconds later by cut-off of power and application of the brakes, if by then the driver has not responded. The driver also has to acknowledge signals, and this automatically resets the pedal-bleep mechanism, so that the driver does not have to respond to signals and the driver's safety device more or less simultaneously, which could result in unnecessary and embarrassing brake application.

Figure 18.19.4 Control console arrangement.

The disc brakes are applied by compressed air, and the wheel-tread-conditioning parking brakes are applied by spring and held off by compressed air. The manual brake control lever has six positions which adjust the hold-off air pressure in steps, through an electrically supervised compressed air control system. If this system should fail, the lever can be pushed past its normal full-on brake position, when it releases the hold-off air pressure through a direct mechanical compressed air valve system. A separate direct-acting push-down emergency knob is also available in the event of any failure of the lever system, and is situated adjacent to the full-on brake operation position. A parking brake is also situated on this wing of the panel, as are gauges displaying the storage and operating cycle compressed air pressures.

The driving controls on the right hand wing include 'start' and 'stop' push-buttons for the diesel engine, reverse and forward controller handle, and the power control lever with a number of steps, working in a quadrant similar to the manual brake system. Instrumentation includes a large, well-sited speedometer, instruments measuring power output and a fault indicator. Railway practice is to omit instruments unless they are clearly essential, in order to reduce to a minimum the complication created by a display of less important information. Based on this philosophy, the fault alarm does not identify the type of fault, since the driver cannot take corrective action himself. More elaborate fault identification to help with maintenance is relegated to other compartments on the power car.

The central area of the desk also provides a panel for the driver's operating timetable and other documents, and a flat space to hold beverages. All the panel horizontal areas are designed to prevent damage to the control system in the event of liquid being spilt on them.

The cab also contains a change-over controller to select the designated control cab, which cuts off control functions from the trailing power car.

The instrumentation has been unified in graphic style and is clear to read, but long-standing design requirements for maintenance have prevented the adoption of modern control panel practice, both as regards detail appearance and miniaturisation based on electronic techniques.

This cab design has created a good operating environment for the control function through ergonomic study, simplicity, clarity and comfort, which all contribute to safety, and its external proportions (Figure 18.19.5) lend support to the visual impression created by the train as a whole.

References

1. Burden, E. S. 'Into service with British Rail's Intercity 125.' Paper presented to Studiengesellschaft '*Leichtbau Der Verkehrsfahrzeuge*', Frankfurt Am Main, 4 October 1977. (Not published in UK, but text has been used in lectures given to many societies, discussion groups and institutions through UK.)
2. Powell, A. J. and Cartwright, A., 'The design of driver's cab.' *Institution of Mechanical Engineers, Proceedings*, vol 191 33/77, 1977.

Figure 18.19.5 Exterior of the British Rail HST, design by Kenneth Grange.

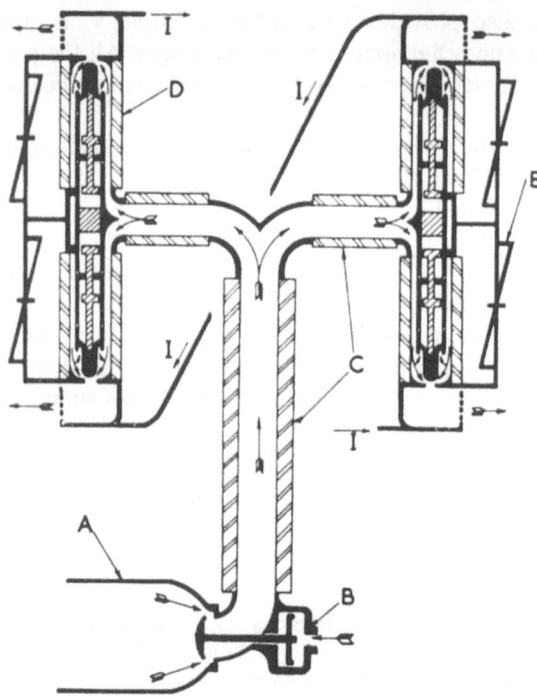

Figure 18.20.1 Schematic arrangement of Type GA high-voltage air-blast circuit-breaker employing four interrupting elements with identical synchronised air supplies.

A—air reservoir
B—blast valve at earth potential
C—porcelain blast tubes
D—porcelain interrupting chamber
E—non-linear resistor

Figure 18.20.2 General view of Type GA installation.

18.20 Type GA high-voltage air-blast circuit-breaker

Manufactured by: Associated Electrical Industries Ltd
                   (now GEC) Trafford Park, Manchester
Designed by: AEI Design Team

Objectives. The GA range of air-blast circuit-breakers, of which over 1000 were constructed in the years up to 1965, used compressed air to drive the mechanism and as the arc interrupting medium, with an automatic sequence switch in air to provide isolation. They were developed to meet a wide range of short circuit ratings and service voltages from 66 kV to 440 kV in modular format using from 1 to 12 interrupters in series per phase.

Design. Contemporary practice was to arrange interrupters in series one above the other in vertical columns mounted on supporting insulators. The GA design had the interrupters electrically in series, but mechanically in parallel (Figure 18.20.1) in order to ensure that the air supply paths through the porcelain assemblies to the interrupters were aerodynamically identical; that the heat generated by the power arc in one interrupter did not interfere with the air-flow to the interrupters in series with it [18.20.1]; and to provide better contact accessibility. Figure 18.20.2 shows one phase of an eight-break assembly for 330 kV duty.

The modular interrupters were designed as pairs except when odd numbers were required, enabling the fixed contacts to be extracted by removing the end covers to the interrupters. Moving contacts, designed as a single cartridge assembly for each pair of interrupters, could then be removed by undoing a central locating bolt, the complete contact system being exposed for inspection in a few minutes, compared with as many hours for designs in which the interrupters were mounted in a vertical array. The blast valves were mounted on the end of the air receiver, and were accessible without the need to dismantle any other components.

High-voltage equipment does not usually offer much latitude in terms of design for appearance, but in this concept the requirements of the electrical, mechanical and appearance criteria were harmonised. In the original designs the support insulation for the interrupters, which also acts as compressed air supply pipes, was mounted horizontally, but in order to reduce turbulent loss of pressure at the bends and to improve the elegance of the structure, this was inclined to the horizontal, reducing the change in direction of airflow by 30°. The ceramic vertical support-insulator columns for the interrupter and isolator assemblies were then designed with a slight taper, 20 per cent in diameter, permissible structurally because of progressive reduction in stress in the insulation with increase in height, which contributed to a useful reduction in live dimensions of the circuit-breaker and also to improved visual balance of form compared with the use of parallel cylinders.

Reference

1   Flurscheim, C. H. 'Switchgear: a review of progress.' *IEE Proceedings*, vol 103 part A, June 1956.

Figures 18.21.1 and 18.21.2 General views of a Brown Boveri steam turbine generator installation.

18.21 Power station steam turbine generator sets

Designed and manufactured by: Brown Boveri Co Ltd,
                                 Baden, Switzerland

Objectives. The design problems experienced with large turbine generator sets running at up to 3600 rpm, with turbine blade tip speeds in the supersonic range, with four or more shafts coupled in line, associated with steam temperatures of the order of 540°C and hydrogen-cooled generators, are immense. In such an environment industrial design represents a very small part of the total design effort but, if integrated with the engineering, it can have a significant influence on the quality of the result.

Design. The 600 megawatt set (Figures 18.21.1, 2 and 3) installed at the Cardinal power station of Buckeye Power Incorporated, Ohio, USA, comprises a combined high-pressure/intermediate-pressure turbine, two low-pressure turbines taking their steam in parallel, a hydrogen-cooled generator and static excitation. The components are necessarily designed in separate specialist departments, and in Brown Boveri co-ordination of their industrial design is made the responsibility of the Chief Designer of the Turbine Division. The philosophy used is to express function, careful detail design being used to make visual disguise of components unnecessary, with covers provided only when they are justified by protective or safety reasons, the design being optimised for maintenance by open access. Examples are the HP-LP steam pipes, which are sometimes encased in a single manifold in the interest of tidiness, but are here exposed, yet the undisguised system contributes to the interest of the machine's appearance. Another practice sometimes used is to enclose the complete high-pressure turbine in a sheet steel housing. Although this looks tidy and reduces noise, it distorts the natural

Figure 18.21.3 Detail of the 600 MW installation shown on the opposite page.

Figure 18.21.4 A cast steel high-pressure turbine casing during manufacture.

Figure 18.21.5 Model of a Brown Boveri 260 MW set, showing steel foundation structure and services.

appearance of a turbine, increases cost, creates a difficult and hot environment for work on the many auxiliaries inside the enclosure, and has to be removed completely for major maintenance. The high-pressure turbine here is not enclosed but is tidy, and the set as a whole gains from the functional impression created.

On the other hand, the excitation equipment, which used to comprise an array of small generators, gearboxes, fans and ducts and access covers, was often untidy and incompatible in size and appearance with the rest of the set. This justified enclosing it in a single housing, which improved the relative scale and tidiness, reduced noise and did not increase the difficulties of maintenance. The recent change in practice to static excitation makes enclosure essential for electrical safety, and the excitation equipment here has been built on to the generator it serves, with which its housing is visually compatible.

The cast steel high-pressure turbine casings (Figure 18.21.4) present an interesting form, derived in part, as with the aircraft engine (Figure 1.18) from a long history of specialist development.

Low-pressure turbine and generator shells used to be curved in three dimensions, taking advantage of cast iron techniques. When welded steel plate fabrication replaced cast iron these rounded shapes were at first imitated, using a number of formed sheets with smoothed-over welded joints. This approach was gradually simplified, and the BBC set here carries rationalisation forward by using single curvature sheets for the shells, and flat sheets reinforced with radial ribs for the ends, all the plate edges being fully exposed emphasising definition of structural form. Although not visible in the illustration, turbine and generator are similar in concept—a logical cost-effective approach that does not interfere with steam path efficiency and contributes to the functional impression.

An access platform emphasises unity of concept by surrounding the complete machine all at one level and provides accommodation (with access doors for maintenance) for pipework and multicore cables, which so often spoil the design of heavy engineering equipment. Turning to detail, the platform guard rails (Figure 18.21.3) are fabricated from flat sections, with rounded edges for the top rail, which harmonise with the style of the turbine better than would a more conventional tubular construction.

The foundation areas of turbine generators house steam and condensate services, cooling water, lubrication and hydrogen cooling systems, and the recent development of steel foundations has created opportunities for improving the layout of these complicated systems in the interest of maintenance and appearance. Because steel structures can be designed to have a natural frequency lower than those produced by the turbo generator, they are light in construction as compared with the established concrete structures, which are designed for frequencies above the highest level that can be so excited, the consequential massive construction restricting space available for auxiliaries. The model of a BBC 260 MW set (Figure 18.21.5) demonstrates the impression of order that can be achieved by attention to layout and detail design using a steel structure.

The co-ordinated approach to engineering, ergonomic and aesthetic factors in the design of these turbine generators has made a valuable contribution to ease of maintenance, the visual impression of order presented, and to the working environment of the power stations in which they operate.

18.22 Type 20 pneumatic road breaker

Designed and
manufactured by: Compair Construction and Mining Ltd
Product development: D. G. Evans, General Manager
D. Hough, Engineering Manager
R. Evans, Manufacturing Manager
M. J. Davidson, Chief Production Engineer
B. R. Kirby, Product Manager

Objectives. The Zitec 20 road breaker [18.22.1] was developed to meet a growing international market for a lightweight, high efficiency machine. A major objective was to improve the man/machine interface relationships in terms of operation, noise, health, safety, the environment and maintenance.

Design. The basic design (Figure 18.22.1) was guided by two properties of shock transmission; that the effectiveness of a road breaker depends not only on the energy of the blows delivered, but upon optimisation of the balance between stress produced in the impact shank and the energy transmitted to it; and that the shape of the stress wave, that is stress in the steel shank versus time, should be such that most of the energy delivered is associated with a high stress level.

Consideration of these criteria together with the optimisation of ergonomic handling features led to the use of a one-inch hexagon steel shank, which is significantly smaller than is usual. By careful detail design the anvil, which is normally required to transmit the blow from the air-operated piston to the steel shank, was dispensed with, and efficiency was improved as a result because this component converts much of the energy which should be transmitted to the shank into heat and vibration, and reduces the steepness of the stress wave front. This basic concept made possible a lighter construction because of its greater efficiency, and design effort was then directed at the man/machine interface problems.

A major problem with road breakers is noise, which arises from the explosive air discharge after each power pulse, and from vibration initiated internally by the piston impact. In this design exhaust noise is silenced by provision of a large expansion chamber formed within a non-conducting polyurethane casing that surrounds the complete operating mechanism, providing two-stage expansion in combination with baffles which are formed as part of the inner components and a long exhaust path of adequate cross-section to prevent exhaust condensation freezing and blocking up in cold climates. This construction provides an integral and effective silencer system, compared with the less effective and separate external mufflers commonly used. Noise generated by vibration has been suppressed by composite construction of the cylinder body, the steel barrel of which has a moulded-on bonded polyurethane sleeve, which damps vibration and noise at its source. This moulding also incorporates integral compressed air passages for returning the piston after the power stroke, and provides the silencer baffles referred to above. Residual high-frequency noise that may be reflected from the outer housing is again damped by the plastics materials used for this component. The combined effect of these measures is to reduce the external noise level by at least 3 dBA, or about one half of that usual with traditional designs.

Following ergonomic study, the conventional T arrangement of handles was retained, but the modular cast-iron handgrips incorporate replaceable soft

18.22 Type 20 pneumatic road breaker

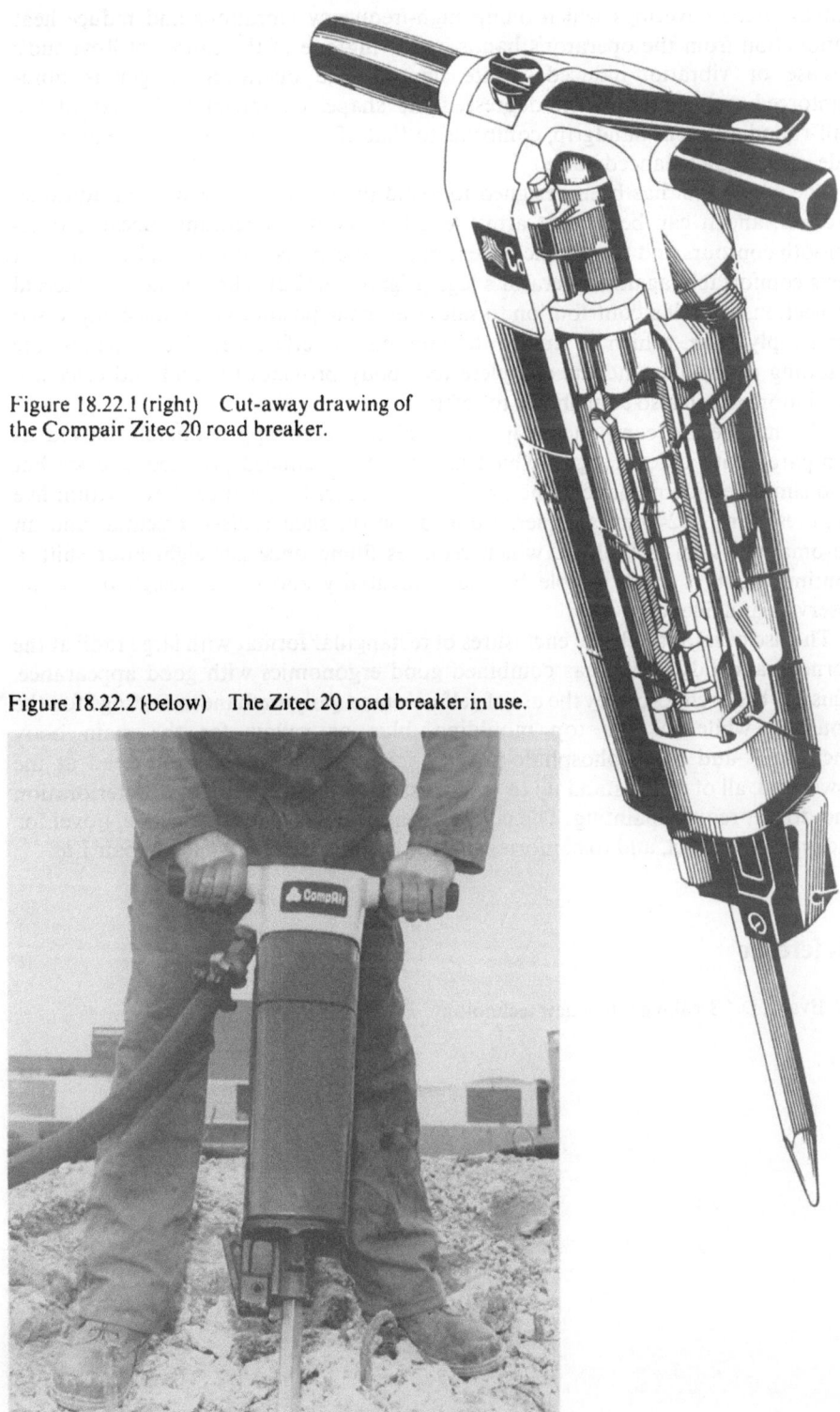

Figure 18.22.1 (right) Cut-away drawing of the Compair Zitec 20 road breaker.

Figure 18.22.2 (below) The Zitec 20 road breaker in use.

polyurethane coverings which damp high-frequency vibrations and reduce heat conduction from the operator's hands, removing one of the causes of Reynaud's disease or vibration-induced 'white finger'. The operating trigger is fibre-reinforced nylon, and when depressed, its shape, combined with that of the half-round adjacent handgrip, conforms to that of the round grip on the opposite side, ensuring a balanced hold.

The whole tool has been designed to avoid the top-heaviness of the traditional breaker, and it can be held at arm's length while it is operating. Because of its smooth contours and the absence of external projections such as silencers, the tool rests comfortably against operator's legs (Figure 18.22.2). The absence of external projections is itself a contribution to safety, as is the positioning of the compressed air supply hose, which is angled to minimise interference. The polyurethane cladding of handles and the complete tool body provides thermal and electrical insulation, which also contributes to safety.

The number of components in this tool has been approximately halved as compared with older designs, which has not only reduced production costs but also simplified maintenance. The breaker can in fact be stripped down within five minutes using a 24 mm spanner. Lubrication on such tools is essential and an automatic system is built in, which requires filling once per eight-hour shift, a routine which is made simple by the accessibility and detail design of the oil reservoir and filler cap.

The use of smooth plastic enclosures of rectangular format with large radii at the corners, and tidy detail, has combined good ergonomics with good appearance. This has been reinforced by the use of self-coloured polyurethane mouldings for the housing—white for the top moulding, blue or yellow for the main body enclosure—and black phosphate-finished forged steel for the front head at the lower end, all of which stand up to rough usage with the minimum of deterioration and do not require painting. The colours were selected to be distinctive, novel for the type of product, and to conform with the corporate colours of Compair Ltd.

Reference

1   Evans, D. 'Breaking into a new technology.' *Engineering*, April 1980.

Subject Index

ABS (acrylonitrile butadiene
    styrene) 183, 188, 192–193, 202,
    204
acceleration and the human body 95–96,
    97–98
accident prevention 127–128, 196
  see also safety
acclimatisation
  to noise 93
  to temperature 83
accuracy of instruments 65
achromatic colours 135
acrylics 183–184
acrylonitrile butadiene styrene
    (ABS) 183, 188, 192–193, 202,
    204
aesthetic design
  definition 13
  history 2–12
aesthetic functions of colour 142–143
ageing in man *see* human ageing
agricultural drill 351–354
air-blast circuit-breaker 366–367
airborne contaminants 100–101
airborne pollutants 100–101
air pressure and human
    performance 98–100
altitude and human performance 99–100
aluminium and aluminium alloys 179,
    189–190, 202–203
analogue displays 64
animals and materials selection 177
anthropometric data 21–24
  adjustments 23
  limitations 21
  selection 22–23
  sources 24
anthropometric models 24
anthropometric values 22–23
appearance
  and materials selection 176
  importance 16–17
  in specifications 293
Articulation Index 94
atmospheric effects
  and human performance 98–102
  countering 101–102
audiograms 93
auditory displays 69
auditory environment 91–95

auditory protection 94–95
auditory signals 41, 62, 69, 70
auditory system *see* hearing
automated systems 34
automatic flight control system 323–327
automatic lathe 346–349
automation and ergonomics 20

balance and proportion 108–116
barrier rails 253–255
Bauhaus movement 110
bi-metal laminates 188
biorhythms 32–33
  guidelines for designers 33
bodily heat exchange 82–83
body metabolic rate 82–83
body movement 24
body size 20–24
  available data 21–22
  guidelines for designers 24
body strength 25–31
  factors influencing 28–29
  guidelines for designers 31
  standard data 29–31
boom crane 354–356
boring machine 349–351
brass 178
British Standards
  cable connections 142
  colour 135, 142, 242
  colour coding 170, 171
  flooring 142
  guards 249, 250, 262
  paint finishes 198–199
  paints 135, 142
  pipelines 142
  plastics 142
  plating 201–202
  safety 142
  vitreous enamel 142
bronze 178
bus 356–360
buttons 236–238

CAB (cellulose acetate butyrate) 182
cable connections, colour standards 142
CAD (computer-aided design) and
    models 274–275

carbon fibres in substrates   186
cash dispenser   310–313
cast iron, painting   199
cast steel *see* steel
cathode ray tubes (CRTs)   239–240
C/D ratio (control/display ratio)   71, 73
cellulose acetate   182
cellulose acetate butyrate (CAB)   182
ceramics
   sintered   187
   see also glass
chipboard   188
chroma   135
chromium plating   201–202
CIE system of colour
      specification   137–138
circadian rhythms   32–33
clad materials   188, 201–204
clad metals   188
cleaning   207–208
   and materials selection   177
   of finishes   196–197
climate
   effects on man   81–87
   see also temperature
climatic comfort   83
climatic descriptors   83–84
cloth, resin-bonded   187
clothing
   effect on anthropometric data   23
   effect on workload   28
CLO units of thermal insulation   84, 87
coating, plastics   201
cold environments
   human response   28, 86
   reduction of effects   86–87
colour   13, 133–157
   aesthetic functions   142–143, 144–145
   altering apparent physical
      properties   140, 154
   and environment, checklist   155–156
   and ergonomics, checklist   155
   and maintenance   142, 207–208
   and marketing   143–144
      checklist   155
   and production   143–144
      checklist   156
   and safety   141, 152
   centres of attention   139
   compatibility   141–142
   connotations   145–146
   data collection   147–148
   design strategies   143–150
   durability of appearance   141
   ergonomic functions   139–142, 144
   finishes   198
   functions   138–143
   function weightings   144–145
   identification   141
   natural lawfulness   146–147
   number used on product   143
   practical problems in product
      design   143–144
   prediction of future trends   148–150
   product examples   150–154
   proportion and orientation   139
   psychological factors   217
   separation/association   139
   standards   135, 142, 242
   structural functions   138–139, 144
   visibility   141
colour blindness   39, 89, 142, 170
colour coding   141
   of machine graphics   170, 171
   of services   142
   product examples   153
   standards   142, 170, 171
colour desirability   147
colour differences   137–138
colour perception   133–134
colour rendering and illumination   90
colour schemes
   apposite   147, 148–149
   monochromatic   146, 147, 148–149
   selection strategies   148–149
   triadic   147, 148–149
colour selection   133, 146
   checklist   150–156
colour specification   134–138
colour tolerances   138, 144
colour value   135
commercial viability and form   116
Commission Internationale de l'Eclairage
      see CIE system of colour
      specification
communication between man and
      machine   219–221
   see also information transmission
communication systems   93–94
company standards   283–284
compressed air processing
      equipment   302–305
computer-aided bio-mechanical
      analysis   275
computer-aided models   274–275
computer system   334–336
confidence in product/system   215–216
conspicuous consumption and
      colour   146
contaminants, airborne   100–101
contrast, luminance   89–90
control-display compatibility   75
control/display ratio (C/D ratio)   71, 73
control panels
   colour standards   242
   psychological factors   221–222
   surface production
      techniques   241–242
   typeface   141

Subject Index

controls
    and form   129
    and materials selection   176
    and safety   238
    backlash   74
    classification   70–71
    combination   231
    comparison of characteristics   72
    deadspace   74
    design   71–75, 227–243
    force applied to   29–30
    grouping   240–241
    inertia   73
    layout   75–76, 76–77
    maintenance reduction   206
    misuse   238
    resistance   73
    response lag   74
    rotary   234–236
    selection   71
    symbols   168–169
    systems approach   55–58
    tactile/heavy   228–233
    tactile/light   233–238
    tactile/light with heavy   238
    two-hand   238, 253
    viscous damping   73
    visual/display aspect   239–240
    visual/graphics aspect   240–243
co-ordination in functional design   119–120
copper and copper-based alloys   178, 192–193
corrosion and materials selection   176
cost effectiveness   1, 17
costs
    in finishes selection   197
    in materials selection   175
    in specifications   294–295
crane   354–356
CRTs (cathode ray tubes)   239–240

danger *see* safety
deafness *see* hearing loss
deceleration and the human body   97–98
decompression sickness   100
design *see* functional design; industrial design in engineering
design checklists   285–287
design reviews   285–287
detail design issues and form   128–130
development specifications *see* specifications
development time-scale   295
digital displays   64, 65
dimension ratios, history   108–111
DIN (Deutsche Institut für Normung), barrier rails   253–254
dirt *see* cleaning

disability   32, 33, 39
    see also colour blindness; hearing loss
discrimination in man   38, 42–43
displays
    alpha-numeric   239
    analogue   64
    auditory   69
        see also auditory signals
    choice of channel   61–62
    digital   64, 65
    electronic   67–68
    hybrid   63
    lamp   170, 171
    layout   75–76
    qualitative   67
    quantitative   64–67
    representational   66, 67
    symbols   168–169
    tabular   170–173
    visual   62–68
distortion analysis, models for   274
DMC (dough-moulding compound)   181, 190–191
domestic press   313–316
dough-moulding compound (DMC)   181, 190–191
driver's cab, high speed train   361–365
durability
    economic   195–196
    of appearance   141

Effective Temperature (ET)   83–84, 85
electrical cable connections, colour standards   142
electronic displays   67–68
electroplating   201–202
emergency conditions   23
    see also safety
emergency stop devices   252
emergency warning signals   41, 62, 69
end-grain wood sandwich composites   188
energy expenditure   25, 27–28
engineering, integration with industrial design   1–12, 297–301
    successful examples   302–375
engineering product design *see* industrial design in engineering
environment
    and colour   155–156
    and materials selection   176–177
    and psychological factors   224
    and specification content   292
    and system design   81–103
    and work capacity   28, 29
    and workplace siting   77
    auditory   91–95
    visual   87–91
environmental load, checklist   52

environmental philosophy   127
epoxy resin   186
equipment, effect on anthropometric
      data   23
ergonomic functions of colour   139–142
ergonomic models,
      computer-aided   274–275
ergonomic problems
   checklist   49–52
   guidelines for designers   49
   identification   45–47
   strategy for tackling   45–52
ergonomics
   and automation   20
   and colour, checklist   155
   and safety   20
   control applications   55–79
   definition   13, 19–20
   display applications   55–79
   environmental factors   81–103
   in specifications   292
   introduction to   19–53
   methodology   55–79
   of guards   253–255
ergonomic strategy for allocation of
      functions   57–58
ET (Effective Temperature)   83–84, 85
ethics   17

Factories Act 1961   246–247
failure anticipation, design for   206
fashion
   and colour   148–150
   and form   107–108, 122–125
fatigue in man   40, 44
fault identification   207
fear of machine/environment   215–216
feedback
   from user experience   288
   to design   287–288
finish
   and functional efficiency   195–196
   and safety   196–197
   cleaning   196–197, 207–208
   colour   198
   cost   197
   market appeal   195
   reflectivity   198–199
   selection   195–204
   standardisation   199
   terminology   198–199
   texture and pattern   198
   see also colour
fire and materials selection   177
fire risks   196
   see also safety
Fitts list of relative abilities of men and
      machines   33, 34–35
fixed scales   64

flame spraying   187
flat screen devices   67–68
flight control system   323–327
floor coverings
   colour standards   142
   materials selection   185
foam plastics
   finishes   204
   soft   184–185
   structural   185
folded steel components   128–129
form
   and commercial viability   116
   and fashion   107–108, 122–125
   and function   107
   and safety   127–128
   and size   108
   and style   122–125
   and visual environment   125–127
   concealing function   123
   definition   13, 105–106
   design guide   117–131
   detail design issues   128–130
   development   106–107
   obtrusiveness   125–126
   psychology   106, 216–217
   suggesting function   122–123
   theory   105–116
form design   105–131
   checklist   130–131
   danger of isolation   125
   special shapes   122–123
French Standards (NF), colour   242
function
   allocation   55–58
   and form   107
   and materials selection   175
   separation   55–58
functional design   117–121, 123
   and size   121
   co-ordination   119–120
   expressing purpose   117
   history   2–6
   purest form   117–119
   style in   117
functionalism, influence on design
      theory   143

German Standards (DIN), barrier
      rails   253–254
German Standards (VDE), colour   242
glare   90
glass   179–180
glass-reinforced cement (GRC)   186
glass-reinforced compounds   187
   see also DMC; GRC; GRP
glass-reinforced plastic (GRP)   177, 186,
      204
graphic patch technique   240–241

graphics *see* machine graphics
GRC (glass-reinforced cement)   186
GRP (glass-reinforced plastic)   177, 186, 204
guards
   adjustable   252–253
   automatic   251
   basic design   248–255
   ergonomics   253–255
   fixed   249–250
   for electrical circuits   253–255
   interlocked   250–251
   self-adjusting   252, 253
   standards   249, 250

Hand Skin Temperature (HST)   86
harnesses   129
Health and Safety at Work Act 1974   47, 246–247, 260–261
hearing
   and displays   61–62
   basis   91, 92
   in complex stimulation   41–42
   working range   36
   see also auditory signals; human sensing system
hearing loss   91–93
   see also sensory disability
heat exchange in man   82–83
heat stress on human body   84–85, 86
high speed train   361–365
honeycomb structures   188
horizontal boring machine   349–351
hot environments
   human response   84–85
   reduction of effects   85–86
HST (Hand Skin Temperature)   86
hue   135, 146
human ability compared with machines   56–57
human ageing
   and body strength   28–29
   and performance   31–32
   guidelines for designers   33
human body
   and acceleration   95–96
   and vibration   95–97
   motion effect on   95–98
   response to climate   81–87
   see also body size; body strength *etc*
human discriminative ability   38, 42–43
human factor problems *see* ergonomic problems
human factors
   experimentation   49
   resolving conflicts   47–49
   systems approach   33–45
   see also ergonomics
human fatigue   40, 44

human information processing
   system   35–37, 40–43
   guidelines for designers   45
human perceptual processes   81
   see also hearing; touch; vision; visual perception
human performance
   and atmospheric effects   98–102
   and product characteristics   47
   effect of ageing   31–32
   mental   40–41
   see also visual performance
human response
   to cold environments   86
   to hot environments   84–85
   to humidity   84–85
human response system   36, 43–45
   guidelines for designers   45
human sensing system   36, 37–40
   guidelines for designers   45
   working ranges   36
   see also hearing; touch; vision
humidity
   and materials selection   177
   human response   84–85
hybrid displays   63
hypoxia   99–100

IEC publications   242
illness   32, 100, 216
illumination
   and perception   89–90
   and visual displays   63, 66
   measurement   88–89
illusions   39
Index of Thermal Strain (ITS)   83–84
industrial design in engineering   297–301
   history   2–12
   introduction   297–298
   methods of integration   298–300
   objectives   1–2
   successful examples   302–375
   techniques   13–18
inertia in controls   73
information analysis   59–60, 61
information feedback
   from user experience   288
   to design   287–288
information flows   59–60, 61, 69–70
information gathering in man   81
information processing in man *see* human information processing system
information transmission   40–45, 60–75, 90–91, 93–94, 213–214, 218
   see also communication between man and machine
instrument accuracy   65
interactive display terminal   330–333
interlocks   250–251

International Organisation for
    Standardisation *see* ISO
    recommendations
ionising radiation   101
iron *see* cast iron
isometric strength   25, 26
ISO recommendations
    graphic symbols   166–168
    plating   201–202
    vibration   96, 103
ITS (Index of Thermal Strain)   83–84

knobs, control   234–236

labels   242–243
laminates, bi-metal   188
lamp indicators, colour coding   170, 171
lathe   346–349
law and safety   47, 246–247
layout   75–79
leakage, measurement   209
levers
    combination controls   231
    feedback   230–231
    foot   231–232
    hand   228–230
    on/off   230
    rotary   232–233
    space considerations   231
    with other controls   235
lever switches   233–234
light levels
    and visual adaptation   89
    and visual displays   63
linear accelerator   337–342
link analysis in workplace layout   77–79
lubrication   208
luminance contrast, definition   89–90
luminance ratio, definition   90

machine abilities compared with
    human   56–57
machine graphics   159–173
    colour coding   170, 171
    definition   13
    in specifications   293
    see also controls, visual/graphics aspect
machine system design *see* system design
maintenance
    accessibility   209
    and climate   87
    and workplace layout   76
    checklist for design   211
    design for   205–211
    in specifications   293
    philosophy of design for   205
    prevention of assembly error   209–210

prevention of pre-commissioning
    troubles   206
reduction   205–206
safety in   210
see also cleaning
man/machine interface
    as a display   160–161
    as an information device   159
    finishes for   195–204
    function   159–161
    information analysis at   59–60
    materials selection for   175–193
man/machine interface design,
    psychological factors   213–226
man/machine interface layout   75–79
man/machine interface
    specification   60–75
manual systems   34
marketing
    and colour   143–144, 155
    and finish   195
    and materials selection   176
material requirements, defining   175–177
materials
    and form   129–130
    clad   188, 201–204
    composite   186–188
    compound   186–188
    history   6, 9
materials selection
    basic design factors   175–176
    environmental factors   176–177
    for the man/machine
        interface   175–193
    philosophy   175
    product examples   189–193
    see also under specific materials
mechanical systems   34
melamine   181, 188
mental performance   40–41
mental stress   215–216
metabolic rate of human body   82–83
metal cladding of non metals   188
metals   178–179
    clad   188
    painting   199–200
    sheet   199–200
    sintered   187
    see also under specific metals
microelectronics   121
miniaturisation   121
misuse and materials selection   177
mock-ups
    full-scale   264–267
    in workplace layout   79
models   263–276
    anthropometric   24
    computer-aided   274–275
    for distortion analysis   274
    for exhibitions   265–267, 267–268

for stress analysis 274
for work station design 270–273
full-scale 264–267
in human factors design 48
introduction to use in design 263
in workplace layout 79
of details 269–270
scale 267–268, 271
system 270, 271
modular systems
   core and dressing 280–283
   dimensional and functional 278–280
motion effect on the human body 95–98
motor performance in man 98
motor responses in man 43–45
moving scales 64
muscular contraction 25–27
muscular power 25–28
Munsell system of colour
   specification 135–137, 146–147, 148–149

name plates 242–243
national grid control centre 327–330
neon indicators 239
NF (Norme Française), colour 242
nickel coating 201–202
noise
   and materials selection 177
   human response 28, 91–94
noise control 94–95
noise measurement 91
non-ferrous castings, painting 200
non-ferrous extrusions, painting 200
nylon 182, 201

obsolescence in specifications 294
operating theatre 342–345
optical illusions *see* visual illusions
ornamentation
   history 2
   styled 124

paint finishes, standards 198–199
painting of metals 199–200
paints, colour standards 142
panels, control *see* control panels
panels, sandwich 187–188
paper, resin-bonded 187
pedals 231–232
perceptual processes in man 81
   see also hearing; touch; vision; visual perception
performance
   assessment 14
   definition 1
   in specifications 292

measurement 208
   see also human performance; visual performance
phobias 216
physiological load 25
pilots 239, 240
pipelines, colour standards 142
pipe systems and form 129
plastic cladding 188
plastics
   coating 201
   colour standards 142
   finishes 203–204
   foam 184–185
   thermosetting 180–181
   see also thermoplastics
plating finishes 201–202
pneumatic road breaker 372–375
pollutants, airborne 100–101
polyacetal 181
polycarbonates 184
polyester 181
polyethylene 181–182
polyphenylene oxide (PPO) 184
polypropylene 182
polystyrene 183
polysulphones 184
polytetrafluoroethylene (PTFE) 182, 201
polyurethane 183
polyvinyl chloride (PVC) 183, 192–193
polyvinylidenefluoride (PVDF) 201
posture 24, 28, 29–30, 76
potentiometers 235–236
potentiometric chart recorder 319–323
PPO (polyphenylene oxide) 184
press, domestic 313–316
prestige 214
printed circuits 237
product characteristics and human performance 47
production and colour 143–144
   checklist 156
production levels
   and materials selection 175
   in specifications 293
product parameter reviews 286
proportion and balance 108–116
protection
   auditory 94–95
   by use of form 128
psychological factors
   design guide 219–226
   effect on work capacity 28, 29
   fear, confidence, and stress 215–216
   in man/machine interface design 213–226
   laziness 218
   love, hate, and indifference 216–218
   misunderstanding 213–214
   phobias and sickness 216

pride and prestige 214
secondhand understanding 218
psychological principles in symbol
        design 165–166, 167
psychological processes
  and noise 93
  in machine control 160
psychological stress 215–216
psychology of form 106, 216–217
PTFE (polytetrafluoroethylene) 182, 201
purpose in functional design 117
push-buttons 236–238
PVC (polyvinyl chloride) 183, 192–193
PVDF (polyvinylidenefluoride) 201

qualitative displays 67
quality 284–288
  and design 284–285
  and standardisation 277
  definition 1
quality assurance in specifications 293
quality levels, requirements for
        ensuring 288
quality manual 285–287
quality requirements in
        specifications 292–293
quantitative displays 64–67
quantitative scales 66

radiation, ionising 101
rating plates 242–243
ratios of dimensions, history 108–111
reach 24
reach values 254
receptor cells in colour
        perception 133–134
reflectivity of finishes 198–199
reliability
  assessment 14
  in specifications 292
repetition of form 114–116
representational displays 66, 67
resin-bonded cloth 187
resin-bonded paper 187
resin-impregnated wood 187
response
  system 71–75
  see also human response
risk assessment 255–257
risk reduction 255–257
risks see safety
road breaker 372–375
rotary controls 234–236
rotary levers 232–233
rubber 180

safety 245–262
  and anthropometric data 22–23
  and colour 141, 142, 152
  and controls 238
  and ergonomics 20
  and finish 196–197
  and form 127–128
  and materials selection 176, 177
  and sensory disability 39
  case studies 257–260
  checklist for designers 261
  design procedures 260–261
  emergency stop devices 252
  emergency warning signals 41, 62, 69
  factors reducing 16
  identification of dangers 247–248
  in maintenance 210
  in specifications 293
  legal requirements 246–247
safety guards see guards
SAN (styrene acrylonitrile) 183
sandwich panels 187–188
scaffolding system 307–310
scale markers 65
scale pointers 67
scale progression 65, 67
scales
  fixed 64
  moving 64
  quantitative 66
  see also displays
scale units 65, 66
senses in man see human sensing system
sensory channels 42–43
sensory disability 39
  see also colour blindness; hearing loss
services and form 129
servicing skills 210
shape
  psychological factors 216–217
  see also form
shape detailing 129–130
shapes in form design 122–123
sheet metals
  painting 199–200
  see also under specific metals
shock forces and materials selection 176
sickness 32, 100, 216
sight see vision
signals
  auditory 41, 62, 69, 70
  visual 62
  see also human information processing
        system
sintered ceramics 187
sintered metals 187
size
  and form 108
  and functional design 121
soft foams 184–185

solvents and materials selection 177
sound *see* noise
specification content 290–295
    component standards 291
    component variants 291
    costs of production and
        development 294–295
    design basis 292
    development objective 290
    development time-scale 295
    future requirements 291
    product definition 291
    production levels 293
    product obsolescence 294
    quality requirements 292–293
    special legal requirements 291
    technical standards 291
    work in progress 294
specifications 289–296
    for type 162, 163, 164
    history 6, 9
    purpose for development 289–290
speech in a communication
        system 93–94
Speech Interference Level 94
stability 111–116
stainless steel 178–179
stainless steel finishes 203
standardisation 278–284
    and design 278
    and quality 277
    and specification content 291
    of construction 283–284
    of design procedures 284
    of finish 199
standard manuals 283–284
standards
    technical 291
    see also British Standards; French
        Standards; German Standards;
        ISO
static workload 25, 28, 31
steam turbine generator sets 368–371
steel 178, 189–190, 191
    painting 199
    see also stainless steel
steps, materials selection 185
sterile operating theatre 342–345
stimuli, human sensitivity to 37–40
storage and specification content 293
strength, isometric 25, 26
strength in man *see* body strength
stress analysis, models for 274
stress in man 215–216
structural foams 185
structural functions of colour 138–139
structural requirements and materials
        selection 176
style 13
    and form 122–125

    in functional design 117
    styled ornamentation 124
styrene acrylonitrile (SAN) 183
superplastic alloys 186
surface colours, coding 170, 171
sweat 84
switches
    lever 233–234
    rotary 234, 236
symbol design, psychological
        principles 165–166, 167
symbols
    and control/display elements 168–169
    for machine functions 165–169
symbol standards 166–168
symmetry 113
synthetic rubber 180
system design
    and environment 81–103
    decision-making process 58
    definition 56
    history 55
system design process 34–35
system models 270, 271
system objectives, definition 56
system response 71–75
systems
    automated 34
    manual 34
    mechanical 34
systems approach
    to control 55–58
    to human sensing system 33–45

tabular displays 170–173
task analysis 161
telescopic boom crane 354–356
temperature
    acclimatisation 83
    and materials selection 177
    working range 36
    see also cold environments; hot
        environments
texture
    maintenance 207–208
    of finishes 198
thermal indices 83–84
thermoplastics 181–184
thermosetting plastics 180–181
threshold limit values (TLVs) 100–101
timber 180
time-scale of development 295
TLVs (threshold limit values) 100–101
touch
    in communication 62
    working range 36
    see also human sensing system
train 361–365
training and workplace layout 76

transportation in specifications 293
trip devices as guards 252
two-hand controls 238, 253
typeface for control panels 241
typeface specification 162, 163, 164
type measures
    metric 164–165
    traditional 163–164
typographic form, principles 163
typography
    fundamentals 161–165
    geometric specification 162, 163
    in tabular displays 171, 173

urea 181
user
    captive 220, 221–222
    non-captive 220, 222–224
    see also psychological factors
user experience, information feedback 288
user identification 220–224

vandalism 216–218
    and materials selection 177
VDE (Verband Deutscher Elektrotechnika) standards, colour 242
VDUs (visual display units) 67, 91
velocity change and the human body 97–98
ventilator, medical 316–319
vibration
    and materials selection 177
    and the human body 95–97
    working range 36
viscous damping in controls 73
vision
    and displays 61–62
    defects 39
        see also colour blindness
    in complex stimulation 41–42
    working range 36
    see also human sensing system
visual acuity 89
visual adaptation to light levels 89
visual displays 62–68
visual display units (VDUs) 67, 91

visual environment 87–91
    and form 125–127
    measurement 88
visual illusions 39
visual perception, basis 87–88
visual performance
    and acceleration 98
    factors affecting 89–90
    improvement 90–91
visual signals 62
vitreous enamel 142, 180

warning signals 41, 62, 69
WBGT (Wet Bulb Globe Temperature) 83–84
wear, measurement 209
weathering and materials selection 177
weight and materials selection 176
weightlessness 98
Wet Bulb Globe Temperature (WBGT) 83–84
wheels, control 232–233, 235
Wind Chill Index 84
women's strength 28–29
wood 180
    resin-impregnated 187
wood sandwich composites 188
work capacity 28–29
work duration, psychological factors 224
work in progress and specification content 294
workload 27–28
    static 25, 28, 31
work method
    mental demands, checklist 51–52
    physical demands, checklist 51
workplace
    mental demands, checklist 50–51
    physical demands, checklist 50
workplace design, models in 270–273
workplace layout 75–79
work rate 25, 26

yarn top detector for textile machines 305–307

zinc alloys 178

Name Index

Adam, Robert 109
Adams, George 2, 5
A J Seward and Co 223
Albers, J 139, 156
Albert, Prince Consort 6
Alexander Gibb and Partners, Sir 126
American Conference of Government Industrial Hygienists 84, 102
Ashby, P 23, 26, 53
Ashman, C S 337
Associated Electrical Industries Ltd 367
Astrand, P O 27, 29, 53
Ayling, K R 337

Bailey, A J 351
Baker Perkins Ltd 279, 281
Bang, Jens 192
Bang and Olufsen Ltd 192, 193
Bardaghjy, J 53
Barrett, B 52
Bauhaus 12, 110
Bentley, W O 9
Berrel 6, 8
BHRCA Fluid Engineering Laboratories 153
Bierhoff, A 307
Billmeyer, F W 156
Bill Moggridge Associates 153, 269, 270
Birren, F 139, 156, 157
Black, M 18
Blank, W 307
BMW Ltd 123, 124
Bolton, C B 52
Booth, Richard T 245, 253, 262
Boroughs Ltd 331
Braun Ltd 110, 127
British Leyland Ltd 268
    see also Leyland Vehicles Ltd
British Petroleum Co Ltd 107
British Rail Engineering Ltd 361
British Railways Board 66, 361
British Standards Institution (BSI) see British Standards in subject index
British Steel Corporation 356
Brook, D 351
Brooks, B M 360
Brown Boveri Co Ltd 368, 369, 370
Bruce Peebles Ltd 279, 280
Buckeye Power Inc 369

Bugatti, Ettore 9, 10, 124
Building Centre 185
Burdon, E S 365
Burger, C E 53
Burns, W 103
Buskirk, E R 53
Butler, Dennis, Garland and Partners Ltd 150

Cakir, A 79
Caldecote, Lord 18
Cardano, Gerolamo 2
Cardin, Pierre 108
Carpenter, J 53
Cartwright, A 365
Case, K 276
Cazamian, P 53
Central Electricity Generating Board (CEGB) 278, 327, 329
Chapanis, A 42, 52, 53, 79, 103
Chaplain, M 354
Chapman, B G 316
Charnley, Sir John 342, 343, 345
Chessell, B 319
Chessell Ltd 222, 319, 320
Chevreul, Michel-Eugène 142
Chubb and Sons, Lock and Safe Co Ltd 110, 310, 311
Cleal, C K 319
Colchester Ltd 12, 15
Collins, J B 89, 103
Commission Internationale de l'Eclairage (CIE) 137
Communications Complex Design Ltd 273
Compair Construction and Mining Ltd 372, 373, 375
Compuda Ltd 275
Cook, J S 79
Cook, S C 103
Cosmos Crane Co Ltd 354
Crannell, C W 79
CRE Design Consultants 118

Data SAAB 240
Davall (S) and Sons Ltd 153
Davey, Paxman Ltd 268
Davidson, M J 372

Davies, David G   55
da Vinci, Leonardo   2, 7, 107, 109
Dean, M   351
Defence, Ministry of   126
Department of Employment   95, 262
Department of the Environment   103
Design Council   223, 303, 305, 310, 316, 319, 334, 337, 342, 347, 351, 354, 356, 361
Dewhurst and Partner Ltd   217
Diffrient, N   53
Dillow, J W   330
Dreyfuss, H   243
Driver, B S   337
Dualform Press and Shear Manufacturing Co   152
Dunnette, M D   53

Easterby, Ronald S   159
Edgerley, David   118
Edholm, O G   103
Electrical Research Association Ltd   330
Elna S A   313, 314
EMI Ltd   189, 190
EMI-MEC Ltd   346, 347
Employment, Department of   95, 262
English Electric Ltd (GEC)   338
Environment, Department of the   103
Euclid   107, 109
European Organisation for Quality Control   277
Evans, D G   372, 375
Evans, R   372
Ewen, Alistair   153

Factory Inspectorate   253, 255
Fanger, P O   103
Farina   108
Fielden, G B R   277
Fitts, P M   33, 79, 243
Flurscheim, Charles H   1, 18, 105, 205, 289, 297, 303, 367
Fox, J G   53
Fox, W F   103
Franks, P E   79
Frauenhofer, Josef von   6, 7
Frister and Rossman Ltd   120
Fry, Roger   279
Fryars, R A   356

Galileo   107
Gantz, C M   157
Gaudi y Cornet, Antoni   113
Gebhard, J W   36, 38, 53
General Electric Company Ltd (GEC)   115, 323, 367
George, R A   327
Ghia   108

Gibbs, Antony   205, 297, 330
Gloag, H L   139, 142, 156, 157
Goethe, Johann Wolfgang von   142
Gold, M J   142, 157
Goldsmith, S   53
Goulder Ltd   12, 16
Grandjean, E   44, 53
Grange, Kenneth   120, 361, 365
Graves, Rod J   81, 102
Gray, R   316
Greenberg, S   52
Gregory, R L   53, 103
Grien, S   53
Gropius, Walter   12
Grundy and Partners Ltd   330
Gucci, Aldo   108
Gunov, M B   79

Hale, A   262
Hale, M   262
Hall, J   157
Hardy, A C   139, 156
Harris, D V   53
Hart, D J   79
Hassall, D   354
Hatchman, P   347
Health and Safety Commission   247, 262
Health and Safety Executive   255, 262
Height, Frank   18, 305
Henderson, C I   337
Henneman, R H   79
Hilton, Jack   263
Holmes, G   351
Honda Ltd   122
Hopkinson, R G   89, 103
Hopkinson, R R   139, 157
Hough, D   372
Howard, R W   327
Howe, Jack   310, 312
Howell Killick Partridge and Amis   126
Howells, R   52
Howorth, F H   342, 343, 345
Howorth Air Engineering Ltd   342, 344
Hunsicker, P A   243
Hunt, D P   79
Hunter, S   53
Hunting Engineering Ltd   130
Hutchinson, A   347
Hyclamet Ltd   327

Illuminating Engineering Society   90, 103
Imperial Chemical Industries Ltd (ICI)   274
Industrial Fatigue Research Board   12
International Computers Ltd (ICL)   154, 282, 334, 335
International Ergonomics Association   49

Name Index

International Organisation for Standardisation *see* ISO recommendations *in subject index*
Issigonis, Sir Alec 108

Jarvis, T R 337
JCB Ltd 191
Jenkins, W L 79
Jenkins, W O 243
Johnson, J W 354
Jordan, K A 305
Judd, D B 156

Kay, R M S 211
Kearns-Richards Ltd 349, 350
Kenward, M 52
Kerslake, D McC 103
Keyte, M J 139, 157
Kinkade, R G 36, 37, 43, 53, 79, 89, 243
Kirby, B R 372
Kirkbride, Alec B 175
Kletz, T A 260, 262
Kreitler, H 156
Kreitler, S 156

Lancer Boss Ltd 282, 283
Landers, D M 53
Lansing Ltd 229
Lauru, L 243
Lawley, H G 262
Leamon, T B 102
Leeuw, J de 307
Leithead, C S 85, 102
Leyland Vehicles Ltd 113, 152, 356, 357
  see also British Leyland Ltd
Lind, A R 85, 102
Lindsay, H P 156
Loewy, Raymond 12
London, Noël 195, 227
London and Upjohn 334
Love, D 347
L S Starrett Co Ltd 119
Lund, M W 79, 103

McCormick, E J 18, 27, 28, 53, 58, 64, 73, 77, 79, 85, 96, 97, 99, 103, 243
McMeekin, R R 53
McWhirter, N 52
Marconi Avionics Ltd 285, 287, 323, 324, 326
Marconi Instruments Ltd 241
Massey Ferguson Ltd 351
Math, F C 107
Mather and Platt Ltd 264, 265, 266, 267, 268, 271, 282
Maudslay, Henry 6, 7

May 254
Mayall, W H 18, 277
Medical Research Council 53
MEL 337
Mellor, David 16
Meyer, R P 52, 53
Ministry of Defence 126
Montecatini Ltd 317
Moore and Wright Ltd 65
Morgan, C T 79, 103
Morgan, R 316
Morris, L A 102
Moulton, M 327
Mowbray, C H 36, 38, 53
Munsell, Albert H 135, 142
Murdoch, Peter 105, 213
Murrell, K F H 53, 79, 103

Nadler, G 53
National Research Development Corporation 275
Negretti and Zambra Ltd 190, 191
Newell, C A E 319
Nicholl, A G McK 102
Norgren Ltd 302, 303
Norman, D A 156
Nottingham University 275

Odell, A 316

Palladio, Andrea 109
Payne, John 150
Penlon Ltd 110, 316
Perkins Ltd 280, 282
Pevsner, Nikolaus 18
Philips Electronic and Associated Industries Ltd 337
Pickford, R W 157
Plato 107
Porsche Ltd 124, 125
Porter, M 276
Post Office 328, 329
Poulton, E C 79
Powell, A J 365
Pritchard, R 347
Prolect Ltd 112
PSA, Ministry of Defence 126
Pye Radio Ltd 279

Rams, Dieter 110, 127
Reed, David 223
Rhodes, Zandra 108
Rockway, M R 79
Rodahl, K 27, 29, 53
Rogers, H B 351
Rolfe, J M 41, 53

Rolls-Royce Ltd 11, 14
Royal College of Art 154
Royal Navy Personnel Research Committee, Medical Research Council 53
Russell-Smith, R 316
Ruston Diesels Ltd 114

SAAB 124
Saloman et Fils S A 17
Saltzman, M 156
Sargeant, R S 351
Satherby, R 349
Satherby Design 349
Scaffolding (Great Britain) Ltd 307
Schraer, R 53
S. Davall and Sons Ltd 153
Seward (A J) and Co 223
SFENA et Cie 323
Shackel, B 35, 53
Simpson, G C 102
Simpson, R E 52
Sinaiko, H W 53
Singleton, W T 31, 33, 34, 35, 52, 53, 79
Sir Alexander Gibb and Partners 126
Slatter, P E 157
Smallhorn, A P 316
Smiths Industries Ltd 310
Smits, B 307
Socrates 107
Somerville, M J 319
Sparshott, F E 157
Starrett (L S) Co Ltd 119
Staveley Machine Tools Ltd 349
Stewart, T F M 79
Sugg, P R 316
Sullivan, Louis 107, 143
Superform Metals Ltd 333
Surry, J 255, 262

Talbot, C F 102
Tasker, B 305
Thomas, S N 307

Thompson, D T 253, 262
Thorn Ericsson Telecommunications Ltd 121
Thorp Modelmakers Ltd 268
Tilley, A R 53
Toshiba Ltd 237
Triplex Ltd 361
Triplite Ltd 305, 306
Trussler, P 337
Tucker, W A 53
Turner, G M 52
Tustin, D E R 319

Upjohn, Howard 195, 227

Van Cott, H P 36, 37, 43, 53, 79, 89, 243
Vauxhall Motors Ltd 229
Veblen, Thorstein 146
Vickers Instruments Ltd 235, 241, 282
Volvo Ltd 224

Ward, F 351
Ward, J W 327
Warren Spring Laboratory 52
Welford, A T 53
Whitfield, D 53
Whitfield, T W Allan 133, 157
Wilkes, E G M 327
Wilkes and Ward 327
Wilkie, D R 53
Williams, C R 319
Wilson, J 354
Wiltshire, John T 133
Wolcott, J M 53
Wond, G I 316
Wood, J 19, 53, 273, 276
Woudhuysen Ltd 316
Wyszecki, G 156

Zeff, C 79